Millets, Rice and Farmers

Phytoliths as indicators of agricultural, social and ecological change in Neolithic and Bronze Age Central China

Alison Ruth Weisskopf

BAR International Series 2589
2014

Published in 2016 by
BAR Publishing, Oxford

BAR International Series 2589

Millets, Rice and Farmers

ISBN 978 1 4073 1217 0

BAR Publishing is the trading name of British Archaeological Reports (Oxford) Ltd. British Archaeological Reports was first incorporated in 1974 to publish the BAR Series, International and British. In 1992 Hadrian Books Ltd became part of the BAR group. This volume was originally published by Archaeopress in conjunction with British Archaeological Reports (Oxford) Ltd / Hadrian Books Ltd, the Series principal publisher, in 2014. This present volume is published by BAR Publishing, 2016.

Printed in England

BAR
PUBLISHING

BAR titles are available from:

BAR Publishing
122 Banbury Rd, Oxford, OX2 7BP, UK
EMAIL info@barpublishing.com
PHONE +44 (0)1865 310431
FAX +44 (0)1865 316916
www.barpublishing.com

ABSTRACT

Late Neolithic and Early Bronze Age Central China was the scene of important cultural developments, which impacted on agricultural practices and local vegetation. Using phytolith data from 4 archaeological-sites and a survey of a further 11 sites in Henan, this project investigates changing crop choices, developments in crop processing and, to a lesser extent, differing local environments both temporally and spatially. Comparisons of relative levels of phytoliths from crop husks, *Panicum*, *Setaria* and *Oryza*, from each site and period were compared to provide evidence of changes in crop repertoire. Results from these investigations show crop variation both over time and in different parts of the region. Millet farming predominated in the more Northern sites in the Yellow River Valley, while rice was clearly the main crop at Baligang, the southernmost site within the Yangtze catchment. However, rice became more important in the Yellow River valley during the Late Neolithic and interestingly despite a changing climate making rice farming more challenging in the Early Bronze Age farmers continued rice cultivation. Crop processing stages were interpreted by examination of differing proportions of phytoliths from crop husks, weed husks and crop and weed leaves which can illustrate differing cultivation systems, harvesting and processing practices. These can be seen especially clearly in the rice data from Baligang suggesting more successful agricultural practices and possible change in social organisation in the Late Neolithic. Evidence of local environmental variation was more challenging to unpick as all the phytolith samples available were from cultural contexts. However, the results of the investigation into the changing local environment reflect other proxy data. Differences in occurrence of specific key phytolith short cell morphotypes and changes in the levels of bulliforms, and cone shaped phytolith morphotypes from Cyperaceae indicating wetland were used to interpret local vegetational change, again both spatially and temporally. Rondel and bilobe shaped short cells represent Pooid and Panicoid grasses respectively and short cell morphotypes, such as rondels, bilobes, saddles, can also be used to track variation in levels of C3 and C4 grasses, so changes in proportions of these morphotypes can indicate larger vegetational change. A comparison between the dataset from Neolithic Central China and one from Neolithic India highlighted possible variations in arable systems with millet and wet rice farming in China contrasting with dry rice in India.

LIST OF CONTENTS

Please note that the DVD referred to above has now been replaced with a download available at www.barpublishing.com/additional-downloads.html

FIGURES

TABLES

ACKNOWLEDGEMENTS

I would like to thank the Arts and Humanities Research Council for providing financial support for this research.

Special thanks are due to Dr. Sue Colledge, who has been a constant mentor. My thanks to Professor Dorian Fuller and Professor Arlene Rosen, Sandra Bond, Dr. Norah Maloney, Professor Ken Thomas, Dr Emma Jenkins also Professor Marco Madella and Professor Gina Barnes for their helpful comments and suggestions.

In China: Professor Li Liu, and Professor Xingcan Chen for allowing me to work at Huizui and for their help throughout the research, Dr Li Xinwei, Dr Ma Xaolin, for allowing me to work and stay at Xipo, Dr Jixiang Song, for her help sampling and translating at Xipo, Dr Gyoung-Ah Lee, for collecting the Erlitou, survey and Huizui, 2003, samples and for her help and advice all the way through the project, Professor Xu Hong, for the survey and Erlitou samples, Dr Qin Ling and Peking University for the Baligang samples.

Thanks to CASS and PKU

At UCL Rhiannon Jackson, Dr Katie Manning, Rebecca Beardmore, Dr Emma Harvey, Mary Larkum, Pat Brolley, Dr Mary Anne Murray, Dr Stuart Brookes, Dr Dominique de Moulin, Lefteris Zorzos, Dr Meriel McClatchie, Professor Yvonne Edwards, Charlotte Veysey, Edgar Samarasundera, and Harriet Martin,

Shelly Cleeve at the Royal Botanical Gardens at Kew for helping collect grasses for my reference collection. Dr Freea Itzstein-Davey, for her grass collection from the Yangtze.

My husband, Alan Michels, my son Max Weisskopf Michels, my parents Anne and Hans Weisskopf and friends June James and Jane Morton.

CHAPTER 1

INTRODUCTION TO ISSUES IN THE ECONOMIC ARCHAEOLOGY OF CENTRAL CHINA

1.1 INTRODUCTION

The Mid/Late Neolithic and Early Bronze Age in north central China saw great socio-economic change, from egalitarian villages during the Yangshao period (4900 - 3000BC) to more hierarchical social organisation during the Longshan (3000-1850BC) to the rise of complex societies and the emergence of the first city state at Erlitou in the Early Bronze Age Erlitou Period (1850-1100BC) (Chang, 1986, Underhill, 2002, Liu, 2004, Ma, 2005, Liu and Chen, 2003, 2006). The main focus of this book is to examine agricultural and social change in Henan, North Central China over the course of these three cultural periods using archaeobotanical evidence from phytolith data. The study is based on 4 sites in Henan, north central China: Xipo on the south eastern Loess Plateau, Huizui and Erlitou on the broad lowland area along the Yellow River valley and Baligang in the Nanyang Basin on the northern margins of the northern subtropical zone, in the northern catchment of the Yangtze River. In addition to these sites samples have been taken from 7 small settlements in the Yilou River Valley.

This chapter will give brief overviews of past and current attitudes towards archaeology, archaeobotany and phytolith analysis in China and the aims and objectives of the research will be set out.

1.2 APPROACHES TO ARCHAEOLOGY IN MODERN CHINA

Traditionally, archaeology in China has been oriented primarily towards historical text, historiography, and is far more closely tied to written history than in the West. For traditional Chinese scholars everything from the Neolithic onwards has potential for text-based investigation (von Falkenhausen, 1993:841). For many in China archaeology is regarded as a sub discipline of history (Chang, 1981, von Falkenhausen, 1993, Olsen, 1987:285, Shelach, 2000:369, Lee, 2002: 20). Xia describes Chinese history and archaeology as 'two wheels on a cart or two wings on a bird', (1990:62-3 in von Falkenhausen, 1993:841). This is in keeping with attitudes in European archaeology in the 1930s-1950s when Chinese archaeology first developed (e.g. Woolley, 1930). This view has in many ways articulated the methodological approaches taken towards archaeology even today with an emphasis placed on individual sites and artefacts (Shi, 2001: 55-7, Shelach, 2000:369).

1.3 APPROACHES TO ARCHAEOBOTANY AND PHYTOLITH ANALYSIS

Palaeoecology and Pleistocene archaeology are not situated in the Academy of Social Sciences (Institute of Archaeology) (Olsen, 1987:286) but in the Academy of Science at the IVPP (Institute of Vertebrate Palaeontology and Paleoanthropologist). The Academy of Science has taken a more multi disciplinary approach and there is a leaning towards investigation of human environment interaction. Phytolith analysis in China has often come from the soil science and geology departments at Universities or the Science Academy, or departments of agriculture rather than archaeology or social science, even when the analysis is based on cultural deposits, for example, Li *et al*, 2007 on the expansion of early agriculture, Lu *et al*, 2005 on ancient millet noodles, Jiang, 1995, 1997, on rice cultivation. Exceptions are Zhang and Wang, 1998 and Wang *et al* 1999 on the origins of cultivated rice both collaborations between Agriculture and archaeology. Phytolith analysis seems to have been used more frequently alongside pollen as a palaeoecological tool and more commonly used to discuss ecological than archaeological change, for example, Lu *et al* 2006, 2007. This division may account for the paucity of Chinese-led archaeobotanical work until recently. Another exception is Zhao Zhijun from the Institute of Archaeology, Chinese Academy of Social Sciences, who has undertaken floatation at more than 30 Chinese sites since 2000, including Xinglonggu, Inner Mongolia (Zhao, 2004) and Zhouyan, Shaanxi (Zhao and Xue, 2004). However, as the bulk of his work is published in Chinese only, it is not readily accessible to western archaeologists.

1.4 AN OVERVIEW OF RESEARCH USING PHYTOLITH DATA IN CHINA

Phytolith research was introduced to China during the 1980s. Most research using phytolith analysis has concentrated on the origins and spread of rice agriculture and domestication (Jiang, 1995, Chen and Jiang, 1997, Zhao *et al* 1998, Zhao, 1998, Lu, 1999:5, Huang and Zhang, 2000, Lu *et al*, 2002, Jin *et al*, 2007, Li *et al*, 2007a, Li *et al* 2007b,) or broad environmental reconstruction from non-archaeological phytoliths (Lu, H-Y *et al*, 2007, Lu, H-Y *et al*, 2006, Li *et al*, 2001, Zhao and Piperno, 2000). There has as yet been comparatively little fine resolution site-based analysis of phytolith data to consider contextual variation and changes through time that relate to agricultural economy and organisation. Past studies in this field mostly focus on *Oryza* and include Jiang and Piperno, (1994) who used a dataset from 7 samples from a Longshan period trench at Yangzhuan, Henan and from 5 samples from a neighbouring ancient lake deposit containing Longshan pottery sherds, to identify *Oryza* cultivation during the Longshan. Jiang, (1995) also used the 7-sample phytolith dataset from Yangzhuan to investigate the origins and dispersal of *Oryza* agriculture. The datasets are small and counts range from 59 to 320 per sample. Jiang was building on earlier work at the Peligang site of Jiahu, Wuyang,

Henan, where she used a dataset from 9 samples to discuss early *Oryza* cultivation (Chen and Jiang, 1997, Jiang 1997). Huang and Zhang (2000) looked at pollen and 7 phytolith samples from two Neolithic phases at the site of Longquizhuang Guayu, Jiangsu, southern China in order to confirm *Oryza* cultivation. Most of these studies use the presence of specific morphotypes to establish presence of *Oryza* but there is little contextual consideration and many of the datasets are statistically small.

An exception is Itzstein Davey *et al* (2007a, 2007b) and Atahan *et al* (2008), who use phytoliths alongside pollen and charcoal to assess human influence on environmental change during the Neolithic and Bronze Age at Qingpu in the Yangtze delta and investigate *Oryza* cultivation, possible domestication and intensification. A larger dataset from 100 samples was analysed for both pollen and phytoliths (Itzstein Davey *et al*, 2007a: 877, 2007b). 26 of those samples were investigated further to determine whether it was possible to distinguish wild from domesticated *Oryza* using the method outlined by Pearsall *et al* (1995) and Zhao *et al* (1998). The samples were taken contiguously from sections from a trench wall so analysed according to sedimentary zone rather than archaeological context. Nevertheless, they are able to suggest the cultivation of wild rice continued to be an important feature at Qingpu until around 1800 bp suggesting rice based agriculture arrived comparatively late in this area.

However, the disappearance of smaller rice keystone phytoliths from this region after c4000 has been noted (Fuller and Qin, 2009: 95, Fuller *et al* 2007, Zheng *et al* 2003). Fuller and Qin suggest directional evolution by selection during domestication and changes to when and how the plant was harvested may be responsible and that this combined with evidence of changes in rice grain size suggests that rice was morphologically wild but cultivated from 6000 until around 4500 when it became fully domesticated (2009:95).

Another example of use of a larger dataset is Li *et al* (2007 a and b) who investigated phytolith, pollen and macro archaeobotanical remains at Xishanping in order to reconstruct the origins and spread of rice cultivation, the development of complex agriculture and the impact of agriculture on the surrounding vegetation. They collected 64 pollen and 65 phytolith samples at 10cm intervals from a 650cm continuous undisturbed sediment profile, but the sequence cuts through cultural layers and the C14 dates are not in stratigraphic order so there are some chronological problems. 15 phytolith types including phytoliths from *Setaria, Panicum* and *Oryza* were identified (Li *et al*, 2007b: 1709). The phytoliths were processed and analysed according to Lentfer and Boyd (1998). There seems to be little data for the phytolith material, only *Panicum, Setaria* and *Oryza* appear in the tables (Li *et al*, 2007b: 1711, 2007a 1375) and despite processing a large number of samples the count seems very small and is difficult to read.

Gong *et al*, (2007) use unspecified phytolith data gathered from published reports with other proxy data, such as, pollen, carbonised grains, and ancient paddy soils, to understand the temporal and spatial distribution of rice in ancient China. They conclude that evidence of temporal continuity of rice in Central China suggests a centre of origin from which rice spread in three waves to the rest of China and beyond to Korea and Japan. They also see a relationship between the past distribution of rice farming and environmental change. However, there are only five references to phytoliths and the role phytolith analysis plays in their interpretation is unclear. The archaeological records on which they found their conclusions are based on somewhat uneven sampling methods, little discussion of crops and none of crop weeds.

Ruddiman *et al* (2008), use a broader, but uncritical, rice database, including phytolith data, from 311 sites to evaluate the effects of human activity, in particular rice farming, on climate change. Although this database seems based entirely on rice remains rather than the more generalised proxy data used by Gong (2007), similar criticisms can be levelled. The archaeobotanical records on which they found their conclusions are not all systematically sampled, there are no tables to show exactly what those records were, the supplementary material gives a list of publications many of which are in Chinese and unavailable in the west so the data are difficult to assess. As with Gong there seems little concern with other crops or crop weeds. There also seems to be an assumption that all early rice farming was wet but it is possible that the earliest farmed rice especially north of the Yangtze may have been dry farmed.

1.5 HOW ARCHAEOBOTANICAL EVIDENCE FROM PHYTOLITHS CAN ADDRESS ISSUES OF SOCIO-ECONOMIC AND CULTURAL CHANGE

Major socio-economic cultural changes in the Yellow River Valley from the Yangshao to Longshan to Erlitou cultural periods have been identified and discussed (Chang, 1986, Underhill, 2002, Liu, 2004, Ma, 2005, Liu and Chen, 2003, 2006). Archaeological evidence points to the rise of social complexity and the emergence of a possible early state at Erlitou during this period (Liu, 2004, Liu and Chen, 2003, Shao, 2000, Chang, 1989, 1986, Pearson and Underhill, 1987, Keightley, 1983), beginning with the appearance of well-established, although still mostly egalitarian, agricultural communities during the Early Yangshao period. Evidence for increasing social differentiation during the Mid and Late Yangshao continuing and expanding into the Longshan can be inferred from the increasing settlement size, density and stratification within and between settlements (Ma, 2005, Liu, 2004). Additional evidence comes from mortuary practices, burial arrangements and grave goods (Underhill, 2002:45). Multiple factors affect social change; environmental and ecological conditions, changes in population, competition between groups, religious or ritual activities, specialisation, and regional interaction (Halstead and O'Shea, 1989, Liu, 2004:239). How do these elements manifest themselves in the archaeobotanical data? Economies are based on the procurement, processing, exchange and consumption of natural resources (Hastorf, 1988, Simmons, 1996, Fuller and Stevens, 2009: 2). Vegetative material acquired by gathering and cultivation has long provided humans with food, fuel and technology. Plant remains are the direct results of these activities so can be linked to economic systems. Data from archaeobotanical remains can be applied to economic and cultural questions (Hastorf, 1988: 119, Johannessen, 1988:145).

In this study I shall, in particular, argue that phytoliths tell us about variation in crop choices and the strategies people in the past may have used to adapt to developments in the agricultural economy. The phytolith assemblages studied here will be used to articulate discussion on changes in crop-processing patterns over time, differences in crop choice in different regions, and key changes especially from the Yangshao to Longshan period, including the relative significance of rice spatially and temporally.

One of the strengths of phytolith data is demonstrating broad patterning in plant remains. Comparing these patterns within samples from different time periods and sites should identify trends within the data that mirror changes in plant populations and human plant interactions (Johannessen, 1988:149). Variables over time, such as crop choices and who is producing, processing and consuming can indicate environmental and/or social change. These can then be correlated with evidence of changing agricultural economies and other cultural developments, such as escalating population, craft specialisation and regional land clearance.

The first place to look is at differences in the crop repertoire across the three cultural periods Yangshao, Longshan and Erlitou. Phytolith samples can demonstrate changes in plant input to site deposits, including evidence for new crops. Identified crop changes can then be queried for their relationship to social or environmental changes. We can also document differences in plant repertoire between contemporaneous sites in different sub-regions, specifically between Xipo, Baligang and Huizui in the Yangshao period, between Longshan period Baligang and Huizui, and between Huizui and Erlitou samples in the Bronze Age. I shall describe differences in phytolith assemblage composition between these sites and periods and use these as a basis for discussion of agricultural variability and change. I shall attempt to disentangle the extent to which crop choices relate to the position, size and type of settlement, or mainly to chronology and geographical position.

Such data can be used to assess existing hypotheses. For example, Ma proposes that changes in settlement strategies in the Yangshao are associated with alteration of natural resources and social landscape especially from after the Middle Yangshao (4000-3000 BC) (2005:26). Larger more integrated settlements may have been a response to changes in agricultural production due in part to the effects of land clearance on the local environment and partly an attempt to alleviate stress placed on resources by a burgeoning population. This might be supported by the decline in wild-gathered fruits, including probable wild soybeans, from Yangshao to Longshan periods documented in Ying Valley (Fuller and Zhang, 2007:946, figure 3). This may be paralleled in phytolith data by higher proportions of rice leaves and stems pointing to the adoption of new crops more suitable to a changed local environment and expansion of the agricultural economy. Also a decline in *Setaria italica* combined with an increase in *Panicum miliaceum* might suggest expansion to more marginal agricultural land.
Charred seed evidence suggests millets, in particular Foxtail, *Setaria italica,* and to a lesser extent Broomcorn, *Panicum miliaceum*, were the primary crops during the Yangshao in north-central China (Lee and Bestel 2007, 51, Underhill, 1997, Lu, 1999). According to archaeobotanical evidence for Middle Neolithic crop distribution *Panicum miliaceum* was more common in the western arid interior, while *Setaria italica* was predominant in wetter eastern areas (Chang 1986; Lee *et al.* 2007, 1088). *Panicum miliaceum* may be a useful insurance against drought as it is more tolerant of drier conditions, and has a shorter growing season, than *Setaria italica* or *Oryza sativa* (Jones, 2004, Underhill 1997, 117). The introduction of new crop plants, such as *Oryza sativa*, which needs a more consistent water supply, may provide insight into variation in local environment, both natural and human induced and human responses to the results of changing environmental conditions (Rosen, 2008:305). Rice cultivation is also labour-intensive and this probably implies social organization able to mobilize larger agricultural work forces (cf. Fuller and Qin 2009). However, the production of rice may also have been facilitated by environmental changes and the creation of alluvial floodplains (Rosen 2008: 305).

Evidence for shifts towards *Panicum* or rice, might be related to changes in demographic pressure and social competition. For example, Underhill (1997) proposes increasing competition over basic resources such as land and labour influenced the ability of people to produce a food surplus causing changes in the nature of interactions concerning food and status (2002:248). If prime land for traditional *Setaria* cultivation became scarce we might expect to see a shift towards increasing marginal land used for *Panicum miliaceum.* In addition growing social groups might be able to organize labour to create rice paddy fields in wetlands. Phytolith evidence for increasing rice and increasing *Panicum* relative to *Setaria* during the Longshan period suggest that both trends might have been present. The extent to which these different strategies were adopted on different sites or different kinds of sites will be explored.

CHANGING AGRICULTURAL STRATEGIES: INTENSIFICATION AND EXTENSIFICATION

The introduction of wet rice cultivation would have necessitated changes in labour organisation and mobilisation for initial construction and maintenance of rice paddies. This would also have involved extensification, expansion of land used for cultivation, and landesque capital intensification, for example, investment of labour in physical evidence of intensification in long term improvements to the land, such as paddies, terraces, dams and irrigation systems (Allen and Ballard 2001:158, Brookfield, 2001:184, Leach, 1999:315). Increased proportions of distinctive phytoliths might indicate the expansion of rice cultivation from the husks (double peaked glume cells) and leaves (scalloped edged keystones) as well as macro remains. The presence of different parts of rice crop weeds such as *Zizania* or Cyperaceae can also indicate cultivation and crop processing stages.

Agricultural intensification has been related to socio-political development (Leach, 2003:31, Brookfield, 1972:38). As Morrison (1996:583) states the underlying causes of intensification are under debate. Five elements seem predominant; demography (Boserup, 1965, Brown and Podolevsky, 1976, Cohen, 1977, Johnson and Earle, 1987 in Morrison, 1999: 583, Netting 1993), socially generated demands for produce (Bender, 1985, Leach 1999, Gilman, 1981, Sand 1999, Brookfield, 1972:38), market forces (Netting, 1993), risk buffering (Halstead, 1989) or another specific cause or causes (Morrison, 1994, 1996, Leach, 1999).

Boserup's (1965) model understands agricultural intensification as a unilineal path from long fallow swiddening to permanent multicropping (Morrison, 1994:137), as a of result of demographic growth and related increasing pressure on limited resources leading to innovation, adoption of new crops, technological advances and a higher labour input (Allen, 2001:204, Brookfield and Hart, 1977:332, Netting, 1974:36).

In contrast, others see multiple pathways (Kirch, 1994 in Leach, 1999). Morrison (1994, 1996) suggests intensification is made up of diverse productive strategies that can be separated into a number of components: intensification, specialisation, diversification and expansion at certain social and spatial scales. Agriculture is practised by social means (Stone, 2001:168,), food production of all kinds requires technology and the social ability to mobilise labour (Stone, 2001:169, Bender, 1985:53, Athens, 1999:323) pointing to central political and social organisation in the emergence of intensive agricultural economic systems (Sand, Comments on Morrison, 1999). This highlights the question of whether social coercion or population pressure are driving forces behind intensive systems (Bayliss-Smith, 1999: 323, Sand 1999:330) or as Brookfield (1972: 38) asserts, intensification is a result of production of goods produced for social purposes, such as prestation. All these factors almost certainly have a bearing on the issue and as Leach (1999:332) proposes, investigation would benefit from focussing case by case on environmental and social elements. Identifying intensification in the archaeological record could articulate social and cultural change (Schurr, 1995), but how achievable is this?

The introduction or expansion of non-indigenous crops, such as rice to the Yellow River Valley may well have meant developing new cultivation strategies. Rice and millet farming involve different agricultural practices, requiring differing degrees of capital (landesque) labour investment, and recurrent labour inputs. So the ability of particular communities to undertake this may have as much to do with social organization and changes therein, as to available land. Changes in weed flora and crop-processing remains will be explored to assess the extent to which new cultivated environments and mobilization of agricultural labour might correlate. Some preliminary indications from the macro-remains analysis in the Ying Valley suggest diversification between settlements in the labour organization inferred from crop-processing but also diversification in the arable environments inferred from weed flora with more wetland weed taxa and more marginal dryland/upland weeds. Similar patterns are found in the contrasts between Baligang and the sites to the north and west of the region, Huizui, Xipo and the survey sites.

Demographic growth has been established in the region through regional survey of settlement patterns (Liu *et al*, 2004). Since cereal crops should be visible in the phytolith record, it may be possible to discern diversification of production or risk buffering by adopting a new crop (Halstead and O'Shea, 1989:5). Expansion or modification of cultivated land possibly showing specialisation in the use of land resources and changes in technology, for example, the development of rice paddies, might point to intensification and expansion (Netting 1974:39). Intensification and a higher incidence of sugars and starches coupled with associated vitamin deficiencies in the diet risks declining nutritional standards which can be seen by examining skeletal remains and stable isotope signatures (Netting, 1974:38, Pechenkina *et al*, 2005:1176, Fuller, 2007: 393). Intensification can also lead to environmental degradation (Netting, 1974:38). This may be evident in the phytolith record and does show in the geomorphological record (Rosen, 2008, 305) and in the phytoliths from Neolithic paddy fields close to Huizui (Rosen *et al*, in preparation).

NON-CROP PHYTOLITHS: INPUTS FROM CROP-PROCESSING AND SOCIAL ORGANIZATION

Another issue to be addressed is the extent to which non-diagnostic phytoliths and those from non-crop plants correlate with the crops. In macro-remains archaeobotany, it is often inferred that most of the non-crop seeds reflect weed seeds, harvested with the crop and lost during crop-processing (e.g. M. Jones 1985; G. Jones 1987, Fuller and Stevens, 2009). The same may apply to non-crop phytolith remains in specific contexts such as ash middens. Harvey and Fuller (2005) argued that most non-crop phytoliths from on-site deposits could be interpreted as also deriving from crop-processing waste of weeds, and straw, as well as identifiable crop husk. They used a small dataset (10 samples) of predominantly midden material from two sites. I propose to test this model on a larger dataset (68 samples) from ash midden and laminated pit fill material from 11 sites and three cultural periods, Yangshao, Longshan and Erlitou.

Both Liu (2004:249-251) and Underhill (2002:5) highlight the dichotomy between corporate, shared power and network, individual personal power, or household versus centralised, hierarchies. Tracking varying levels of production and activities involved from harvest to consumption might suggest a distinction between household and centralised organisation (Stevens, 2003:61). Storage is central to redistribution and trading (Netting, 1974:44). Large bell shaped storage pits are ubiquitous in the region (Chang, 1986) and the remains of their contents may provide insights to organisation, for example how far along the processing path the crops are when stored, and from this infer information about relative quantities of labour needed to process and store the crop in that condition.

The dividing line between bulk and daily processing can suggest the level of organisation and labour mobilisation during the harvest period whether it is small scale e.g., a family unit, or points to large-scale mobilisation. (Stevens, 2003: 72). As harvest is a time of great labour demand (Stone *et al*, 1990 in Fuller and Stevens, 2009: 8) fully processed grain might suggest large groups mobilised to process the crop fully, whereas partially processed grain might point to smaller household use where processing is completed as the grain is needed (Fuller and Stevens: 2009: 8). Assemblages reflecting changing densities of parts of crop plants leaf/culm versus husk, crop husk versus crop weed, can be used to understand where crop processing was taking place and if it was occurring at a household level possibly within kinship groups, or whether it took place on a grander scale requiring labour mobilisation. This might suggest changes in scales of social organisation from household to more centralised systems and also a contrast in differentiation between settlements to within settlements. This may be achieved by comparing ratios of husk to leaf/stem and identifying the proportions of phytoliths from a particular plant part within samples (Harvey and Fuller, 2005: 742).

Another important issue is whether any stages are missing from the crop processing chain of activities. Waste from bulk processing may only be present, if at all, for a short period of time so would only show in high densities in particular samples, whereas waste from routine processing throughout the year is likely to show in broadly similar ratios across many samples (Stevens, 2003:64). The potential of such patterns are explored in the Henan assemblages studied here and used to relate phytolith patterns to those from macro-remains. Fuller and Zhang (2007:955) propose that regularly occurring processes, such as crop processing, can be expected to leave numerically dominant patterns in charred seed remains. During the Yangshao the crop-to-crop weed ratio suggests a consistent input of waste from primary winnowing pointing to uniform organisation of crop processing. And they find a shift from uniformity amongst Yangshao settlements of a large scale (semi-communal) processing pattern to a diversified Longshan pattern in which some communities continue to mobilize on a large scale, while others mobilize on a small, household scale, which Fuller and Stevens (2009) term a "focused" scale of labour mobilization. This diversification between settlements can be seen as part of the increase in social complexity, similar to the emergence of Chiefdoms in Iron Age Britain recognized by Fuller and Stevens (2009) and chiefdom level social organisation on the Chinese Neolithic Central Plains described by Liu (2004:189).

NON-CROP PHYTOLITHS: EVIDENCE FOR ENVIRONMENTAL CHANGE?

In addition, I have asked whether non-crop morphotypes might suggest anything about the broader, non-agricultural vegetation around sites. Palaeoecological studies have demonstrated considerable climatic and environmental change between the Yangshao and Erlitou periods (An *et al*. 2006; An *et al*. 2004; Dykoski *et al*. 2004; Herzschuh *et al,* 2004; Maher and Hu, 2006; Rosen 2007; Wu and Liu 2004). The climate gradually warmed and ameliorated to a peak during the Mid-Holocene Climatic Optimum, followed by an abrupt cooling and drying and a subsequent unstable period of climatic fluctuations. Therefore, we might expect phytoliths, from plants brought onto the site from non-agricultural vegetation, to reflect this, while also considering cultural selection. Unlike pollen, which is usually windblown, phytolith data typically represent *in situ* deposition so can illustrate the vegetation present in a specific place and accordingly can be good indicators of local environment, in particular from offsite natural sediments (Delhon *et al*, 2003:11). Exceptions are windblown phytoliths which often have identifiable pits, small indentations, on their surfaces and more rarely, phytoliths carried in by water, they are silt sized so can be carried some distance in rivers depending on the flow. Even if phytoliths are brought into a site by human or animal activity the flora that produced them is likely to be local.

Various indices (ratios) can be proposed to reflect broader patterns of vegetation change. Most classically, phytoliths can provide a morphological base to distinguish grass subfamilies (Bremond *et al*, 2004:215, Twiss, 1992, Fredlund and Tieszen, 1997). Ratios between these grass subfamilies can be used to infer broad patterns of vegetation. Phytolith analysis can be used to suggest patterns in grass subfamilies that often relate to whether species with a C3 or C4 photosynthetic pathway dominate grasslands (Iriate, 2006, Parker *et al* 2004, Fredlund and Tieszen, 1997). For example, Iriate (2006) suggests that a change from mostly C3, Pooid grasses, to predominantly C4, Panicoid grasses during the mid Holocene in Uruguay indicates a climatic change to substantially more arid conditions. Wetland plants such as Cyperaceae (Ollendorf, 1992) and *Phragmites* produce identifiable abundant phytoliths, which can be used to point to a damp habitat. The challenge, however, is to tease apart the changes in non-crop phytoliths that are due to changes in agricultural practices (changing weed flora or changing crop-processing) from broader climate-driven changes in vegetation. Thus, for example, the balance of C4 grasses to wetland species might reflect the relative inputs of rice and millet agriculture as much as any climatic change.

1.6 ORGANISATION

Chapter 2 will deal with issues surrounding economy and social organisation through the Yangshao, Longshan and Erlitou periods in the Yellow River Valley.

Chapter 3 is an introduction to the geography and palaeoenvironmental background to the project area. The subsequent Chapter 4 is an outline of crops and agricultural production in early north central China.

The next three chapters are background to this project. Chapter 5 is a detailed discussion of the sites and sampling methods used. Chapter 6 gives a background to phytolith research and the methodology used to bring the phytoliths from the field to analysis, including the reference collection and identification criteria. Chapter 7 is a discussion of analytical approaches used with the dataset collected for this project, including quantification, context related variation, crop-processing and comparative datasets.

The results and their implications are presented in the next group of Chapters; Chapter 8 summarizes general results, while Chapter 9 contains the results for specific issues such as crop changes and changing agricultural processes, leading to the discussion of how the results relate to the questions raised in the introduction and some general conclusions in Chapter 10.

CHAPTER 2

ISSUES SURROUNDING ECONOMY AND SOCIAL ORGANISATION DURING THE EMERGENCE OF THE FIRST POLITIES IN CENTRAL CHINA

2.1 INTRODUCTION

This chapter presents a brief description of approaches to archaeology in China over the past one hundred years. There follows an overview of the changing Neolithic and Early Bronze Age cultures in Henan, north central China; Peiligang, which immediately preceded the periods covered in this project, Yangshao, Longshan and Erlitou. Agricultural subsistence will be dealt with in the following chapter.

Following conventions in the majority of both English and Chinese publications on Chinese archaeology, I use BC when referring to dates. However, I have provided both BP and cal. BC dates for the major cultural phases (Figure 2.1)
I use Pinyin for the Chinese spellings.

2.2 PIONEERS

Chinese specialists such as, Li Ji, Liang Siyong, Feng Hanyi and Xia Nai, trained in modern methods in the United States and Britain, were the pioneers at the foundation of Chinese archaeology (von Falkenhausen, 1993: 841-2). Unlike other places with physical remains of ancient civilisations, archaeology in China did not become monopolised by foreign expeditions. However, western archaeologists also played a dominant role from the onset of scientific archaeological research in China at the beginning of the 20th Century (Shi, 2002: 55, Shelach, 2000: 367). The most notable of these is the Swedish mining engineer and geologist J. G. Andersson who, while working for the Chinese Geological survey, was the first to identify the Neolithic remains at Yangshao, Henan, in the Yellow River (Huang He) Valley, (Andersson, 1929, Chang, 1968:5, 198614-15, Shelach, 2000:366).

2.3 ARCHAEOLOGY SINCE 1949

After the establishment of the People's Republic of China in 1949 archaeology became state controlled and interpreted theoretically via Marx's historical materialism (Chang, 1986:18, Olsen, 1987 288). Until the late 1970's a primary goal was to explain the origins and trajectories of cultures found in early historical texts, thereby illustrating the indigenous emergence of Chinese civilisation in contrast to diffusionist ideas of western archaeologists, such as Andersson, focussing on the central and lower Yellow River Basin where these cultures, the traditional dynasties, Xia and Shang were believed to have evolved (Underwood, 2002:19-20, Shelach, 2000:366-7, von Falkenhausen, 1993: 839, Ho, 1977,Linduff *et al*, 2004:46, Chang1986:414). Chinese archaeology has been criticised in the past for an emphasis on typology and chronology (Olsen, 1987: 288). However, recent work would suggest this is not always the case today. The Origins of Chinese civilisation project, commissioned by the Chinese government in 2004, takes an interdisciplinary empirical approach to the investigation of the technological and economic development of early Chinese state level polities in the Central Chinese Plain. The emphasis is on archaeological sciences, palaeoecology, archaeobotany, zooarchaeology and staple isotope analysis (Jing and Campbell, 2009). Far more so than the earlier controversial Three Dynasties Chronology project, commissioned by the People's Republic of China in 1996 to establish a time frame for the three ancient kingdoms, Xia, Shang and Zhou, which focused on tying the historigraphic tradition to an archaeological chronology (Jing and Campbell, 2009, Lee, 2002). Other examples of a broader approach are the collaborative systematic survey undertaken in south eastern Shandong by Underhill *et al* (1998) to investigate settlement patterns and hierarchy in the Longshan period, and the Yilou River valley research project, an interdisciplinary Sino-international collaborative research project which aims to test theoretical proposals on the emergence of social complexity through regional survey, geoarchaeology, archaeobotany and lithics analysis in the Yilou region (Liu *et al*, 2005).

2.4 INTRODUCTION TO THE YANGSHAO, LONGSHAN, ERLITOU SEQUENCE

The geographic and cultural heart of this project is the central Yellow River Valley in Henan. This region saw a number of major cultural shifts throughout the Neolithic and Early Bronze Age (Figures 2.1 and 2.2). The focus here is the transition from the Middle Neolithic Yangshao to the Late Neolithic Longshan and then to the Early Bronze Age Erlitou in the central Yellow River Valley. The traditional view accepted by pre 1980's Chinese archaeologists regarded the Central Plains as the place where civilisation originated and from where cultural elements spread outwards (Underwood, 2002:19-20, Shelach, 2000:366-7, von Falkenhausen, 1993: 839, Chang, 1986, Liu and Chen, 2006:151).However, current research suggests prehistoric cultures emerged here separately from those that developed in other river valleys (Underhill, 2002:20, Barnes, 1999:100, Murrowchick, 1994, Chang, 1989). Numerous recent archaeological studies demonstrate social complexity emerged in areas previously considered peripheral and interpretations changed from mono-centric to multi-regional independent development and inter-regional interaction (Su and Yin, 1981, Chang 1986, 234-42, Linduff *et al*, 2002-2004, Shelach, 2000, 1999, 1994, Underhill and Habu, 2006:122) (Figure 2.1).

Figure 2.1 Early Chinese culture

<table>
<tr><th></th><th>Upper Yellow River</th><th colspan="4">Middle and Lower Yellow River and Huai River</th><th colspan="2">Middle and Lower Yangtze River</th></tr>
<tr><th>Date BC</th><th>Gansu</th><th>Shaanxi</th><th>Shanxi</th><th>Henan</th><th>Shandong</th><th>Hubei</th><th>Jiangsu</th></tr>
<tr><td>6500</td><td>Dadiwan I</td><td>Laoguanti</td><td></td><td>Peligang</td><td>Houli</td><td>Cengbeixi</td><td>Kuahuqiao</td></tr>
<tr><td>5000</td><td colspan="6"></td><td>Hemudu</td></tr>
<tr><td></td><td colspan="4">Yangshao</td><td>Beixin</td><td>Daxi</td><td>Majiabang</td></tr>
<tr><td>3000</td><td></td><td colspan="4">Early Longshan</td><td>Quijialing</td><td></td></tr>
<tr><td>2500</td><td>Majiayao</td><td colspan="3">Late Longshan</td><td>Dawenkou</td><td>Shijiahe</td><td>Liangzhu</td></tr>
<tr><td>2000</td><td>Qijia
Machang
Banshan</td><td colspan="2">Erlitou regional cultures</td><td>Erlitou
Xiaqiyuan</td><td>Yueshi</td><td colspan="2">Erlitou regional cultures</td></tr>
<tr><td>1100
1500</td><td>Regional Bronze cultures</td><td colspan="2">Shang and Proto - Zhou</td><td colspan="2">Shang</td><td colspan="2">Shang and regional Bronze cultures</td></tr>
</table>

After Liu, 2004: Table 1.1

The transition from the Neolithic to Bronze Age and the accompanying emergence of a state level society at Erlitou in the Yilou River Valley during the early second millennium BC has been thought to correspond with the rise of the Xia in ancient texts. Much archaeology in China is focused on finding relationships between archaeological remains and historical events, in particular the origins of the Xia dynasty (Liu and Chen, 2006:149, Lee 2002, Thorp, 1991, Underhill 2002:20, Linduff *et al*, 2002-2004:61, Lee 2002:23). However, there is no direct evidence from Erlitou linking it to the Xia (Liu and Chen, 2006:148, Lee, 2002:25). Liu and Xu argue it is time to change focus from texts to empirical data (2005:887).
In recent years focus has changed from the search for the Xia to identifying trajectories to early states and the development of social complexity (Liu, 2004:9).

MODELS FOR SOCIAL CHANGE

There are numerous models for social change in Neolithic and Early Bronze Age China. In the past there has been a division between western evolutionary views and those within and China that are based on classical Marxism (Linduff *et al*, 2004, Liu, 2004). Liu (2004:10-13) describes evolutionary approaches to emerging social complexity within China. The Western neo-evolutionary model (e.g. Service 1962) of bands, tribes, chiefdoms, and state was introduced to China in the 1980's but did not immediately gain widespread recognition (Liu, 2004:11). Up until twenty-five years ago theoretical thinking was dominated by Marxism, which resulted in a predominately unilinear view of social evolution. There was a general perspective of the progression of early societies from 'matrilocal/matrilineal/matriarchal clan organisation to patrilocal/ patrilineal/ patriarchal society' corresponding with the transformation from egalitarian to stratified society as a result of changes in the means of economic production leading to a divergence of craft from agriculture (Liu, 2004:10). According to Shi (1983 in Liu, 2004:10) this resulted in a clan-based class society founded on private ownership and exploitation. More recent research including osteoarchaeological and DNA evidence from Neolithic human remains from the Yangshao period does not support the proposed matrilocal burial pattern (Gao and Lee, 1993, Jilin University, 2001 in Liu 2004:11). In addition metal tools do not seemed to been produced for functional use in either the Longshan or early Bronze Age (Liu, 2004:11).

More recently Chinese archaeologists began to develop models based indigenous, historical traditions. For example, some recognize a phase of '*gucheng guguo guwenhua shidai',* the period of archaic towns, archaic states and archaic cultures and look for a link between cultural developments, such as jade manufacture, and the origins of civilisation (Su, 1988, 1997 in Liu, 2004:12). Until recent years there has been little distinction made between civilisation, as in cultural tradition, and the state, representing political organisation (Cowgill, 1988:256, Service 1975:8, Yoffee, 1991:15 in Liu, 204:12). Although stating it is better to define Chinese social evolution with indigenous terminology, Yan (1997 in Liu, 2004:13) turns to Earle's (1991:1) understanding of chiefdom as a regionally controlled polity with economic stratification and heritable social ranking to describe the trajectory to early states taking place during Longshan period (Liu, 2004:13). Liu (2004:13-15) proposes an anthropological approach to understanding the social processes leading to emergence of social complexity and development of complex society in China. Like Yan, she prefers the term chiefdom in the sense defined by Earle (1991:1), to describe a centrally organised polity with economic stratification and heritable ranking, externally but not internally specialised (Wright, 1977:381) in contrast to a state society which is also organised centrally, has a minimum two class system, ruling and commoner, but is both externally and internally specialised (Wright, 1977:383).

Linduff *et al* (2004:45) divide western models for the emergence of social complexity into two groups, one emphasising the effect of beneficial environmental conditions and technological development on population growth and surplus production, the other stressing how responses to conditions of risk and scarcity can lead to technological and organisational complexity (cf. Clarke and Blake, 1994, Spencer 1993, Drennan, 2000 in Linduff, 2004:45).Both types of model are applicable to early China, particularly in the contrasting trajectories to complexity between the Yellow River Valley and the Chifeng region in northern China (Linduff *et al*, 2004:45, 69).

DEFINING NEOLITHIC AND BRONZE AGE

In China the term Neolithic is used to refer to Holocene communities that subsist on farming and exhibit at least one specific cultural characteristic, for instance, particular ceramic styles, animal or crop husbandry, ground stone tools, or sedentism (Underhill and Habu, 2006:122). Here I use the terms Yangshao, Longshan, and Erlitou more to describe the temporal period or phase more as a cultural description although both apply.

The cultures discussed were originally, and to a large extent still are, defined by distinctive material culture assemblages, or *'leixing',* in particular ceramic types (Liu and Chen, 2006:3, Underhill, 2002:20, Barnes, 1999:98), typically Yangshao black on red polychrome pottery and Longshan grey ceramics or black eggshell, (Ma, 2005:1, Barnes, 1999:102).The Bronze Age is defined by bronze production and state level polities (Underhill, 1997:105).

SETTING

The geographical areas covered here will be discussed in detail in chapter 3. While the focus is on the central Yellow River Valley in Henan, neighbouring areas are also discussed. The central Yellow River Valley includes Eastern Gansu, Shaanxi and southern Shaanxi to the east, and Henan and southern Hebei to the west. The lower Yellow River valley covers Shandong and parts of Anhui and Jiangsu. Parts of northeast China and the Yangtze River valley will also be discussed.

Peligang	
ca.7000/6500/ 6000 – 5000 BC (9000/9500/8000 –7000 BP/ 5900 – 4500 cal BC)	
Yangshao	
4900 and 3000 BC (6900 – 5000 BP/ 4500 –2500 cal BC) (painted pottery)	
Early Yangshao	5000/4900 – 4000 BC (6900 –6000 BP)
Mid Yangshao	4000 – 3500 BC (6000 – 5500 BP)
Late Yangshao	3500 – 3000 BC (5500- 5000 BP)
Miaodigou (Xipo)	ca.4000 –3500 BC (6000 – 5500 BP)
Early Qujialing (Baligang)	ca 3500 BC (5500 BP)
Qujialing	ca 3500 – 3000 BC (5500-5000 BP)
Longshan	
ca. 3000 –2000 BC (5000- 4000 BP) (Black pottery)	
Early Longshan	ca. 3000 - 2500 BC (5000 – 4500 BP)
Miaodigou II	ca. 3000/2800 –2600/2500 BC (5000 – 4500 BP)
Shijiahe (Baligang)	ca. 3000-2500 BC (5000 – 4500 BP)
Late Longshan	ca. 2600/2500 – 2000 BC (4500 – 4000 BP)
Xinzhai	ca. 2000-1900BC (4000-3900BP)
Erlitou	
ca 1900 –1500 BC (3900 – 3500 BP) Phases based on ceramic styles	
Erlitou phase I	ca. 1900 –1800 BC (3900 – 3800 BP)
Erlitou phase II	ca. 1800 – 1700 BC (3800 – 3700 BP)
Erlitou phase III	ca. 1700 – 1600 BC (3700 – 3600 BP)
Erlitou phase IV	ca. 1600 –1500 BC (3600 –3500 BP)

Figure 2.2 Timeline of cultures considered in this study (Ma, 2005:2-3, Liu, 2004:16, Institute of Archaeology, 1999:392 in Liu and Chen, 2003:29, Xia-Shang-Zhou, 2000, An, 1991:137-9, Underhill, 2002)

2.5 EARLY NEOLITHIC – LATE PEILIGANG (6000 – 5000 B.C.)

Peiligang, (Xinzheng, Henan) one of the earliest Neolithic cultures with evidence of millet cultivation in the eastern middle Yellow River Valley (Henan and southern Hebei), developed at the beginning of the postglacial climatic optimum (Liu *et al*, 2004:81, Chang, 1986:87-95). Characterised by the domestication of millets (Underhill and Habu, 2006:126, Zhao, 2006, Yan, 1992, 1997:120,Ren, 1995 in Underhill, Barnes, 1999:93 First Henan Team, 1984,) rice, (Lu *et al*, 2004:81, Chang, 1986:87-95), the domestication of animals such as, dog, pig and chicken, well polished stone tools, for example, grinding stones, spades, sickles, and kiln fired pottery ranging from thin fragile orange paste to finely made black ware, the highest density of Peiligang sites is found in central Henan (Underhill and Habu, 2006:126, Chang, 1986:87-95, Henan Institute, 1999 in Liu *et al,* 2004:81-82, Flad *et al,* 2007:168). There appears to have been a mixed economy combining hunting, gathering and millet cultivation with some seasonal settlements (Underhill and Habu, 2006:126, Liu *et al* 2005:84). At the contemporary site of Cishan, (Wuan, Hebei) situated on a river terrace at the base of the Taiheng Mountains, there are pit buildings, burials and more than 80 storage pits containing millet remains, possibly *Setaria italica.* Bone adzes and polished stone axes as well as domesticated dog, the earliest evidence of domestic pig chicken bones and ox, alongside wild fauna suggest a fully-fledged village economy supplemented by hunting, fishing and gathering (Ren, 1995:41 in Underhill, 1997:121, Barnes, 1999:93, Yan, 1992:123, Yuan and Flad, 725). At Jiahu, (Wuyang, Henan) in the Huai river valley, the boundary between the cooler more arid north and warm humid south, house foundations, cellars, and kilns have been

excavated along with bone flutes (Zhang *et al*, 1999). There is evidence of rice cultivation from charred grain, phytoliths, tools (spades, reaphooks and knives) and carbon isotopes, as well as gathered plants such as wild beans and water chestnuts (*Trapa natans*). Faunal remains from pig, dog, deer, rabbit, crocodile and fish reflect the mixed village economy at Cishan (Hu *et al*, 2006:1320). Peiligang settlements also have cemeteries with burials, generally 75-100 of equal size and containing similar grave goods, pottery vessels and stone tools suggesting familial links to the land through generations (Underhill and Habu, 2006:126, Henan Team, 1995 in Underhill and Habu, 2006:126).
Contemporary early Neolithic cultures in the western Yellow River Valley with similar cultural features including dry farming are Laoguantai, Dadiwan and Lijiacun or Baijia in Gansu and eastern Shaanxi (Underhill, 1997:120). Lithic evidence from Dadiwan points to southward population migration from the Upper Yellow River region during periods of climate deterioration. Bettinger *et al* (2009:5) and Barton *et al* (2009:4) raise the possibility that intensive plant use patterns brought by northern hunter gatherers were the origins of early millet agriculture during Dadiwan I (8000-7000 BP).

In Shandong the Houli culture (ca.6200-5500BC) shares many similar characteristics but so far there is no evidence of millet agriculture (Underhill, 1997:122).

In northeast China sites, including some settlements surrounded by moats, from the Xinglonggua culture (6200- 5300 BC) have provided evidence of *Panicum miliaceum* (Xinglongguo, Xinle) and domesticated pig as well as acorns, hazelnuts, hawthorns, wild animal and fish remains and stone and bone hunting tools, pointing to a mixed economy (Underhill, 1997:122, An, 1988 a: 757).

2.6 YANGSHAO (4900-3000 B.C.)

CHARACTERISATION

The first Yangshao artefacts, hundreds of axes, knives and other well preserved objects, were discovered in 1920 by villagers in Yangshao, northwest Henan and bought by Andersson's collector Liu Ch'ang. In 1921 Andersson excavated at Yangshao unearthing the characteristic Yangshao black painted red pottery and stone tools (Andersson, 1923 in Chang, 1986:109, An, 1988a: 754, Ma, 2005:1). Numerous finds along the Huanghe River resulted in the theory of Western cultural diffusion; this was subsequently negated by new finds (An, 1988a: 754). Since then many hundreds of sites have been discovered and although Yangshao is no longer the type-site, due to problems with understanding the stratigraphy, the culture bears its name (Chang, 1986:109).

The culture is distributed over a vast geographical area and is described in regional phases that share common distinguishing cultural and material characteristics, such as architecture, tools, ceramics and social organisation (Chang, 1986:109).

This project focuses on the Yangshao culture in Henan. The Yangshao spanned nearly 2000 years (4900-3000 BC) and can be subdivided into temporal phases (Table 2.2).

DATING

With advances in radio carbon dating and stratigraphic excavation by the late 1970's and early 80's the Yangshao culture was dated to the seventh millennium BC and its local origins and path to development identified alongside numerous other Neolithic sites on the Central Plain and beyond, (von Falkenhausen, 1993, Shi, 1986 in Shelach, 2000:366, Chang, 1986: 87-88).

SETTLEMENT DISTRIBUTION

The core of the Yangshao culture was concentrated in the middle Yellow River valley in the same way as the earlier Peiligang with a wider distribution of sites in central Henan (Chang, 1986:107). The Yangshao culture thrived during the climatic optimum of the northern Chinese postglacial in the mid Holocene (See chapter 2, Shi *et al.* 1993, Chang, 1986:112), suggesting abundant vegetation and wild game as well as the availability of other natural river and mountain resources. The majority of Yangshao sites are the remains of densely distributed Neolithic villages on the lower loess terraces, often the second mesa, along three major rivers, the Yellow River, Weishui and Fenho and their tributaries (Chang, 1986:112, Zhang and Zhao, 2007:766).

The Yangshao is characterized by an amplification of population as seen by the steady increase in number of sites. Close to 800 Yangshao sites were found in Henan before the 1990's, in contrast to around 40 Peiligang settlements (Yang, 1991 in Liu *et al* 2005: 84, Chang, 1986:410). A correlation between demographic expansion and emerging social complexity has been widely discussed (Carniero, 1967).

Settlement distribution patterns provide an insight to the organisational complexity of a society. Liu (2004:159) reiterating, Carniero, (1981) and Earle, (1987, 1991) describes a complex society as characterised by central organisation of a regional population and following Drennan and Uribe (1987:60) illustrates the development of complex societies in the Central Plains by reconstruction of settlement patterns (Liu, 2004, Liu *et al*, 2005, Liu and Chen, 2003).

The early Yangshao people lived in villages that seem to have been repeatedly occupied by shifting settlements. Village deposits are usually thick and the remains of multi discontinuous occupations suggest repetitive shifting occupation episodes (Chang, 1986:114, Barnes, 1999:106). This may have resulted from a slash and burn based cultivation method (Underhill, 1997:126, Barnes,

1999:106, Chang, 1986). However, there is no direct evidence of this type of agricultural system and Barnes (1999:106) questions whether people would habitually abandon villages where they had invested enough labour to build substantial ditches, for example of 5metres width and depth as at Banpo.

The Middle and Late Yangshao saw the beginning of two-tiered and three tiered settlement hierarchies although communities were still comparatively egalitarian (Liu *et al*, 2004:85, Liu, 2004: 239, Ma, 2005:28). The large (20ha) site of Zhaocheng (Gan'gou, Henan) occupied during the mid and Late Yangshao shows the development of a two-tiered hierarchy. However, despite preservation of extensive cultural remains, there is no evidence of elite groups (Liu *et al*, 2005: 85).

In the early Yangshao site sizes seem uniform, ca. 2-5 ha with some exceptions, for example Xiaojiawan, (Lingbao, Henan) ca. 10 ha (Ma, 2005:17). By the mid and late Yangshao there is considerable variation in settlement size, for example, in Zhudingyuan, among the 20 mid Yangshao sites areas ranged from Beiyangping ca. 85 ha, to Xipo ca. 40 ha and Dongchang ca. 12 ha (Ma, 2005, 19-20). By the Late Yangshao in this area the number of sites had decreased to 9 and, apart from the 30 ha site at Qiaoying, the villages were small ranging from 8-4 ha (Ma, 2005:16).

In some areas the sites are distributed over an extensive area yet there are few significant changes in the typologies of the material culture over time (Chang, 1986:114). The remains of numerous storage pits and house foundations suggest that by the Mid Yangshao a sedentary lifestyle had been adopted with year round settlement occupation (Ma, 2005:100).

Ma (2005:24) notes that site distribution and arable land availability are closely linked; there is a close relationship between the location of sites and soil quality (Adler, 1996, de Montmollin, 1989). Although he also states social factors could have been just as important (2005:100).

In Zhudingyuan, a loess tableland, 75km by 5km, between the Sha and Yanping rivers in central Lingbao 11 early Yangshao sites were located on lowlands at or below an elevation of 410m, only 2 above at 440m and 445m. In contrast, of the middle Yangshao sites 11 sites were higher than 410m including the settlement at Wupozhai at 655m and 7 sites below. In the Late Yangshao, 6 of the 8 sites were higher than 410m Ma, 2005:16, table 2.1). Today the lower reaches of the Sha and Yanping river valleys are unsuitable for farming (Ma, 2005:24). Of the 18 multi component sites in the area 10 were occupied in the early Yangshao, (elevations ranging from 355-440m). Ma (2005:25) suggests the sites were chosen for their proximity to fertile agricultural land. The settlements were regularly spaced suggesting sufficient resources were available to support each village. This distribution pattern changed in the mid Yangshao with new small sites in high or remote locations. Ma (2005:25) suggests the larger sites occupied the more productive areas, while the infertile mountain sites may have been specialist sites for stone tool production or hunting. If this is the case the site distribution pattern may demonstrate early social differentiation between groups. Evidence from isotopic analysis of faunal remains points to increased land clearance in Zhudingyuan in order to farm millets.

The Late Yangshao saw a sharp decrease in settlement density and size suggesting intensive exploitation and depletion of natural resources during the previous period followed by abandonment (Ma, 2005:25).

SETTLEMENT ORGANISATION

Settlements often consist of three distinct areas, dwelling, cemetery and kilns. These villages are typically oval and around 50,000 – 60,000 square metres in area and in many, for example, Banpo, Jiangzhai (Lintong, Shaanxi) and Beishouling, (Baoji, Shaanxi); large circular ditches surround the dwelling areas, storage pits and animal pens with kilns and cemeteries positioned outside (Flad *et al*, 2007: 184, Underhill and Habu, 2006:129, Chang, 1986:116, Barnes, 1999:104-6).

Early Yangshao houses tended to be semi-subterranean both round and square and constructed from perishable materials, wattle and daub and thatch. Mid and Late Yangshao saw an increase in ground level houses (Underhill and Habu, 2006:129). A variety of house types and size can be distinguished including large structures, which may represent communal public buildings (Pearson and Underhill, 1987: 812, Ma, 2005: 46, Liu, 2004:239). Houses commonly seemed to house single families (Underhill, 2002:142) and tend to be arranged in groups facing a central plaza, for instance at Jiangzhai (Barnes, 1999:104), possibly suggesting social organisation according to membership in unilinear groups or kin based clans. The remains of late Yangshao longhouses, for example at Baligang, (Jia, 1996 in Underhill, 2002:144) point to corporate organisation (Blanton *et al,* 1996, in Underhill, 2002:142). The large buildings found at Yangshao sites appear to be public centres rather than private houses so do not illustrate a distinction between elite and non-elite (Liu, 2004:241).
A rammed earth wall surrounded Xishan, (Zhengzhou Henan), (National Bureau 1999 in Underhill and Habu, 2006:132) a feature that became more common in the Longshan.

SUBSISTENCE

Local environment may have determined the position of Yangshao settlements, which are usually sited along the banks and river terraces of the valleys of major rivers and their tributaries, overlooking the fertile alluvial floodplain (Chang, 1986:112, Barnes, 1999:103, Ma, 2005:100). These lower terraces would have been rich in exploitable floral and faunal resources. The rivers themselves and the mountains beyond would have meant ready provision of supplementary resources to agricultural subsistence as demonstrated by the faunal remains ranging from Chinese bamboo rat, badger, elaphure, deer, turtles, wild horse, and rhinoceros to fish, snails and molluscs. Gathered fruit and nuts; chestnut, hazelnut, pine and Chinese hackberry, have been found at Banpo (Xian, Shaanxi) highlighting the importance of uncultivated supplementary foodstuffs during the early Yangshao (Chang, 1986:112) and suggesting a broad spectrum diet.

Cultivated food also played an important role. Stable isotope analyses of human remains suggest millets made up over half the subsistence diet during the Yangshao period (Cai and Qu, 1984: 949 in Pechenkina *et al*, 2002:16).

There were established agricultural economies (see Chapter 4) millets, domesticated dogs, pigs and chickens in north and central Henan and rice and possible water buffalo in the south and west (Underhill, 2002:39, 1997:128). Part of a silkworm cocoon has been reported at Xiyin, (Xiaxian, Shaanxi) raising the possibility of silkworm husbandry in the Late Yangshao or Early Longshan (Chang, 1986:113). Bovid bones, possibly water buffalo, have been reported at Xiawanggang, (Xichuan, Henan) (Ren, 1995:42 in Underhill 1997:128). The evidence for water buffalo is weak as the now extinct aurochsen and wild cattle were common and as yet there is no distinguishing evidence suggesting the bovid remains belong to a particular species (Fuller pers. comm., Yang, 2008:2783).

Pigs, the principal domesticated animal during the Neolithic, were important for dietary and ritual purposes. Kim, (1994:120) relates control over intensive pig production to the emergence of political hegemony during the middle Neolithic in the Lower Yellow River Valley. Pig husbandry is closely linked to agriculture. Pigs do not graze, although they do forage if woodland is available, but can be fed on the by-products of crop cultivation so pig domestication and cereal surplus are likely to be connected (Yuan and Flad, 2002: 726, Lee, 1994:135). Pig and *Setaria italica* remains dated to ca. 2000BC have been found together at Taosi, (Xiangfen, Shaanxi). The carbon isotope evidence from these bones and also from pig bones from the same period at Xipo (Lingbao, Henan) suggests pigs were being fed millet grain (Cai *et al*, 1984 in Yuan and Flad, 2002:726, Pechenkina *et al*, 2005)

Subsistence practices concerning cereal crops will be discussed in more detail in chapter 4.

MORTUARY PRACTICES

A variety of burial customs have been encountered, some single, others multiple, some primary and others secondary. Children were buried in urns among the dwellings in some villages but at others in the cemetery. All adults were buried in rectangular earthen pits, usually with head west, frequently flat and face-up, although on occasion flexed. Graves were furnished with pottery, bone and stone tools and ornaments. While Chang (1986:119) believes individual graves frequently stand out as especially richly furnished, Underhill (2002:248) considers little effort is made to show individual status during the Late Yangshao. Social organisation reflecting the kinship groups described above is suggested by the layouts of the cemeteries such as that at Yuanjunmiao and Dawenkou in Shandong (Zhang, 1982:21, Chang, 1986:117-119, Barnes, 1999:106, Kim, 1994:124).

Ranking can be seen in burials. Grave goods, such as pottery drinking vessels, stone and bone tools and pig mandibles, include items made from prestige materials such as jade, turquoise and elephant ivory (Fung, 2002:75, Kim, 1994:123 Underhill, 2002, 111, 116).

FEASTING

Social and ideological functions of food became important during the Yangshao period, for example dog and pig burials might represent food consumed by mourners or offerings to the dead. The practice of placing pig mandibles in graves was widespread throughout the Yangshao, Longshan and Erlitou (Underhill, 1997:147, 2002:1). Underhill (2002:4) outlines how gifts of food can enhance hierarchical or integrative relations at intra and intercommunity levels and are a key element in mortuary ritual, provisioning the deceased with food and drink for the next world and providing other mourners with food and fermented beverages (McGovern *et al*, 2005:249). Feasting is an important element of social life for initiating and enhancing social relations (Hayden, 2001, Dietler and Hayden, 2001). Feasts can be integrative, exclusionary or competitive and used to demonstrate social stratification based on rights over food surpluses, labour for food production and rights of access to prestigious foods and containers.

At Xipo, where there is evidence of feasting, isotope analysis indicates pigs were being fed millet grain rather than just the crop processing residues, discarded human food and waste, highlighting their potential value for meat (Pechenkina *et al*, 2005:1185, Ma, 2006, Jing and Campbell, 2009:101).

Fermented beverages prepared from rice, honey and a fruit have been suggested by evidence from residues in jars and basins at the early Neolithic village of Jiahu (McGovern, *et al*, 2004).

ECONOMIC ACTIVITIES

There was already a range of economic activities during the early Yangshao, including evidence of spinning from ceramic spindle whorls (Underhill and Habu, 2006:128) and fibre and oil from hemp (Chang, 1986:113-114). Some early Yangshao sites show clusters of kilns or animal pens close to the houses possibly suggesting household specialisation (Yan, 1999:136-137). Two kinds of pottery were produced, a course grey earthenware and fine- grained ware coated in a kaolin based white slip and decorated with characteristic red and black. There is evidence of fast wheel manufacture on sherds from Houzhuangwang and Yiquanma (Yudong, Henan) and the colour variation suggests a range of firing temperatures (Wang and Andrews, 2002: 241-246).

Regional interaction was made possible by links along and between river systems, the circulation of goods, the concentration of wealth leading to production of surplus, but also the threat of external violence. Competition for arable land may also have been a source of conflict (Ma, 2005:24). The exchange of unique resources and scarce goods may have been a motive for expansion (Chang, 1986: 244, 1986:410).

2.7 LONGSHAN (3000-1900 B.C.)

Like the Yangshao, the Longshan culture describes groups of cultures sharing common features. The Longshan culture, as defined by the thin bodied, lustrous black ceramic assemblage at Longshan in Licheng, (Chenziya, Zhangqiu) Shandong (Wu, 1930, in Liu, 2004:1, An 1988a:754), originated from different cultural settings, Dawenkou in Shandong and Yangshao via Miaodigou II in Henan (Liu, 2004:2). Sites have been found in Henan, Hebei, Shanxi, Shaanxi and Shandong. Because of this regional diversity the Longshan is more usually described as a cultural period than a single culture (Yan, 1981 in Underhill, 1992:173. At first Longshan was considered a contemporary eastern version of Yangshao but this was disproved at Hougang (Anyang, Henan) where a stratigraphic sequence of Yangshao, Longshan and Shang was discovered in 1931 (An, 1988a: 754). There is a question of exogamous influence from the south as seen by ceramic styles (Huber 1981:3).

CHARACTERISATION

There are several cultural divisions within the Longshan. This study looks at samples from the Early Longshan, Shijiahe, defined by distinctive jade and painted spindle whorls, succeeding the Qujialing at Baligang (from the Middle Yangtze) and the Late Longshan (Table 2.1). During this period the Neolithic cultures of the region become increasingly complex, shifting from mainly egalitarian towards more stratified societies (Chang, 1986:234, Liu, 1996a: 243, Barnes, 1999:117).There are distinctive cultural traits characteristic of the increasing cultural complexity of the Longshan (Liu, 2004:1):increase in population density and settlement size (Gong and Jiang, 1993 in Underhill, 2002: 32 Liu *et al,* 2005:85), early writing systems (Liu, 2004:1, Chang, 1999:64-65, Postgate *et al*, 1995:647-648), distinctive forms of ceramic and jade production, exotic trade and exchange items such as ivories and turtle shells (Barnes, 1999:111), small copper and bronze implements and ornaments (Liu, 2004:1, Linduff *et al*, 2000), construction of town walls, widespread violence and warfare (Liu, 2000:1, 2000, Underhill, 1992:174,1994), a heavier dependence on domestic animals, introduction of sheep and cattle(Barnes, 1999:113), organisation of cemeteries and graves suggest increasing social integration and stratified clans (Li, 2004:1, Liu, 1996a, Pearson, 1981, Underhill, 2000, Barnes, 1999:112-113), and increased interaction between more widely distributed regional cultures (Li, 2004:1, Chang, 1986:234). Regional cultures were increasingly extensively distributed and experienced more intensive contact, for instance, Dawenkou and Qujialing type ceramics and cultural characteristics, such as tooth extraction, in sites in western Henan during the late Yangshao and early Longshan periods (Du, 1992, Sun, 2000, Liu *et al*, 2005: 85), at Huachenghe, Gan'gou river, an early Longshan burial contained Dawenkou type ceramic vessels. The presence of these cultural traits may also indicate large-scale immigration from southeast regions into Henan (Liu *et al*, 2005:85, 89).

SETTLEMENT DISTRIBUTION

There was a rapid decline in population from the Late Yangshao (Ma, 2005:28, Liu 2004:241) marked by a decrease in the number of sites followed by a clear increase in site number and size throughout the Longshan (Liu *et al*, 2005:86, Gong and Jiang in Underhill, 2002:32). By the third millennium BC the Longshan culture had spread throughout hundreds of sites the Lower and Middle Yellow River Valley (Underhill, 2002:31).Catchment areas expanded from a few hundred km^2 to $3000km^2$ in the Taosi area during the Longshan (Liu, 2004:240)

As in the Yangshao period, Longshan sites seem to cluster in areas of arable land close to rivers (Liu and Chen, 2006:154), for example, Nanshi, sited at the confluence of the Yilou and Wulou Rivers and Luokou NE, situated in a large catchment of alluvial plains in the Wulou valley (Liu *et al*, 2005:86).There is evidence of increasing variability and differing regional settlement patterns (Liu, 2004:241). Liu and Chen (2006:154) have identified two predominant distribution patterns in the Middle and Lower Yellow River valley. The first is mono centred; each group of settlements has one large centre, for example the walled site of Taosi and surrounding settlements in the environmentally circumscribed Linfen Basin, Shanxi (Liu and Chen, 2006:154).

The other multi centred, where evenly spread medium sized (>100ha) sites co-exist (Zhao, 2001:142-144 in Liu and Chen, 2006:157). These are generally in less circumscribed situations. This group can be further divided into linear multi-centred and scattered multi-centred. Linear usually consist of 2 to 3 tiered settlement hierarchies in clusters of regularly spaced walled centres (ca 40km apart) found in the semi circumscribed regions of north and west Shandong (Liu and Chen, 2006:157). The scattered multi centred systems in North and Central Henan are also regularly spaced but less clustered. The settlement hierarchies are largely 2 tiered. Some sites are walled, Haojiatai, Wanchenggang, Pingliangtai in the

Ying Valley, and there is evidence of warfare (Liu and Chen, 2006:157, Liu, 1996:249).

SUBSISTENCE

The early Longshan period was marked by rapid climatic fluctuations (see Chapter 3). The climate gradually dried and cooled, causing less favourable conditions for agriculture. However, this was combined with population expansion, so may have lead to greater land use for agriculture, especially rice and millet cultivation (Rosen, 2008: 11, Liu *et al*, 2004:86, Pechenkina *et al*, 2002:16). For example, Luokou NE may have expanded to a large settlement due to the availability of cultivatable land (Liu *et al*, 2004:86).Access to wild foods is likely to have decreased as people aggregated to larger settlements (Pechenkina *et al*, 2002:16), for example in the Ying Valley fruit was present in the Yangshao samples but not in the early Longshan (Fuller and Zhang, 2007:946 Figure 3), in turn leading to a greater reliance on domestic animals (Liu, 1996, Underhill, 1997).An increase in bovid bones has been found at Jiangou, southern Hebei and Keshengzhuang in central Shaanxi (Ren, 1995:42 in Underhill, 1997:128). It is possible cattle, possibly water buffalo, could have been used for paddy preparation but there is no direct evidence of this (Underhill, 1997:128). The earliest finding of domestic goat is from the early Longshan Miaodigou II site (Ho, 1977:474 in Underhill, 1997:128). Sheep (*Ovis* sp.) and goat (*Capra* sp.) have been identified at Kanjia (Flad *et al*, 2007:193).

At Wanchenggang, Xinzhai and Taosi pigs were still the predominant domesticate but both cattle (*Bos* sp.) and sheep appear in the faunal assemblages. The arrival of the non-native sheep suggests East-West contact (Jing and Campbell, 2007:99).

SOCIAL STRUCTURE

Settlement patterns were marked by the emergence of chiefdom level social organisation with many small competing polities coexisting in the Yellow River Valley (Liu and Chen, 2003, Liu, 1996, Underhill 1994). Settlement hierarchy could be noted in Western and Central Henan with two or three tiered systems in Late Longshan (Liu and Chen, 2003:29, Shao, 2000: 197, Liu 1996a).

Rammed earth town walls appeared during the Late Yangshao and developed during the Longshan (Underhill *et al*, 2002:745, Barnes, 1999:116). These would have required considerable labour mobilisation. Collective labour forces from many communities would have been needed suggesting elite groups had the ability to organize the population from a considerable area, indicating a chiefdom level polity (Liu, 2004:115). Building walls has been understood as a response to increased violence and warfare (Liu, L. 2000a: 268-9, Underhill, 1994, Shao, 2004: 197, Liu: 2004:115). Underhill (2002:32) highlights the debate as to whether these walled sites represent political centres in chiefdoms (Underhill, 2002, 1994, Liu, 1996b), or city-states (Dematte, 1999).

SETTLEMENT ORGANISATION

Residential areas in Late Longshan settlements generally consist of elite houses built on large earthen mounds or *hangtu* platforms built using adobe and with lime plaster walls and floors (Underhill and Habu, 2006:133, Wiesheu, 1997:100, An 1988: 758) and simpler non elite houses, refuse pits, ritual pits, walled enclosures surrounded by ditches (Crawford, 2005:309, Chang, 1986:248).Houses at sites such as Hougang and Baiyang in Henan were predominantly round and built with mud, wattle and daub, and adobe. Some floors and walls were covered in lime plaster and there is evidence of house decoration in the geometric patterns cut into the plaster at Taosi (An, 1988a: 757). Underhill, (2002:194) suggests a greater range of differences in range of size across two earlier phases than later could indicate greater social competition earlier in the Longshan, particularly as this is also suggested by the ceramics. But, a household could have used more than one structure so it may be better to compare compounds or look at the building materials. Adobe is a relatively expensive and time consuming material to work with and is also associated with larger floor areas in the Mid Longshan at Hougang although at Baiyang there is only one small adobe house and from the Late Phase. Larger sites tend to have more variation in both size and building materials (Underhill, 2002:195). At Pingliantai two adobe brick houses guard the southern gate in the wall (Barnes, 1999:116, Underhill, 1992:175). At Kanjia (Lintong, Shaanxi) there are rows of connected houses (Liu, 2004:49) and large palace like structures mark the beginning of residential segregation at Taosi during the Late Longshan (2500-2000BC) (Liu, 2004, 109-110).

MORTUARY PRACTICES

Longshan cemeteries provide clues to a changing social order. Variables in burials suggest social hierarchies (Liu, 2004:120, Fung 2000; Underhill, 2000, Liu, 1996a: 243, Liu, 1996b, Chang, 1986:249, Pearson 1981). Liu (2004:120) describes three classes of material remains; grave size and structure, quantity and quality of grave furnishings, the occurrence of prestige grave goods, such as jades, high quality pottery, musical instruments and pig skulls or mandibles.For example, the burials at Chengzi (Shandong) can be separated into four rankings according to the grave furnishings. There are two parts of the cemetery occupied by kinship groups of different rankings (Liu, 2004:144, Liu, 1999:603). Since grave size indicates energy expended in labour, large graves demonstrate high status. Liu (2004:244) makes the point that the evidence from burials gives a clearer indication of different social groupings than the evidence from the houses. During the Longshan there was a marked decrease in burial sizes and grave goods, such as pig skulls and long distance prestige items (Kim *et al*, 1994:133). This may represent a change in economic systems from wealth finance; horizontal exchange associated with a prestige goods based economy to staple finance and tributary systems possibly founded on subsistence intensification (Kim *et al*, 1994: 133).

FEASTING

There is evidence for feasting and social display with a variety of domestic animals in the Longshan Kanjia in Shaanxi (Liu, 1995:24). Underhill (2002) proposes that increased feasting activities produced demand for prestige food, drink and vessels. High numbers of the '*gui'* tripod vessel, associated with heating fermented beverages, have been recovered from Longshan contexts at Liangchengzhen (Shandong), and tests on cups from this site have demonstrated they contained fermented beverages made from rice, honey and fruit, but perhaps surprisingly not millet (McGovern *et al*, 2005: 256,260, 262).

CRAFT PRODUCTION AND TRADE

During the Longshan a higher percentage and variety of prestige and utilitarian ceramics were being produced by the fast wheel and hard fired at higher temperatures indicating increased specialisation (Underwood, 2002:34, Chang, 1986:251). However, there is little evidence for change in the mode of production over time. Most pottery seems to have been produced in what Underwood defines as 'complex household industry', where ceramic production undertaken in the home by males and is a primary source of income, rather than in specialised workshops (1999:23), suggesting possible labour organisation as a part time specialisation.

During the late Longshan new technologies involving metallurgy were developed such as, bronze casting, for example, at Wanchenggang and Meishan (Underhill, 1992:176), the manufacture of elite drinking vessels for ritual purposes (Barnes, 1999:118, Chang, 1986:287, Underhill, 1992:176) and the use of copper and bronze for small tools and ornaments (An, 1988 a:758, Linduff *et al* 2000).

Some settlements became specialised craft centres associated with the manufacture of specific items, for example spade production at Huizui (Liu *et al*, 2007:96). The evidence for production or circulation of prestige goods in Henan at this time is poor (Liu, 2006:161). However, in contrast, elite goods, jade and fine ceramics were mainly found outside the central Yellow River Valley in Shandong and southern Shanxi although some items with characteristic designs were widespread suggesting interregional interaction (Liu, 2003, Liu and Chen, 2006:159, Liu, 1996b, Shao, 2000:199).

Specialised craft production was often based on a local resource so elite control over resources and production was within a limited range, for example half a day's journey. (Liu, 2004:245).

The remains of possible stone workshops and lithic evidence consisting of stone tools, blanks, semi finished and finished tools made from 15 types of material from Huizui (Yanshi, Henan), a multicomponent site, suggests regional specialisation in stone tool manufacture during the Longshan and Erlitou periods (Henan 1st team, 2003, 2005 in Chen, 2007).

Another innovation was the development of possible writing systems (Chang 1999:64, Postgate *et al* 1995: 467-468, Shao, 2000: 214)

SOCIAL STRATIFICATION

Social stratification became more apparent throughout the Longshan. All the aspects described above, in particular, the variation between and within settlements, and mortuary practices demonstrate a more integrated society. Importantly, the differential access to resources, both agricultural (Fried, 1976 in Barnes, 1999:117, Liu, 1993:117) and to the procurement of raw materials for bronze ritual drinking vessels (Liu and Chen, 2003, Barnes, 1999:117), allowed dominant social groups to emerge (Underhill, 2002: 204, Barnes, 1999:118).

2.8 ERLITOU (1900 –1500 B.C.)

There is a gap in the archaeological record of around 100 years between the end of the Longshan and beginning of Erlitou (ca. 2000-1900BC), which Zhao has identified as the transitional Xinzhai phase (2002 in Liu and Chen, 2006:168). The Erlitou period can be divided into four phases each lasting around 100 years (Kaogu Yanjiusuo, 1999: 1 in Lee, 2004:177, Institute of Archaeology, 1999:392 in Liu and Chen, 2003:29).

CHARACTERISATION

From phases II – III (1800-1600 BC) rapid population growth and expansion of material culture to neighbouring regions, increased stratification within the social system, elaborate burials often contained high status prestige goods fabricated from bronze, jade and white kaolin ceramics, specialisation and craft production, production monopoly of bronze prestige items by a ruling class, are all characteristics of the Erlitou period (Erlitou Team, 1984, 1992, Institute of Archaeology, 1999 in Liu *et al* 2005:74).

Characteristics at the Erlitou site itself are; considerable expansion for phase II, the construction of palatial structures, state controlled bronze production and a highly stratified society (Liu, 1996:275). The bronze *jue* (wine vessel) and *jia* (tripod with hollow legs) appeared (An, 1988 a, 758). Social polarisation is illustrated by differentiation in burial practices, (Liu and Chen, 2006:166).

DATING

The Erlitou cultural period existed from 1900 –1500BC based on calibrated radiocarbon dates from 38 Erlitou sites in Henan (Institute of Archaeology, 1991 in Liu and Chen, 2003:29) and the 'Xia-Shang-Zhou Chronology Project' (Xia Shang Zhou, 2000 in Liu and Chen, 2003:29). There are four consecutive phases of the Erlitou culture based on changes in distinctive ceramic styles, each lasting around 100 years (Kaogu Yanjiusuo, 1999: 1 in Lee, 2004:177, Institute of Archaeology, 1999:392 in Liu and Chen, 2003:29). The Erlitou culture is centred on the type-site Erlitou, in the Yilou River basin near the modern city of Louyang. The site was occupied on a small scale during the middle Yangshao (ca. 4000-3500BC) and early Longshan (ca. 3000-2500 BC). After a 500-year hiatus at the end of the Longshan, Erlitou became a large urban centre during Erlitou Phases II and III, declining in Phase IV to a small village in the Erligang (ca. 1500-1400) (Lee *et al*, 2007:1087, Liu and Chen, 2003:58).

SETTLEMENT DISTRIBUTION AND SETTING

Sites containing Erlitou cultural assemblages are spread over a wide region encompassing Henan, southern Shanxi, eastern Shanxi and Hubei. However, there is some question as to whether the site distribution indicates the extent of Erlitou political sphere (Liu and Chen 2006:163). Over 120 Erlitou sites had been identified by 1996 (Zhao, 1987, Kaoguxue Jikan, 1989, NBCR, 1991 in Liu, 1996:274).

The core of the Erlitou culture is the semi circumscribed alluvial plain of Yilou River Basin in Henan, rich agricultural land but with few non- agricultural resources (Liu and Chen, 2006:166, Lee, 2003:173). Assemblages of Erlitou material culture have been found in peripheral regions rich in timber, kaolin, metals and salt. Liu and Chen suggest this is as a result of state managed population movement in order to secure these resources (2003, 2006:166).

Sites are densely distributed throughout the region. The Yilou valley research project identified a total of 53 Erlitou sites in the Yilou River Basin (Liu *et al*, 2005: Table 2).

Overall settlement patterns in the Yilou valley during the Erlitou phase became more complex and showed a clear hierarchy (Lee, 2004:190, Liu and Chen, 2006:164). There is evidence of political centralisation with the primary urban centre at Erlitou. Settlement nucleation can be seen in the Yilou region (Liu and Chen, 2003:33) and a four tiered settlement hierarchy in the Yilou River Valley (Liu *et al* 2005: 95). Three medium sized settlements of 20 – 60 ha within 25km of Erlitou have been excavated in detail: Shouchai in Gongyi, Nanzhai in Yichaun, and Huizui in Yanshi. All are located on rivers with access to resources in mountains (Liu and Chen, 2003 chap 4) illustrating a connection between demand for raw materials and the location of large urban centres (Chang, 1976, 1985 in Liu and Chen, 2003:14). According to Liu and Chen (2003:133) settlement patterns and their links to control and transport of natural recourses such as those associated with processing bronze prestige items, copper and salt, are key to territorial expansion during the Erlitou. Resources such as metal and salt were only found more than 200km from Erlitou (Liu and Chen 2001, 2003, Liu *et al* 2004: 79). Settlement patterns throughout the core and periphery show key central settlements in areas of fertile land with easy access to water for transport to the periphery. Erlitou itself is situated on an alluvial floodplain on the banks of the Luo River (Lee, 2004: fig1). Regional centres were placed on transport channels (Liu and Chen, 2003:132).

SUBSISTENCE

The Erlitou period agricultural economy was based predominantly on the staples millets and rice, (Chen *et al* 2003 in Lee, 2004, 291).There is also evidence for wheat (*Triticum* sp.) at Huizui as well as soybeans (*Glycine max*) and fruit, plums and brambles (Lee and Bestel, 2007:52).

Isotope analyses of human and pig bone from Taosi, Xinzhai and Erlitou demonstrate a predominance of C_4 plants in the diet pointing to millet-based economies in both Longshan and Erlitou Periods. However, while results from the northernmost site Taosi suggested only cereals with a C4 pathway were consumed, human remains from both Xinzhai and Erlitou contained evidence of consumption of small percentages of C3 plants, probably rice. Of the 22 human samples from Erlitou two individuals had consumed a diet consisting of principally C3 food (Zhang *et al*, 2007, Wu *et al* 2007, in Jing and Campbell, 2009:101).

SOCIAL STRUCTURE

Erlitou has been at the centre of debate on the development of early state level polities (see above). Many scholars, particularly those educated in China, believe the archaeological Erlitou culture is the textual Xia dynasty (Chang 1999: 71-73, Du, 1991, in Liu and Chen, 2006:163, Zhao, 1987 in Liu and Chen, 2006:163). Others both in China and the West consider Erlitou as a state but are not convinced by the reliability of the historical connection between Erlitou and the Xia presented by the textual information (Bagley, 1999: 130-131, Liu and Chen, 2006:163).

Liu and Chen (2003:131) consider Erlitou the earliest state in China. Once developed, it expanded rapidly in all directions and displayed many of the characteristics of a territorial state as defined by Trigger, (1993, 10-11, 1999), a political entity with control of a large area through a hierarchy of local administrative centres, a two tiered economy where food in urban centres is procured from local communities (Liu and Chen, 2006:19). The settlement data from the Yilou River Valley survey demonstrates a correlation between 'political and economic integration, between urban and rural areas, the appearance of an early state, and demographic expansion during Erlitou II'. By the peak of Erlitou, phase III the urban population had increased to an estimated 18,000 – 30,000(Liu *et al* 2005: 96).

SETTLEMENT ORGANISATION

Early phases of the Erlitou site may be similar to the Longshan social structure; there is no evidence for construction of large palatial foundations before phase II. Large settlements for early phases of Erlitou culture were settled in short periods (Liu, 1996a: 274). Twenty-three houses from phases II to IV have been excavated, demonstrating various sizes and architectural styles (Lee, 2004:180). Houses were rectangular with evidence of wattle and daub walls and rammed earth floors (Chang, 1968: 197). At the centre of the Erlitou site are the 7.5 ha palatial zone and several dozen rammed earth foundations ranging in size from 600 m^2 to 10,000 m^2, (Liu and Chen: 2003:34, Lee *et al*, 2007: notes) demonstrating the ability to mobilise labour on a grand scale.

Evidence from Erlitou during the peak phase III shows the settlement was partitioned into areas with specialist functions, such as a bronze foundry, pottery kilns, and bone workshops.

MORTUARY PRACTICES

There is a clear mortuary hierarchy. The variation in house size and style is mirrored in the burials. (Lee 2003:180). Members of the elite were buried in richly furnished graves with bronze, jade, and white pottery made of kaolin (Lee *et al* 2007: notes, Erlitou Archaeological Team, 2004: 18-37, Institute of Archaeology, 1999 in Liu *et al* 2005:75). Burials ranged from elaborate with grave goods to simple interments containing nothing or a few pottery vessels. Ceramic grave goods declined as bronze items became more important in graves (Underhill, 2002:238). The remains in more elaborate burials in graves tended to be supine or on one side in contrast to those without grave good found in habitation layers which were in a variety of positions, flexed, squatting or stretched. Up to 20% of the human remains found so far at Erlitou are from either human sacrifices or casual disposal of human skeletal elements, many in ash pits and cultural layers, suggesting some form of control exercised by institutional violence (Lee, 2003:181, Chang 1968:198).The large number mutilated body parts point to violence and conflict (Lee, 2004:186, Chang, 1968:198).

ECONOMY AND CRAFT PRODUCTION

The size of territory and political organisation of Erlitou resemble Trigger's territorial state (1993, 1999), but the political economy is different (Liu *et al*, 2007:101). The urban centre at Erlitou produced prestige goods, such as bronze and turquoise objects (Liu and Xu, 2007:886) and also manufactured utilitarian products, pottery and bone items (Liu and Chen, 2003:146). Chang (1986:412) argues that the development of social complexity was political rather than the result of technological development because bronze was rarely used to cast farm implements or tools. Liu and Chen (2003:147) agree with Chang that the production of bronze ritual objects and weapons are a major component of urbanism. Barnes (1999:119) suggests Erlitou co-opted bronze for production of ritual vessels. During the early Bronze Age there was a clear increase in different kinds of bronze vessels for food and drink, coupled with amplification of other prestige mortuary goods such as jade items, lacquer goods, and highly decorated pottery this might suggest independent craft production played a central role in the economy (Underhill, 2002:239).

Secondary and tertiary regional centres, such as, Huizui, situated close to natural resources and on transport intersections afforded the opportunity of state expansion beyond the single day's journey. This could demonstrate delegated decision making which requires organisation and administration (Liu, 2004:246)

2.9 HOW SOCIAL CHANGE IN THE CENTRAL YELLOW RIVER VALLEY HAS BEEN INVESTIGATED AND INTERPRETED

The development of social change and trajectories to early states in north central China is a common theme of discussion. Here, I am going to briefly examine the interpretations of KC Chang, Li Liu and Anne Underhill who have all produced significant models for understanding how and why societies in these three cultural periods developed social complexity.

CHANG

Exploration of the trajectories of state development reaching from the Neolithic to the Bronze Age is recurring topic in the work of KC Chang. Until the 1980's Chang shared the traditional mono-centric view that Chinese civilisation spread outwards from the Central Plains but archaeological evidence of socially complex Neolithic cultures outside the Central Plains, caused a revision of his earlier interpretation. Writing in 1986, Chang identifies two developmental trends in the Neolithic. As regional cultures spread, interaction between them increased resulting in what he described as a 'sphere of interaction' during the 4th millennium, and progressive stratification and diversification of cultures within each region. Chang sees these as paths to the emergence of early states (1986:234).

Sites from isolated cultural clusters in the early Neolithic in North China yielded millet, sickles, *ting* tripods, and mortar and pestles. The houses were semi subterranean and storage pits common. By 5000BC the number of cultures and sites had increased and clusters joined and expanded to form the Yangshao, a single distinctive culture with local phases. Other neighbouring millet growing but distinct cultures formed, for example, the Dawenkou in Shandong as well cultures with rice agriculture further south (Chang, 1986:236).

Evidence of communication between the cultures of both North and South China appears in the material culture at around 4000BC (Chang 1986:236). Within North China interaction between the Yangshao and Dawenkou can be identified through the presence of typical Dawenkou pottery styles in Henan, while Yangshao influence can be seen on Dawenkou ceramics (Huber, 1981 in Chang 1986:237). Chang also makes links between the cultures of North China and those in the Yangtze valley and then south to the Pearl River and Taiwan through similarities in their material culture in particular the painted ceramics (Chang 1986:239). He uses the term 'interaction sphere' to describe these connections between cultures and also to describe interaction within a region (Chang, 1986:242-3). In this way he links the development of the Chinese interaction sphere at the beginning of the 4th millennium and the emergence of complex societies within individual regions. This work can be applied to the question of changing agricultural economies; does the uptake or expansion of crops, especially rice, or new farming methods reflect Chang's interaction spheres?

LIU

Like Chang, Liu is interested in clarifying the socio political transformation from non-stratified Neolithic villages to the development of early 'complex political entities' during the Longshan and Erlitou periods in the middle and lower Yellow River Valley, North Central China (2004:1). She makes cross-cultural comparisons between China and the emergence of complex societies elsewhere in the world (Liu and Chen, 2006:150, Liu and Chen, 2003). Liu uses analysis of both settlement archaeology (2005, 1996) and the manufacture of craft items (Liu *et al*, 2007), in particular prestige goods (2003), as well as access to raw materials to discuss the emergence of social complexity in the Yellow River Valley (2004, 2000, Liu and Chen, 2003, 2006). Through extensive fieldwork, excavation and survey using multidisciplinary teams from inside and outside China, Liu considers interrelationships between wide ranges of sources (Liu *et al*, 2005). She is one of the few people who have used systematic floatation to consider these issues (Liu and Chen 2007, Lee *et al*, 2007, Lee and Bestel, 2007).

In her approach to understanding the development of social complexity in north central China, Liu uses rank size analysis and the theoretical concepts of Longshan period chiefdoms to create three models of regional settlement systems and relate them to geographic circumscription. The most complex, centripetal or mono centred in geographically circumscribed regions, the next, less stratified, centrifugal systems, site with many centres in semi- circumscribed regions and the third, least integrated system, decentralised, in non-circumscribed regions. Liu also builds on Chang's proposition of a connection between the location of Shang cites and the procurement of raw material resources to discuss early state formation in China (Liu and Chen: 2003). She connects craft specialisation, the acquisition of raw materials, manufacture and distribution of elite items with the development of the political structure of a society (2003:1 2007). Testing Trigger's (1999) model of craft production as a two-tiered entity in territorial states, Liu and Chen (2007) compare community based craft specialisation in the manufacture of stone spades at Huizui in the Erlitou hinterland during the Longshan and Erlitou periods with the production of prestigious white ware and availability of raw materials in relation to settlement position. The result is a complex organisation as opposed to Trigger's two-tiered model (Liu, 2003:101).

Liu asserts the development of societies from simple, relatively egalitarian groups in the early Neolithic to complex, stratified early states was not a unified process but a series of interrelated developments and declines stimulated by a wide range of factors (2004:252).

Some of Liu's work can be applied to questions on changing agricultural economies. The developments in the way the crops are selected, cultivated and processed may reflected the settlement systems where they are grown. A relationship between changes in craft specialisation and in crop organisation may be visible in the archaeological record.

UNDERHILL

Underhill bases her model of social change on how access to food surpluses were controlled, especially prestige foods, the relationship between craft production of labour intensive prestige goods, in particular food vessels, and the development of social stratification in the Yellow River Valley in northern China during the mid to late Neolithic and early Bronze Age (c4100-1046BC) (Underhill, 2002:3). Underhill (2002:242-4) suggests where there is little social inequality and easy access to food surplus small scale feasting accompanying social and economic transactions is common. In societies where there is competition for power, and significant social mobility, large scale feasting to negotiate economic influence will take place at a minority of ambitious households. Increasing competition is likely to restrict access to prestigious food and vessels. If a low level of social mobility accompanies high degrees of social economic inequality and restricted access surplus, large scale feasting will become restricted to a few elite households. Social relations with the dead are crucial. The gift of food at a funeral is fundamental to future prosperity so public display as part of mortuary ritual is key to socio-economic competition among descent groups. This can be linked to competition for scarce agricultural resources, allowing dominant descent groups to emerge. Control of access to food surplus and prestigious goods could be signalled by display of elaborate personal consumption.

According to Underhill's model, social change should be visible in evidence of differences in consumption and intensification of craft production (2002:244). Like Liu,

Underhill starts with systematic survey. Her main focus is on mortuary practices, the layout of cemeteries and analysis of grave goods, principally labour intensive food vessels, to interpret the role various vessels play in the negotiation of social relationships between the living and the dead (2002:17). She tests the model on archaeological data from the Dawenkou and Late Yangshao, the Longshan and early Bronze Age, Erlitou and Shang.

None of these models have a strong evidential basis for the production of agricultural surplus or organisation of agricultural production. In part this is because until recently there has been little systematic archaeobotany in China (see chapter 1). This thesis contributes to the issue of the emergence of social complexity through new hard evidence on aspects of agricultural production and the organisation of labour in Central China between the middle Neolithic and the early Bronze Age.

CHAPTER 3

INTRODUCTION TO THE STUDY REGION AND PALAEOECOLOGY

3.1 INTRODUCTION

In order to answer questions about past cultural activities and changes, an understanding of the environments and ecosystems people lived in and their potential relationship with their surroundings is essential. An idea of the palaeoenvironment can grant an insight to what people in the past had to consider when making choices about the way they lived. Ecological and vegetative history can provide indications to the distribution of plants used for food, why particular crops were selected for cultivation, why certain areas were settled and afford clues to changing agricultural economies and processes and how these can be linked back to cultural transformation and development. However, there are limitations to environmental reconstruction that should be considered. It is not possible to make an analogue of environmental systems as a whole. Proxy indicators should be selected carefully and their limitations understood.

In this chapter I will examine topography, geology, climate and vegetation both regionally and of the principal area covered by this project and how spatial and temporal variations in these key fields may have impacted on the way people lived during the Yangshao, Longshan and Erlitou periods in central north China.

First, a regional overview of present geographical features, topography and soils and examination of the current climate and vegetation will provide a broad ecological context, which will be narrowed by focusing on the core area in Henan Province. Once this has been established I will present an outline of the Holocene Palaeoclimatic and Palaeovegetative sequence and the associated environmental transformations based on proxy evidence from pollen profiles, palaeosols, and oxygen isotope records (An *et al*, 2006, Maher and Hu, 2006, Chen *et al*, 2005, Dykoski *et al*, 2004, An *et al*, 2004, Herzschuh, *et al*, 2004, Wu and Liu, 2004, Yi *et al*, 2003, Li *et al*, 2002, Wang *et al*, 2000, Huang *et al*, 2000, Hong *et al*, 2000, Yu *et al*2000 Hu, *et al*. 1992). The potential problems associated with ecological modelling and environmental reconstruction in China will be considered and finally I shall ask how changing climatic and environmental conditions might affect humans and how this relates to my key questions.

3.2 PHYSICAL GEOGRAPHY

TOPOGRAPHY

China is a vast sub continental area lying on the west coast of the Pacific Ocean, in the east of Asia at approximately 70° - 134° east Longitude, 18° - 53° north Latitude (Li, 1983:28). The land descends from west to east in three broad latitudinal steps echoed by three major rivers, the Yellow River (Huang he), Yangtze (Ch'ang Chian) and Sui Chang. In the southwest the Qinhai-Tibet Plateau is made up of mountains and highlands that are an average of between 4,000 and 5,000 metres above sea level. Below this, between the Western Highlands and Eastern Uplands lie a series of basins and plateaus around 1,000- 2,000 metres above sea level. Finally the land drops to low eastern plains of less than 500 metres above sea level. Among these is the Great North China plain. North and south China are divided by the natural boundary of the Qinling Mountains and the Huai River Valley (Wu and Lu, 2004: 154, Zhou, 1992:3, Robinson, 1976:350).

GEOLOGY

China is made up of three major physiographical structural zones related to the ancient landmasses of Tibetia in the west, Gobia, in the north and Cathaysia in the east. The granitic and metamorphic rocks from these landmasses remained above sea level during marine transgressions, which deposited sedimentary rocks such as, limestones, sandstones and shales. Coal beds were formed during warm moist periods. More recently, geologically, continental formations were laid down over the northwest and extensive alluvium was deposited in the northeast (Robinson, 1976:350-2). Three longitudinal structural areas are associated with these physiographical divisions; the comparatively recent Western Highlands, the three geanticlines east of the Tibetan escarpment and the great central depression which is marked by a series of plains running north east to south west, in particular the Great Plain of North China (Robinson, 1976: 352).

3.3 PRESENT CLIMATE

SEASONALITY AND CLIMATE

China has a characteristically monsoonal continental climate (Wu and Liu, 2004:154). Climatic patterns throughout the Pleistocene and Holocene have been influenced by the uplifting of the Tibetan Plateau, which affects the regionalization of temperature and precipitation by dividing the mid–latitude westerlies (jet stream) and consequently the climatic zones into northern and southern parts (Barry and Chorley 1998: in Schettler *et al*, 2006:1043).

Today China can be divided into 6 climatic zones; equator, tropical, subtropical, intermediate-temperate, warm temperate and cold temperate zones. The temperature in the coldest season ranges from an average of more than 16°C in the tropical zones (latitude 15-23°N) to below -24°C in the cold-temperate north (latitude 50°N) (Zhou, 1992:4). Average yearly temperature varies from north to south. During the coldest part of the winter in China the mean temperature north of the Qinling Mountains and Huaihe River is less than 0°C, in some areas below -20°C (Zhou, 1992:4). In contrast, the hottest part of the summer has average temperatures between 20°C and 28°C with the highest temperatures often above 38°C.

Isotherms run from east to west and in winter in particular have a latitudinal disposition (Zhou, 1994: Robinson, 1976 :). North China winter monsoon winds are strong and very cold and blow from land to sea so the period from November to March is dominated by Polar continental air. Asiatic high pressure wanes after March (Robinson, 1976:353).

China becomes progressively more arid from the east coast to the inland northwest and can be divided into moist, semi-moist, semi-arid an arid regions (Zhou, 1992:4) (Fig. 1). There is higher precipitation in the south than the north, on the coast than inland and in the mountains rather than the plains (Zhou, 1992:4).

MONSOON

Asia, the largest landmass in the world, consists of an arid hinterland edged by humid regions bordering oceans. Air pressure differences are caused by thermal variation between the dry heartland and the Pacific Ocean. Warm humid tropical air from the south and southeast are drawn to the interior by the low-pressure conditions, which develop over Inner Asia in the summer. This is usually the only period of precipitation in North China and the scant rainfall decreases as it goes further north and inland. The resulting seasonal reversal of surface air flow and dominant seasonal winds determine the Asiatic monsoon system (An, 2000:172, Li, 1983:22, Robison, 1976:352)

The Asian monsoon system has three subsystems, the Plateau or winter monsoon and the summer monsoon system, which is made up of the southwestern or Indian monsoon, and the eastern Asian monsoon systems. These are divided along a line from around 105° to 110°E longitude. The summer monsoon brings warm humid air from the ocean to around 40.3°latitude spreading across eastern northwest China (An, 2000:174, An *et al*, 2000: 744). The winter monsoon carries a cold arid northerly continental air mass from Central Asian deserts to around 22°N latitude (Wu and Liu, 2004:15, Li, 1983:22).

The East Asian monsoon is a governing force on the present and past climate and environment in east and central China (An *et al*, 2004: 744, Wu and Liu, 2004:154, An, 2000:172) and as a result a major influence on vegetation, flooding and ecological adaptation of past populations (Liu, 2004:23).

3.4 VEGETATION

China can be divided into three broad phytogeographic zones; the North China, South China and South Asia belts with distinct physiographic and phytogeographic characteristics (Li, 1983: 22). In the East Asian monsoon region of China modern vegetation zones are related to both precipitation and temperature (An *et al*, 2000: 743-762).

The division of the North and South China belts is the Qinling Mountain range and its eastern ranges, which lie along the 34th parallel (Li, 1983:22). The north China belt covers area that includes the Yellow River basin, the loess highlands and north China plains. The southern part of the belt lies between the Yangtze and Huai He rivers (Li, 1983:22). The project area lies on the southern margins of the North China belt so a key factor would be evidence of imported or traded crops and other plant material coming from the south, for example, the spread of rice farming north during the Yangshao period.

3.5 HENAN

LOCATION AND TOPOGRAPHY

The province of Henan ('south of the river' or 'Yu') is situated in the middle and lower reaches of the Yellow River. Covering an area of more than 167,000 square kilometres, Henan lies 110"22 – 116"38 east longitude and 31"23"-36"22 north latitude. The province's generally mid altitude terrain slopes from northwest to southeast. Much of the province is made up of basins and plains (55.7%) including part of the vast North China Plain in eastern Henan (Zhou, 1992:247).

RIVERS, IRRIGATION, SOILS AND ARABLE LAND

Henan has a radiating river system with in excess of 460 rivers with drainage areas of 100km^2 or more. Most rivers flow from the western mountains and follow the terrain northeast, east, southeast and south. The Yellow River, the largest in Henan with a drainage area of 26,000km^2, flows for 700km, west to east through the northern part of the province. It has a number of tributaries including the Yilou River in the south (Zhou, 1992:251), the Liujian River, which flows adjacent to the Huizui site (Liu and Chen, 2003:66) and the Sha River, close to Xipo. The Yellow River and its tributaries carry a heavy load of silt due to severe soil erosion in the loess region (Quine *et al* 1999 in Liu, 2004:20). As the river crosses the plains of central Henan it broadens and slows, depositing gravel and sand on the river bed and, during floods, silt and clay on the banks and beyond. The bed gradually rises above the plain until it seeks a lower bed resulting in dramatic changes in its lower course, switching back and forth north and south of the Shandong peninsula several times during the Neolithic (Murphey, 1972 in Liu, 2004; 20).

Meadow soil covers an extensive part of the Yellow River Plain. These soils are deep, calcareous and stone less, containing large quantities of löess, which has been carried and dispersed by flooding rivers. The land is level and the soil fine textured so drainage is often poor. Sometimes salt accumulation in the low-lying plain leads to saline-alkali soil. However, these azonal, alluvial soils are usually fertile and attractive to agriculturalists.

THE CLIMATE IN HENAN

Located between the subtropics and warm temperate zones, the present climate is continental of the North Temperate Zone, mild with abrupt transitions between seasons. The summers are hot and wet contrasting with cold dry winters. Springs are windy while autumns are fine and clear.

The mean annual temperature is 12°C - 15°C. Within this range the high terrains of the West Henan and Taihang Mountains have an annual mean temperature of less than 13°C in contrast to the low lying Nanyang Basin which has an average annual temperature above 15°C. July is the warmest month in Henan. The average is 27°C to28°C with a maximum of 40.9°C while January, the coldest month, has a mean between 2°C and -2°C, increasing from north to south. The record minimum is -23.4°C (Zhou, 1992:247).

The average annual precipitation is 600-900mm decreasing from south to north and varying from an annual 1,300mm in the south to 600mm in the north. 45-65% of the rainfall occurs during the summer. The province is vulnerable to drought, waterlogging and flooding (Zhou, 1992:247).

3.6 HOLOCENE PALAEOCLIMATE AND PAST ENVIRONMENT, POTENTIAL DIFFICULTIES ASSOCIATED WITH PALAEOENVIRONMENTAL RECONSTRUCTION IN CHINA

In the past few years there have been abundant studies on climate and environmental change in China, many of these have taken place in northeastern China, Inner Mongolia, the Loess Plateau and north central China. Both local and broad scale sequences and climatic events have been discussed (Rosen, 2008: 298).

There are issues that need to be considered when making palaeoenvironmental reconstructions. Reliable analogues and chronologies are needed. Many proxies' bioclimatic dependencies have not been tested or even convincingly established (Feng *et al*, 2006:120).

RELIABILITY OF CHRONOLOGY

There are some potential difficulties associated with C14 dating in the Holocene sequences of arid and semi arid China. Charcoal, the best C14 dating target, can be rare, while bulk organic matter, which is commonly used, can be unreliable due to potential mixing of diachronous organic material by bioturbation such as root penetration and animal activity (Birkeland 1999 in Feng *et al* 2006:120).

Most information on Palaeoclimate is based on pollen profiles. There can be an interval between climate and vegetation change of up to 300 years and another pause between environmental change and evidence of any related social response (Bradley, 1999:365-370, Liu, 2004:23).

The majority of geological chronologies used for Palaeoclimatic reconstruction in China are drawn from uncalibrated radiocarbon dates (BP), while archaeological chronologies are derived from calibrated dates (BP or BC), making the two datasets incomparable without calibrating the geological BP dates (Liu, 2004:23).

TYPES OF EVIDENCE

There are numerous recent studies of climatic and environmental change in China using a range of types of proxy evidence to reconstruct the shifting Holocene palaeoclimate (Appendix 1 Holocene sequence table). Choosing suitable proxies and establishing consistent chronologies are crucial. However, there are problems associated with specific proxies that need to be taken into consideration.

MAGNETIC SUSCEPTIBILITY

The magnetic susceptibility of the Loess-palaeosol sequence has been used as a proxy indicator for effective moisture and monsoon strength. Aeolian deposits since the last glaciation in north central China have been interpreted as representing a strengthening winter and diminishing summer monsoon while the palaeosols reflect the opposite (An *et al*, 2000:752, Maher, 1998: 25). However, Feng *et al* (2006:120) summarise some potential problems; magnetic susceptibility can be influenced by *in situ* pedogenic enhancement, allochthonous material, and dilution by carbonate concentration. Leaching can cause magnetic enhancement and redox cycles can trigger magnetic alterations, (Zhou *et al*, 1990, Sun and Liu, 2000, Heller and Liu 1986, Anderson and Hallet 1995 and Feng and Chen, 1999 in Feng *et al*, 2006: 120).

POLLEN

Pollen is used as a proxy to establish past vegetation patterns and thus palaeoenvironments at a range of spatial scales (Bradshaw, 1991:41). Most pollen found in sediments is from wind-pollinated (anemophilous) plants, which produce vast quantities of pollen to be carried by air. Pollen from insect-dispersed (entomophilous) plants is not well represented (Küster, 1991:19), while self-pollinating plants are virtually invisible. This creates an obvious bias. The distribution of pollen rain may not be uniform. Some pollen collected from fluvial or lacustrine sequences may not be *in situ* and as some are transported long distances there is no clear-cut certainty of chronology (Evans and O'Connor, 1991:134-5, Zhu *et al*, 2001 in Feng *et al*, 2006:120). Survival is dependent on soil conditions and taphonomic factors must be considered when using pollen for environmental reconstruction (Edwards, 1991: 53). There may be issues of identification as very few pollen grains can be identified to species (Küster, 1991:19).

PALYNOLOGICAL DATA RELATED TO CLIMATE/ ANTHROPOGENIC ACTION

Climate and anthropological action in the pollen data need to be disentangled. Human modification has shaped much of China's natural vegetation. To a large extent the natural steppe formed on the plateaus of northwest China is impoverished due to over grazing and soil erosion. Most natural forest cover has gone, either for timber or has been replaced by agriculture (Robinson, 1976:356). Deforestation and agricultural activity add to sediment loads in rivers and the creation of gullies, which in turn results in greater drainage and erosion (Quine *et al*, 1999 in Liu, 2004:19).

Intensive anthropogenic disturbance over prolonged periods means transfer functions articulating the relationship between modern pollen rain and associated vegetation are likely to be distorted by the lack of natural vegetation (Feng *et al* (2006:120).

However, numerous recent studies using pollen data to reconstruct Holocene environmental history in China suggest with consideration of limiting resolution validity pollen can prove a reliable qualitative proxy (Sun and Xia, 2005, Herzschuh *et al,* 2004, Li *et al*, 2004, Yi *et al*, 2003, Yu *et al*, 2000, Li *et al,* 2002, Ren and Zhang, 1998).

3.7 HOLOCENE SEQUENCE ON THE LOESS PLATEAU, A GENERAL OUTLINE

(Appendix 1, Holocene sequence table)

THE LAST GLACIAL MAXIMUM

In contrast to the warmer, wetter more stable Holocene the environment during the cold dry Pleistocene was extremely variable (Barton *et al,* 2005). Pollen records suggest that during the last glacial maximum at 18,000 BP (±2000) steppe and desert vegetation stretched to the modern coastline of east China between 32°N and 40°N, while taiga extended south to 43°N. Broadleaved evergreen /warm mixed forests retreated south as far as 24°N, ca.1000km further than today, bordered on their northern margins by cool mixed forest (Yu *et al*, 2000:658).

THE EARLY HOLOCENE

The Early Holocene experienced general warming and increased precipitation. There was a heightened contrast between seasons and the northern frontal zone retreated (An *et al*, 2000, Lister *et al,* 1991). The ensuing dry phase (Feng *et al,* 2006) was followed by an abrupt transition from dry to humid at around 10,700-10,000 BP (An *et al,* 2000:758). During this period the first sedentary villages appeared on the Central Plains, indicated by early ceramics, domestic features and possibly domesticated dogs and pigs (Liu, 2004:24).

The vegetation between 10,250–7900 cal BP was dominated by *Artemisia* (61.1-75.5%) and other arid halophyllic herbs and shrubs such as Chenopodiaceae and *Ephedra* suggesting a dry environment, while an increasing trend in arboreal pollen from *Pinus, Quercus, Betula* and *Ulmus* (12.8-27.9%) indicate a mild climate with gradually rising temperatures and rainfall (Xiao *et al*, 2004: 1674, Li *et al*, 2004:198).

THE MID HOLOCENE CLIMATIC OPTIMUM

Changing orbital cycles in the Early and Mid Holocene leading to amplified solar radiation influenced climatic events around the globe (Kutzbach and Gallimore, 1988, Rosen 2007: 40, Wang *et al*, 2005: 855), resulting in the warm and humid Mid Holocene Climatic Optimum, a gradual increase in warmth and humidity, to the warmest and, in many places, most humid period of the Holocene (Xiao *et al,* 2006, Feng *et al*, 2006, An *et al*, 2000). The warming Asian continental landmass and low latitude Pacific Ocean temperatures almost certainly led to a stronger East Asian Monsoon and higher precipitation levels (Feng *et al,* 2006:126, Rosen, 2008: 298). However, there is some debate on the intensity and synchronicity of the Climatic Optimum in different regions of northern and north-eastern China (Rosen, 2008: 4, An *et al,* 2004, An *et al,* 2000:744). Early Neolithic cultures emerged in the Yellow River valley (9000-7000 BP). This was followed by a weakened East Asian monsoon episode ca 6000 cal BP. The maximum and subsequent weakening are consistent with the Middle Neolithic Yangshao Culture and a rapid population increase (Liu, 2004).

From 7900-4450 cal BP, arboreal pollen, principally *Pinus* with some *Quercus* and *Ostryopsis*, increased to 69.4%, while *Artemisia* declined pointing to predominantly mixed coniferous and broad-leafed forests and a warm and humid climate. Finer resolution stages can be seen within this, reflecting climatic fluctuations. Between 7900-7250 cal BP Pinus dominates while *Picea* and *Abies* first appear indicating a cool and wet environment.

From 7250-6050 cal BP the proportions of herb, *Artemisia*, and broad-leafed tree, *Quercus*, pollen increased while *Pinus* declined rapidly suggesting a drier and warmer period. At this time the Yangshao Culture was developing in Henan.

6050-5100 cal BP is characterized by a return to the dominance of *Pinus*. *Betula* and *Artemisia* decline indicating a warmer and more humid stage, the Mid Holocene climatic optimum in this region.

By the mid Holocene 6000 BP (± 500) the forest biomes had expanded westwards and moved north, replaced by broadleaf evergreen/ warm mixed forest, which was also present at higher elevations than previously. The northern limit for temperate deciduous forest was 800km north of where it is today. (Yu *et al*, 2000:656).

The warmer wetter environment shaped vegetation patterns creating a denser cover, which in turn may have elevated the Mid Holocene Optimum by feedback mechanisms (An *et al,* 2006, Feng *et al,* 2006:127). Radiocarbon dated pollen cores from taken from Dahai Lake north of the Loess Plateau in the transition between semi humid and semi arid areas of the temperate zone illustrate vegetation change from 14,000 BP to the present day (Li *et al,* 2004:198-200, Xiao *et a,* *l*2004:1671). Although this is north of the key area of this project the general picture presented is relevant as related climate and vegetation changes would have also occurred further south. Rosen (2007:41) draws attention to the Holocene pollen sequence from the southern edge of the Loess Plateau. Their pollen Zone 2 is from 8800-5300 BP, corresponding to the Yangshao culture, and demonstrates an increase in broad-leafed deciduous trees such as, *Quercus*, *Corylus* and *Betula* indicating a warm and humid climate. This could indicate the environment during the Yangshao in northern Henan at that time.

LATER HOLOCENE

The climate cooled and there was a brief cool humid period (4800-3400 BP, Li *et al*, 2004:202, Li *et al*2003) followed by mild arid conditions (3400 –1950 BP) (Xiao *et al*, 2004:1675, Li *et al*, 2004: 198-200 Figure 3.2). Correspondingly the Longshan culture emerged in the Central Plains and the population steadily increased (Liu, 3004:27). This was followed by a series of climatic fluctuations as the weakening monsoon moved north (Xiao *et al*, 2004 Schettler *et al*, 2006 Li *et al*2004 Zhou *et al,* 2004, An *et al,* 2005, Wang *et al*, 2000, Bond *et al*, Wu and Liu, 2004, Jin and Liu, 2002, Liu, 1989). Environmental changes would have impacted on river systems. Heavier vegetation cover during warm moist periods would have stabilized the landscape reducing soil erosion and runoff, leading to rivers with a steady flow. The light sediment flow meant less alluvial deposition and soil formation. In contrast, during cold arid episodes reduced vegetation would have lead to greater erosion, heavier sediment loads and the likelihood of flash flooding (Quine *et al*, 1999, Rosen, *et al*, in preparation).

Much of the following phase can be related to the Longshan period (5000 –4000 BP). From 5100–4800 cal BP there was a rapid decrease in *Pinus* and *Quercus* pollen while *Artemisia* dominated, although there is less pollen overall indicating the cooler drier conditions that characterized much of late Holocene period in northern China (Xiao *et al*, 2004:1675, Li *et al*, 2004: 198-200, Feng and An, 2006).

During the phase from 4800-4450 cal BP *Pinus* increased while *Artemisia* decreased although the pollen concentration remained the same pointing to a slightly warmer and wetter period (Xiao *et al*, 2004:1675).

The next major stage is from 4450-2900 cal BP and begins with a marked decrease in tree pollen with a steady increase in arid herbs and shrubs, indicating forest steppe with increasing grassland and a cooler more arid climate. Fluctuations can be seen in three sub stages, the first 4450-3950 cal BP shows rapidly decreasing arboreal pollen contrasting with increasing herbaceous pollen indicating advancing grassland and cooler more arid environment.

The subsequent period corresponds to the Erlitou Cultural Period, 3900-500BP. From 3950-3500 cal BP arboreal pollen increased, particularly *Pinus*, while herbaceous pollen fell in particular *Artemisia,* suggesting a recovery of mixed conifer and broad-leafed deciduous forests and warmer wetter conditions.

The stage from 3500-2900 cal BP shows another decrease in arboreal pollen and growth in herbaceous plants, principally *Artemisia* and Chenopodiaceae pointing to a further decline in forest as the climate dried (Xiao *et al*, 2004:1676, Li *et al*, 2004:200). The final phase from the sequence at Yaoxian on the southern Loess plateau dates from 5300-2000 BP, which corresponds with the Longshan and Erlitou periods and shows a peak of Chenopodiaceae and *Ephedra*, both arid plants suggesting the climate had become drier (Li *et al*, 2003:3).

3.8 HOW VARIATIONS IN CLIMATE AND ENVIRONMENT MIGHT AFFECT HUMANS

SETTLEMENT AND CULTURAL CHANGE

Changing environments are likely to impact on selection of settlement position and size. During the mid Holocene climatic optimum there was a rapid expansion of settlements in the Yellow River Valley region. 800 Yangshao sites have been recorded in Henan contrasting with 2000 in neighbouring Shaanxi (National Bureau, 1991,1999 in Liu, 2004:26), whereas in the western highlands there is a dense distribution, especially during the Miaodigou phase (mid Yangshao, ca 6000-5500 BC) (Liu, 2004:26).

As the climate cooled and dried, and land previously covered by lakes and marshes became available, lowland central Henan saw a steady increase in Longshan culture settlements. At around 4000BP the climate became cooler and drier. The Longshan culture declined followed by the rise of the Erlitou around 3900BP. While, at present, only 200 Erlitou sites have been found in Henan compared to 2000 Longshan, the social organisation, particularly of the core area, was transformed to nucleated and urbanised settlements.

CHANGING AGRICULTURAL/ ECONOMIC PATTERNS

Increasing temperature and humidity are likely to have encouraged expansion of any established agricultural activity, such as millet (*Panicum* and *Setaria*) farming, northward, while allowing the spread and amplification of rice (*Oryza*) farming from the south of the Yellow river region. Conversely, as the climate cooled and dried the agricultural zones retreated southward (Liu, 2004:19).

3.9 CONCLUSION

The Holocene has seen numerous ecological and cultural changes. It is important to note that human responses to environmental change do not always take the same path and cultural changes are circuitous and multifaceted. While I have briefly correlated the broad environmental and archaeological sequence using calibrated dates where available, having an understanding of the palaeoenvironment does not equate with an understanding of the complex relationship that different groups of people had with their surroundings but it can facilitate a path towards an understanding.

CHAPTER 4

CROPS AND AGRICULTURAL PRODUCTION IN PREHISTORIC CHINA

4.1 Introduction

To date there has been relatively little systematic archaeobotanical investigation or floatation in China. Most has been focussed on the origins of agriculture (Lu 2006, Crawford *et al,* 2005, Crawford, 2006, 1992), and domestication of millets (Song *et al,* 2013,Liu, *et al,* 2009, Lu, 1999, Nasu *et al,* 2007, Zhao, 2006, Barton *et al,* 2009, Lu *et al,* 2009,Yang et al 2012), rice and the spread of rice and cereal cultivation (Nasu *et al*, 2012, Fuller *et al,* 2007, 2008, Liu *et al,* 2007, Zong *et al,* 2007, Cao *et al,* 2006, Lu, 2002, Wang *et al,* 1999, Crawford and Chen, 1998, Higham and Lu, 1998, Pei, 1998b, Zhang and Wang, 1998, Zhao, 1998, Wu, 1996, Zhao and Wu, 1986/1987, Lu, 2006, Underhill, 1997:109).

As long ago as 1997, Underhill (1997: 106) highlighted the dearth of projects using systematic collection, analysis and interpretation of agricultural remains to investigate changes in subsistence on a regional basis. Until recent years a large proportion of archaeobotanical data in China has consisted of hand collected chance finds, with little consideration of systematic floatation (Lee *et al,* 2007:1087, Crawford *et al,* 2005:309) until Zhao (2004). There are exceptions, predominantly studies set in an interdisciplinary framework, Yuezhuang and Liangchengzhen in Shandong, (Crawford *et al,* 2006, Crawford *et al,* 2005) the Yilou Valley project, (Liu *et al,* 2003, Lee *et al,* 2007), the Upper Ying Valley (Fuller and Zhang, 2007), Baligang (Peking University), Tianluoshan (Fuller *et al,* 2009) and Xinglongguo (Zhao, 2006). Most systematic archaeobotanical sampling in China has been carried out over the past 10 years (Zhao 2008). Most phytolith analysis seems to have been motivated from outside archaeology departments and while there has been archaeological investigation, for example Lu *et al,* 2009 at Cishan, again there has been relatively little work within multidisciplinary projects (see Chapter 1).

This chapter first provides an overview of the origins of agricultural systems in north central China. Next, the origins, dispersal and cultivation of the three key cereal crops *Setaria italica*, *Panicum miliaceum*, and *Oryza sativa* will be considered. Although there are many indigenous Chinese plants that could have been and were exploited (for example, soybean (*Glycine max*), jujube (*Zizyphus jujuba*) (Harlan, 1995, Underhill, 1997:109, Li, 1983:33), evidence discussed in the literature is predominantly for the cereals, millets and rice. These also produce the most easily recognisable phytoliths and are central to the questions asked in the present study.

Finally, this chapter will consider what is known about crops and agricultural production in Henan during the Yangshao, Longshan and Erlitou with brief descriptions of the preceding and subsequent periods.

4.2 Origins of agricultural systems in north central China

The origins of agriculture in China have been widely discussed inside China and beyond (Harlan, 1973, 1995, Bellwood, 2005, Barker, 1985). It is commonly believed that agriculture developed indigenously in different regions producing the domesticates; *Oryza sativa, Setaria italica,* and *Panicum miliaceum*, which can be divided into two systems, millet based in the north and rice in the south (Barton *et al,* 2009:1, Crawford, 2006:82-4, Underhill, 1997, Yan, 1992, An, 1988a: 756, Chang, 1986:182). Cohen (1998) and Bellwood (2005) in Barton *et al,* (2009:1) propose an alternative notion that the northern millet based system is a result of a diaspora of southern rice farmers adapting to a cooler arid environment by using their experience of the drought tolerant qualities of *Setaria viridis,* the ancestor of *Setaria italica,* to expand into areas where *Oryza* was less productive, the migration paralleling the distribution of Sino-Tibetan and Austronesian related language families (Bellwood, 2005, 2006:100-101, Sagart, 2008:134). But the evidence so far from northern China demonstrates millet is earlier than rice (Lu *et al,*, 2009, Zhao, 2005, 2006).

Two major river valleys in central China, the cool dry middle and lower Yellow River valley in the north and the warm humid middle and lower Yangtze in the south, have provided archaeological evidence for early crop cultivation (Underhill 1997:107). Their respective climates provided suitable environments for the adoption of different cereal crops, particularly millets in the temperate Yellow River Valley and rice in the warmer Yangtze (Imamura, 1996:444). In both the Yellow and Yangtze valleys agriculture is generally accepted as established by 8000 BP (e.g. Lu, 1999, Crawford 2006), although controversies continue around the actual evidence and dating for the start of cultivation, domestication or agriculture in both valleys.

The limitation of available datasets may call into question common generalizations. For example, Lu (1999:133) dates the emergence of agriculture in the middle Yellow river basin as no later than 8500BP based on AMS C14 dating of an assemblage from the Peligang culture site at Jiahu (on the upper Huai River), although archaeobotanical evidence from this site lacks evidence for the classic Yellow River Neolithic crops, with neither *Panicum miliaceum* nor *Setaria italica* present there (Henan Provincial Institute of Archaeology 1999; cf. Zhao 2008: 234). The Late Peiligang sites, such as Fudian and Wuluoxipo in the Yilou valley, produced the expected *Setaria italica* (Lee *et al,* 2007). Evidence from Xinglongguo in Inner Mongolia (Zhao, 2006) and Dadiwan in Gansu on the upper Yellow River Valley, both suggest early agriculture focused on *Panicum miliaceum* was widespread across north China by shortly after 6000 BC (Barton *et al,* 2006), although phytolith evidence and new radiocarbon dates from the sediment in Cishan storage pits may indicate *Panicum miliaceum* cultivation by ca. 2000 earlier years, with *Setaria italica* added to the repertoire at ca. 8000 BP (Lu *et al,* 2009). Unthreshed inflorences of *Panicum* tied together in bundles were found at Yangshao period Linjia (Yan, 1989, in Crawford, 1995:16).

The Qinling Mountains are a dividing line between the Neolithic northern millet-based economies and southern wet rice agriculture, running along the 34th parallel between the Huai and Yangtze rivers, the North China belt covering the Yellow River basin, Loess Highlands and north China plains and the South China belt (Li, 1983:22, Barnes, 1999:97) Above this line winters are cold and there is limited precipitation. The staple crops are herbaceous, heliophilous, mostly annuals of open grassland (Ames, 1939:8, Li, 1983:22-3). The ecological setting of the Central Plains, dry cold winters, infertile loess soils retaining little water, means farming is better suited to the river valleys where the soil is more productive. Crop choices are limited to plants that can thrive in drought conditions and have a short growth period before harvest (Yan, 1992:115). The earliest main staple cultigens, during the Neolithic in north central China were millets (Li, 1983:53) *Panicum miliaceum* (2 kinds glutinous and non-glutinous) and *Setaria italica,* both meeting these requirements (Lu, 2006: 61,129, Yan, 1992:115 Li, 1983:29). There is a general consensus that the temperate Eurasian millets *Panicum miliaceum* and *Setaria italica* were domesticated indigenously and probably the earliest cereal cultigens in Central China (de Wet *et al,* 1979, Smith, 1995, in Lu, 1999:5) although evidence at Chengtoushan (Hunan) at ca 4500 BC raises the possibility of a separate domestication alongside *Oryza sativa* in south central China (Nasu *et al,* 2007:492), or an early dispersal from the north.

These cereals were supplemented by legumes, in particular, *Glycine max* (Soybean). Like *Setaria* and *Panicum,* soybeans are highly adaptive and drought tolerant, although it is still unclear archaeological when soybeans were cultivated or collected wild prior to the evidence of historical texts (Crawford and Lee 2003; Crawford *et al,* 2005). Fruits, including jujube (*Ziziphus jubjuba*), peach (*Prunus persica*), apricot (*Prunus armeniaca*), persimmon (*Dyospyrus kaki*), and nuts, chestnut (*Castanea sativa*), hazelnut (*Corylus* spp.), as well as green vegetables, *Brassica* spp. and hemp (*Cannabis sativa*) have also been grown since antiquity (Li, 1983:29, Yan, 1992:117).

Rice (*Oryza sativa*) arrived in north China later than local millet cultivation, presumably spreading from origins to the south. Rice remains have been found at Jiahu, in southeast Henan, although whether this was wild-gathered or actually cultivated has been a point of recent debate (Fuller *et al* 2007; Liu *et al,* 2007). Rice from Yuezhuang in Shandong, is of a similar age, ca. 6000 BC although the domestic/wild status is again uncertain (Crawford *et al,* 2006). Two early Neolithic sites, Liajiacun and Hejiacun, in Shaanxi south of the Qinling Mountains also have reported rice, perhaps as early as ca 7000 BP, although this is not a direct date for rice, which may have been introduced at these sites as late as 5800 BP. Lu (1999) suggests the Shaanxi sites may represent a southbound diffusion of the Dadiwan culture. The earliest direct AMS date on a rice grain north of Qinling Mountains comes for the Early Yangshao levels at Nanjiaokou, at 3900-3800 cal, BC (Ling Qin, personal communication). Diffusion northwards after ca. 4000 BC fits with dispersal after a late fixation of domestication traits in Yangtze rice (Fuller and Qin 2009) although whether any of the earlier finds represent a local start to rice cultivation remains unclear. Nevertheless, it is generally supposed that contact between northern dryland farmers and southern wetland farmers is most likely to have occurred in Henan (Li, 1983:54). Recent finds suggest Shandong should also be considered.

4.3 MILLET CROPS

INTRODUCTION TO FOXTAIL AND BROOMCORN MILLET (SETARIA ITALICA AND PANICUM MILIACEUM)

Setaria italica and *Panicum miliaceum* are both well adapted to varied soil types including alkali, infertile and semi arid environments (Chang, 1993:69). *Panicum miliaceum* matures earlier with a shorter life cycle (60 plus days) than *Setaria italica* (90 plus days) and is more drought resistant (Pursglove, 1972:199, Liu *et al,* 2009:89). Neither *Panicum miliaceum* nor *Setaria italica* need irrigation, so pre 5000 BC domesticated millets are likely to have been from rain fed summer crops for instance, dry millet farming (Hunt *et al,* 2008:S14, Zhao, 2006, Yan, 1992:117, Zohary and Hopf, 2000:83).

During the middle Neolithic *Panicum miliaceum,* like *Setaria italica,* was cultivated on the Central Plains but also grown in the northeast and west where the climate is drier and cooler. Today a similar pattern is shown. The centre for *Setaria italica* is in southern Hebei and northern Henan in the temperate forest zone with more than 600mm annual precipitation (Winker and Wang, 1993, Lu *et al,* 2009), while *Panicum miliaceum,* is grown in hilly more arid environments such as the Loess Plateau with a lower annual rainfall of around 400-500 mm (Liu *et al,* 2009:89). A parallel can be drawn with findings in temperate Europe ca. 2000 BC where *Setaria italica* was grown in the south and *Panicum miliaceum* in the north (Brothwell and Brothwell, 1969:99).

FOXTAIL MILLET (SETARIA ITALICA)

The wild progenitor of *Setaria italica,* one of the oldest domesticated cereals in Eurasia (Sakamoto, 1987, de Wet, 1995, Nasu *et al,* 2007:482), is considered to be *Setaria viridis,* green foxtail (Lu, 1999:61, Yan, 1992:115), identified by Le Thierry d'Enniquin *et al,* (2000, in Hunt *et al,* 2008:S13, Li *et al,* 1945, Gao and Chen, 1988 in Lu, 1999:61). The centre of origin for Eurasian wild *Setaria, Setaria viridis,* has been widely discussed (Dekker, 2003a: 94, Rao *et al,* 1987, Werth, 1937).

Setaria viridis has a longer highly heterochronous life cycle compared to *Setaria italica. Setaria viridis* germinates in spring and summer. Plants with summer germination have a shorter life cycle. There is variation in length of time taken for tiller growing, flowering, ripening and shattering on each spikelet so the season can last for up to four months and during this time there would always be some spikelets ready to harvest (Lu, 1999:61). There is also a major change in the architecture of the plant, from many tillers and auxiliary branches ending in a short inflorescence in *Setaria viridis* to few tillers, no auxiliary branches and long inflorescences, typical of domesticates, in *Setaria italica.* This is combined with a decrease in maturation time (Harlan, 1992 in Doust and Kellog, 2006: 136, Doust, *et al,* 2009:138). *Setaria italica* is predominantly self pollinating (Zohary and Hopf, 2000:86). Both *Setaria italica* and *Setaria viridis* have high nutritional values and it is claimed the seeds can survive for up to twenty years in storage (Anonymous 1976:2 in Lu, 1999:63). *Setaria italica* was found in the Central Plains during the mid Neolithic where it remained concentrated, although there is also evidence beyond at Xinglongguo, Chifeng, Inner Mongolia (Zhao, 2006). By the late Neolithic, *Setaria italica* had become diffused all over North China, spreading to Korea and Taiwan (Yan, 1992:117; Crawford and Lee 2003; Fuller *et al,* 2008).

BROOMCORN MILLET (PANICUM MILIACEUM)

There is debate on the progenitor of *Panicum miliaceum* although it is thought to be the *Panicum miliaceum* ssp. *ruderale,* a common maize and millet crop weed growing in Northwest and central China is most likely (Zohary and Hopf 2000, Hunt *et al,* 2008, Li, 1979, Wei, 1986 in Yan, 1992:115, Lu. 1999:61, Zhao, 2006). *Panicum repens* has also been suggested (Zhao, 2006). Panicum miliaceum is nutritious; grains are 10-11 % protein (Zohary and Hopf, 2000:83). It is also highly drought tolerant with an extremely short growing season of just over three months (Hunt *et al,* 2008).

There are several hypotheses on the domestication of *Panicum miliaceum,* Bettinger *et al,* (2007 in Crawford, 2009:7271) suggest rapid domestication took place in the Wei Basin at ca. 8000 cal BP. Charred grain from early *Panicum miliaceum* has been found at Xinglongguo (ca. 80-7500 cal BP) (Zhao, 2005, 2006), Dadiwan ca. 7800-7350 cal BP (Barton *et al,* 2009), and Yuezhuang, ca. 7870 cal BP (Crawford *et al,* 2006), and phytoliths from *Panicum miliaceum* have been identified at Cishan 10,300-7800 cal BP (Lu *et al,* 2009:7369). These widely dispersed finds at ca. 6000 BC could indicate multiple domestications across North China, as implied by Barton *et al* (2009). However, the now earlier dates from Cishan could also support earlier origins at ca. 8000 BC in southern Hebei (cf. Lu *et al,* 2009). Archaeobotanical evidence is still too limited to pinpoint origins or elucidate the process of domestication.

TRADITIONAL PANICUM AND SETARIA CULTIVATION

Reddy (1997) and Fogg (1983) have made ethnographic studies of *Setaria* cultivation and processing in India and Taiwan, respectively. Reddy describes winter flood plain opportunistic cultivation of *Panicum miliare* (=*P. sumatrense*) and *Setaria* spp. (including *S. pumila* and *S. italica*) in Andhra Pradesh, India. (1997, 1994) Extensive and opportunistic cultivation is a communal activity that takes place on the clay soils of the river floodplains is described in India (Reddy 1997:166). The small Indian millets are sown into the largely weed-free, heavy clay soils after the water from the summer monsoon river flooding has receded. *Panicum miliare* has thin stalks so plants are gathered together for harvesting and cut with an iron sickle. Thus selection against weeds is difficult and seeds often incorporated into harvest to be selected out at later stages of processing. The plants are threshed to release the seeds by beating with a stick or rubbing and trampling with feet to optimise grain retrieval, *Panicum miliare* is winnowed by shaking with a winnowing basket again to prevent loss of the small seeds. This is repeated until the gain is clean. Once clean the seeds are pounded on a grinding stone with a wooden staff to separate the grain from the husks and again to produce flour (Reddy, 1997, 166-171). *Setaria* spp. is likely to be treated in the same way as it also has comparatively small seeds.

Fogg (1983) made an ethnographic study on *Setaria italica* swidden cultivation in Taiwan He observed mountain tribes in central Taiwan and a coastal community, the Yami, in the east. The methods of cultivation were quite different. The central mountain people threshed the seeds by rubbing the panicles between their hands or feet and then winnowed them before broadcasting with a handful of *Phaseolus radiatus* (L.) and *Chenopodium album* (L.) per 1-2 L of *Setaria italica* seed into ground prepared by narrow bladed hoes, they also dibbed individual *Coix lachryma jobi* seeds. The Yami sow all varieties together in the same field on the same day. They leave the seed exposed on the surface and if necessary re-sow a few weeks later. Only dibbing sticks are used and although they do not sow any other crops with the *Setaria italica* they often polycrop with plants already in the swidden such as taro. The Yami only weed once. The central mountain people harvest each panicle individually. The stem is broken or occasionally cut with a small bamboo knife, at a node around 40cm below the panicle. The panicles are tied at the ends of the stems into sheaves and after the harvest carried in back baskets to the field hut or home site for seed selection, drying and storage. Seed selection for the following year's crop is started during the harvest. Each panicle is scrutinized when cut and particularly good panicles are selected, woven into a long stemmed sheaf of 'the beautiful millet' and kept for planting the following year. The remaining sheaves are trimmed of their stems, dried and stored outside in raised granaries or inside homes for threshing as needed. The Yami's cultivars have more supple stems so they use harvest knives and cut at the base of the panicle. They gather and make sheaves according to panicle size, starting with the largest. Not all reach maturity at the same time so they return every one or two weeks. Panicles of less than 3cm are left in the field to become mulch. Varieties of *Setaria italica* are seen as items of prestige and exchanged between hamlets and tribal groups (Fogg, 1983, 103-6)

In north China sites with millet remains are often found above river courses between the upland and foothill sediments positioned where rainwater run off can be exploited in free draining plots, (Shi, 1982:131, Tong, 1984, in Liu *et al,* 2009:90). Nevertheless, it is suggested that for successful cultivation *Setaria italica* needs regular rotation or shifting fields to prevent overwhelming increases in weeds, pests and diseases (Lu, 1999:136).

4.4 RICE (*ORYZA SATIVA*)

INTRODUCTION TO THE CROP

Today rice is a productive staple crop for a large proportion of the world's population (Sweeney and McCouch, 2007:1, Tao and Song, 2007:760). Rice is potentially the most productive crop per unit of land known and China has a long history of rice cultivation (Bray, 1984). It is unique among cereal crops in its ability to tolerate standing in water for long periods and it was therefore a good choice for cultivation and subsequent domestication in tropical and warm temperate regions prone to flooding or with abundant standing water (Chang, 1987:408). In general, archaeology has established that rice was earliest in the middle and Lower Yangtze (e.g. Crawford 2006, Fuller and Qin 2009), and then spread out to other parts of China.

RICE ORIGINS, DOMESTICATION, AND SPREAD

Rice is widely accepted as a crop native to southern China (e.g. Crawford, 2006; Fuller, 2007; Vaughan *et al,* 2008). Although originally many scholars believed there was a single Chinese origin for rice (Oka, 1988, Chang, 1995, Higham, 1995, 2005, Glover and Higham, 1996, Bellwood, 2005), genetic studies demonstrate the domesticated annual *Oryza sativa* was domesticated more than once (Chen *et al,* 1993, Cheng, *et al* 2003, Londo *et al,* 2006 in Fuller *et al,* 2007:1). There are two Asian domestic lineages for *Oryza sativa, indica* often related to the wild annual *Oryza nivara* and *japonica* derived from the wild perennial *Oryza rufipogon* (Vaughn, 1994, Fuller and Qin, 2009:89) although some geneticists consider amalgamating *Oryza nivara* and *Oryza rufipogon* into a broader concept of *Oryza rufipogon* (Fuller and Qin, 2009:89). Each ecotype provides different opportunities for exploitation. *Oryza nivara* produces abundant seeds and thrives in wet and dry conditions such as monsoon flooded land that subsequently dries or lake shores between low and high water (White, 1994:48, Vaughn, 1989; 1994, Fuller and Qin, 2009:90). *Oryza rufipogon* produces fewer seeds and production can vary from year to year (White, 1994:48, Vaughn *et al,* 2008; Fuller and Qin, 2009:90), but is perennial, using new tillers and rhizomes to thrive in permanently inundated soils, although more grain productive stands are found when the plant is stressed such as where the water table temporally dries out, which may have provided incentive for early cultivators to sow seeds in similarly water stressed conditions. These growing habits suggest early gathering may have been opportunistic and connected to foraging for other wetland foods, for instance, water chestnuts (*Trapa* spp.) and fox nuts (*Euryale ferox*). Environmental manipulation, such as water control and land clearance through burning, seems key to rice cultivation (Fuller and Qin, 2009:91). In the tropics, rice germination from seeds falling to the ground may have been enough to maintain early rice cultivation but in the more temperate regions of north and central China propagation was more likely to have depended on annual seeding by early cultivators (Chang, 1987:410). Where the dividing line between these regions should be drawn for the Early of Middle Holocene is, however, unclear.

The wild progenitor populations of Yangtze rice domestication(s) are extinct today, a concern when using changes in phytolith husk morphology to consider domestication. Firstly, changes in morphology are not demonstrably connected to the domestication process and secondly any differences may be related to unique characteristics transmitted to the domesticate from the wild progenitor so a result of phylogenetics rather than the domestication process (Fuller *et al,* 2007:4).

Currently, there is debate as to how early and how many times *Oryza* domestication has taken place (Fuller *et al,* 2007, 2008, Liu *et al,* 2007). Traditional thought in China has tended to assume early rice domestication (Ho, 1977, 1987). Evidence for pre domestication cultivation and wild rice foraging has not always been considered (Fuller *et al,* 2008, 2009). Disputed dates for early domesticated rice range from ca 10000 to 4000 BP. In the past the lack of consistent systematic flotation, large quantitative sampling programmes, detailed analyses of weed floras, among other issues means there has been considerable difficulty in developing comprehensive models of rice domestication (Fuller and Qin, 2009:94). However, recent work in the Lower Yangtze has collected and analysed weed and spikelet data, as well as examining husk size and shape (Zheng *et al,* 2007, Fuller and Qin, 2008, Fuller *et al,* 2009).

Wild rice was collected in the Yangtze valley during the terminal Pleistocene and Early Holocene. Phytolith analysis at Xianrendong and Diaotonghuan sites (Zhang, 2002, Zhao, 1998) indicates rice was a wild resource at sites with some of the world's earliest pottery at ca. 15000 years ago. Rice-husk tempered pottery was present in the Yangtze by 10000 – 8000 BP, for example at Shangshan (Jiang and Liu, 2006). Whether this rice is wild, morphologically wild cultivated, or domesticated has been debated (Fuller *et al* 2007, Liu *et al* 2007, Fuller and Qin, 2009, Sweeney and McCouch, 2007). Jiang and Liu (2006) imply that the rice is domesticated, based on bulliform phytolith morphometric evidence or spikelet Length-to-Width ratios (Liu *et al,* 2007). However, systematic studies of bulliform phytoliths call into question their use as domestication criteria (Pearsall *et al,* 1995; also, Harvey, 2006:198-220). Similarly, the range of variation documented in modern wild and domesticated rice populations, as well as the potential presence of immature grains to skew Length-to-Width ratios, removes the possibility of relying solely on grain/spikelet size or shape to determine if small numbers of rice finds were domesticated or wild (Crawford and Chen 1998; Fuller *et al,* 2007; 2008).

Because rice panicles mature over several days the number of ripe grains available at any one time changes. In order to maximise grain recovery, plants need to be targeted early on in grain production, consequently a large proportion of immature spikelets will also be gathered. This has two implications, the reduction of selection pressure for the evolution of non shattering (domesticated) plants and the production of assemblages with high proportions of immature spikelets, some containing long thin immature grain. Increasingly mature harvesting allowed by loss of wild shedding and selection for larger grains while under cultivation both contribute to morphometric change in archaeobotanical assemblages (Fuller *et al,* 2007:10: Figure)

Regardless of the controversies surrounding the earliest rice finds, it seems clear that rice was a major economic resource in the Middle to Lower Yangtze region, as well as the Huai river, by around 9-8000 BP. Analysis of rice remains found in quantity at the sites of Pengtoushan and Bashidang in the Dongting Lake region has been interpreted by some scholars as demonstrating cultivation was taking place by this period. Finds at Jiahu are of similar date (Pei, 1998b: 879, Zhang and Pei, 1997 in Lu, 1999:134). However, detailed analysis of morphological domestication traits is lacking from these sites, and possibilities of intensive wild collection or pre-domestication cultivation are not yet ruled out (cf. Fuller *et al* 2007). In the Lower Yangtze region the site of Kuahuqiao (6000-5400 BC) has produced rice remains, which include some non-shattering spikelet base types (Zheng *et al,* 2007), which can therefore be unequivocally placed in pre-domestication cultivation.

Hemudu (Yuyao, Zhejiang, ca 5000-4600BC) yielded extensive early rice remains including culms, panicles and chaff (Pei, 1998 b: 880, Liu, 1985, Zhao and Wu, 1987, Yan, 1982). Originally these remains were considered to represent fully domesticated rice agriculture (Crawford, 2005:84). However, Fuller *et al* (2007, 2008) have presented evidence based on the numerous immature spikelets that point to a wild or intermediate population, suggesting the rice at Hemudu was not fully domesticated. The finds at Hemudu were accompanied by wooden and bone agricultural tools that suggest possible use for rice cultivation (Yan, 1992:117, Fuller *et al,* 2007).

Recently the Hemudu-culture site of Tianluoshan has been systematically sampled, allowing this issue to be reinvestigated on the basis of substantial waterlogged and charred remains (Fuller *et al,* 2009). Tianluoshan produced vast quantities of wild nuts especially acorns and *Trapa,* which have also been reported from Hemudu and Kuahuqiao (Fuller *et al,* 2007), as well as middle Yangtze sites. The nut evidence, which has in the past been largely ignored, suggests rice was one component of a broader subsistence base focused on wild foods as well as some cultivation. At Tianluoshan the study of rice spikelet bases indicates the gradual increase over a 300 year period in morphologically domesticated (non-shattering) spikelets, and increase in rice remains versus nut remains, and the presence of a plausible arable weed flora of early wet rice fields (Fuller *et al,* 2009). These data indicate that the domestication process was underway at Tianluoshan (4900-4600 BC) and that by 4600 BC domesticated spikelet bases outnumbered wild shattering types. This domestication process was gradual and had begun many centuries earlier, certainly by the time Kuahuqiao was occupied (from 6000 BC) (Zheng *et al* 2007, Fuller and Qin 2009).

Additional evidence for a local domestication process at this time in the Lower Yangtze comes from morphometric data of both grains and phytoliths. A shift in grain size occurred at later levels at Longquizhuang (4800-400BC) and Chuodun (a Majiabang site) suggesting the process of morphological domestication took place during the Majiabang/ Hemudu period, i.e. (Huang and Zhang 2000, Fuller *et al,* 2007).

Another morphometric shift is found in the size of *Oryza* keystone phytoliths. Recent data from the Lower Yangtze indicate significant shift towards larger bulliforms through time. These data agree with the evidence of the grain data, that suggests earlier rice was being harvested at a less mature stage than later, presumably domesticated, rice of the Songze and Liangzhu phases (Fuller, *et al,* 2007, Zheng *et al,* 2004 in Fuller and Qin, 2009). More mature plants can produce larger bulliforms, so this may relate to allowing more rice plant to reach maturity as the populations became predominantly domesticated, although some adaptive shift in the plant population cannot be ruled out.

ORYZA WEEDS

Examining the weed floras associated with crops offers the possibility of indicating varying cultivation practices as has been demonstrated in the Near East (Colledge, 1998, Hillman *et al,* 2001, Wilcox *et al,* 2008, Fuller and Qin, 2009:95). Different weed assemblages should highlight the differences between wet or irrigated and dry farming. For example, many Cyperaceae are found with wet rice as are *Coix lachryma-jobi* and *Echinochloa. Chloris barbata* and *Brachiaria reptans* are weeds of dry rice (Thompson, 1996 Table 3.2), while *Phragmites* is often found alongside *Oryza* (Barnes, 1990:10), but usually grows at the edge of the fields rather than among the crop, so is unlikely to show in crop processing assemblages. A preliminary collation of potential rice weeds of dry-cropped versus wet rice is provided in Table

4.1, with emphasis on those weeds known to produce phytoliths.

Table 4.1 Wet and dry Oryza weeds that produce phytoliths (after Thompson, 1996: Table 32 Sources Kittipong, 1983, Vongsaroj, n.d, Fuller and Qin, 2009) Allard et al: 2007, Arora, 1977) (See chapter 7 for more detailed table)

Wet rice weeds	Dry rice weeds
Echinochloa colona	*Dactyloctenium aegyptium*
Panicum cambogiense	*Digitaria adscendans*
Leptochloa chinensis	*Eleusine indica*
Echinochloa crus-galli	*Eragrostis temella*
Paspalum scrobiculatum	*Panicum repens*
Ischaemum barbatum	*Cynodon dactylon*
Cyperus pulcherrimus	*Chloris barbata*
Fimbristylis littoralis	*Brachiaria reptans*
Cyperus difformis	*Ischaemum* cf. *rugosum*
Scirpus juncoides	*Andropogon sp.*
Cyperus sp	*Cenchrus ciliaris*
Scleria	*Fimbristylis* sp
Carex sp.	*Poa* sp.
Coix lachryma-jobi	*Echinochloa*
	Cyperus rotundus

4.5 TRADITIONAL *ORYZA* CULTIVATION PRACTICES

TRADITIONAL WET FARMING

Traditional rain fed wet rice cultivation in the Ban Chiang region in northeastern Thailand, a tropical plateau with wet and dry seasons, follows the water regime and ecology of the local lake-edge wild rice. Seedbeds are prepared for sowing early in the rainy season, where the seedlings grow for a month before being transplanted to paddy fields bunded with low dikes, where the water level rises to 10-20cm. There is some small-scale water control, occasionally water is channeled into fields or lifted from streams. By the time the dense stands of rice have matured the rains have finished and the fields dried (White, 1989:153-155).

TRADITIONAL DRY FARMING

The Baduy people in West Java farm rice as mixed crops in traditional swidden fields, intercropping the rice with vegetables, often nitrogen fixing legumes (*Cajanus cajan, Vigna sinensis*). Rice straw is commonly used as compost. Harvested rice is stored husked, in bundles in rice barns sometimes for up to 90 years (Iskander and Ellen, 2000:6-10).

In northern Laos rain fed rice has traditionally been grown in long fallow swiddens as a subsistence crop (Saito *et al,* 2008:64, Roder *et al,* 1995:27). Fallow periods of up to 38 years have restored nutrients to the soil and acted as a preventative measure against weed and insect infestation and disease. In the past few years more intensive agriculture has meant shorter fallows of 2-3 years coinciding with lower yields. However, variations in rainfall, especially during the reproductive stage, disease prevalence, weed competition and rodent damage are also considered to have a major impact on yields (Saito *et al,* 2008: 64, Saito *et al,* 2006:175, 182, Roder *et al,* 1997:111, Roder *et al,* 1995:31, 34,).

4.6 *ORYZA* CULTIVATION IN ARCHAEOLOGY

Oryza typically grows in two contrasting micro niches or ecotypes, lowland paddies, wet farming and upland swiddens or dry fields, dry farming. While the cultivation method differs, most varieties will grow in either niche although paddies are more productive. Lowland wet *Oryza* will grow on almost any soil as long as physiographic conditions allowing a suitable water supply are present. Wet farmed rice *Oryza* probably originally grew in more indiscriminate conditions in the past but as it became adapted to agriculture the niches became polarized (Barnes, 1990:1, Glover and Higham, 1996:413). Within these niches *Oryza* field systems can be divided into four categories, lowland irrigated, rain fed, flood prone and upland dry farming (Khush, 1997: 31, Mathews *et al,* 1995:100-1 Glover and Higham, 1996: 413).

WET FARMING, LOWLAND IRRIGATED

Irrigated rice is productive, one or more crops per annum, and grown in bunded puddled fields with guaranteed irrigation. This system is responsible for close to 40% of all rice production, while most of the rest is from rain fed systems (Khush, 1997:31, Barnes, 1990:2). Sufficient water is needed to inundate the seedlings for between 4–6 months of the growing season or at least 3 months for lower yields (Barnes, 1990:2). Landscape modification allowing the field to drain as the plants mature is also necessary (Fuller and Qin, 2009:89).

WET FARMING, RAIN FED

Rain fed rice is also grown in bunded fields, with ditches, dykes and reservoirs, which are flooded for part of the cropping season, ideally not by more than 50cm and not for more than 10 consecutive days. Some areas are more conducive to rain fed rice agriculture than others and some rain fed systems are as productive as irrigated fields. However in others, lack of water control and erratic rainfall can mean problems of both drought and excess waterlogging.

WET FARMING, FLOOD PRONE

Despite low unpredictable yields, flood prone cultivation exploits environments such as the back swamps of flood plains and slopes on natural levees in deltas in South and Southeast Asia. There can be a wide variation in water depth, from 50cm to 8m. Rice may be submerged as long as 10 days, stand in stagnant water for up to 5 months, or cope with daily submergence by tidal flow and seasonal flooding (Stoskopf, 1985:337-8, IRRI, 1993 in Mathews *et al,* 1995:101, Glover and Higham, 1996:413, Khush, 1997:31).

WET FARMING, PADDIES

The paddy base should be impermeable making deep ponding possible (Barnes, 1990:3). Taro is also grown in ponded conditions so can be intercropped with rice, although not in north China. In Japan during the winter a crop of '*igusa'* a grass grown for matting can take the place of rice. However, the majority of winter crops do not require as much water so during the winter months paddies are often ploughed into ridge and furrows to support winter vegetable crops and levelled in the spring in preparation for a summer rice crop (Barnes, 1990:9).

DRY FARMING, UPLAND RICE

Upland rice is cultivated in hilly or mountainous conditions on 0°–30° slopes. It generally forms part of a crop rotation and has lower yields. The fields are usually unbunded (there are exceptions in East India and Bangladesh) so there is little accumulation of surface water. The dry soil is prepared and seed directly sown. Both shifting or swidden cultivation and permanent fields are common. Shifting fields are used for up to 3 years then moved to either a new or previously abandoned site. Some fields are left fallow for permanent cultivation, which often involves intercropping, relay cropping or sequential cropping with other crops (Khush, 1997:31, Mathews *et al,* 1995:100-101, Barnes, 1991:1). Dry soils account for two thirds of arable land in South East Asia (Kawaguchi and Kyuma, 1977:1 in Barnes, 1990:1). Pfeiffer *et al* (2006:3-4) highlight a number of distinctions between upland and paddy rice in present day Indonesia. The crop of choice for upland rice is usually *Oryza sativa* subspecies *indica* as opposed to subspecies *japonica,* which is usually grown in paddies. Upland rice tends to be planted on more marginal stony soils, from seeds drilled into the soil rather than transplanted seedlings. It is rarely grown as a monoculture stand but often interplanted with a range of annual and perennial crops, such as maize. Upland rice is frequently part of a rain fed shifting swidden system with regular periods of fallow.

Swidden or slash and burn agriculture does not necessarily mean frequently moving settlements. It is likely that past swidden cultivation was supplemented by gathering from the wild and possibly also from fallow swidden land which is often rich in wild vegetables (Fu *et al,* 2003:293).

4.7 MODELS FOR AN ORIGINAL SYSTEM

Is there an original agricultural system for rice? Several models have been proposed. Gorman (1971) drawing a parallel with the Near East (Flannery, 1969:77), proposes a shift from tropical, inter-montane broad-spectrum hunting, gathering and possible cultivation of Cucurbits, Legumes, and tree crops (fruit and nuts) to lowland rice farming on the plains brought about by the introduction of cereals. White (1995:62) suggests both wet and dry rice cultivation evolved from a swamp rice cultivation system under natural inundation conditions, which is likely to have originated in subtropical climates with seasonal rainfall distribution. Fuller and Qin (2009:91) agree with White (1995) and also argue that environment management was key to the development of early rice cultivation. Clearing wetland and adjacent vegetation, including trees and drainage manipulation to encourage flooding during the wet season and subsequent drying at the end of the summer would encourage increased grain harvest from perennial wild rice (2009:94). While the latter hypothesis would fit with our current understanding of the evolution of *Oryza rufipogon* into *japonica* rice, for example in the Yangtze, the trajectory from annual wild rice, *Oryza nivara,* to *indica* cultivar, probably taking place in India, would be different.

4.8 SOUTHWEST ASIAN CEREALS: WHEAT AND BARLEY

Today bread wheat (*Triticum aestivum*) has replaced millets as the major crop in the north central China (Hsu and Hsu, 1977:300). Wheat and Barley both have their origins in the early Holocene in Southwest Asia and had reached central Asia (Turkmenistan, Pakistan) by ca. 6000 BC (Zohary and Hopf, 2000). Southwest Asian cereals arrived late in the Neolithic in western areas of the Yellow River region, becoming established between 2500 and 1500 BC. Finds of wheat are generally more widespread and may arrive earlier in some areas. There is evidence for wheat at Donghuishan, (Hexi corridor, Gansu), perhaps as early as at 2800 BC (Li, 2004 in Li 2007:558) and at Xishanping ca. 2600 cal BC (Li *et al,* 2007:558). Wheat has also been recorded in Longshan contexts at Liangchengzhen (Shandong) although not yet AMS dated (Crawford *et al,* 2005:311), and Late Longshan/ Shijiahe samples from Baligang (Ling Qin, personal communication). Wheat grains have been identified in Erlitou period ash pits at Huizui, (Lee and Bestel, 2007: table 2), and from Shang period samples in the Ying Valley (Fuller and Zhang 2007).

Barley (*Hordeum vulgare*) first makes an appearance in the central Plains in the Bronze Age, such as late Shang period grains (from 1500 BC) from the Ying Valley, (Fuller and Zhang 2007) but is present earlier on the north western periphery at Xishanping, (Tianshui, Gansu), perhaps at ca. 4600 cal BP along with wheat but without a direct radiocarbon date (Li *et al,* 2007:555). Other finds at this site included millets, rice, oats (*Avena*), and quantities of buckwheat (*Fagopyrum*) pollen.

4.9 EARLY AGRICULTURE IN NORTH CHINA

Evidence from sites across China shows a north south divide between the crop choices of the earliest farmers, millets in the north and rice to the south (Lee *et al,* 2007:1087, Barnes, 1999:92). The earliest cultivated cereal crops in North China appear to be *Setaria italica* and *Panicum miliaceum* (Lu, 2007, Zhao, 2005, 2006, Lee *et al,* 2007, Fuller *et al,* 2007:11, Underhill, 1997: 109). The cultivation of millets appears to be earlier than undisputed domesticated *Oryza* in the south (Fuller *et al,* 2007). Underhill (1997:120) suggests *Panicum miliaceum* is more common in early western central Yellow River Valley sites and *Setaria* in the east (Underhill, 1997:120). Apart from northern Xinglongguo and recent early finds at Dadiwan (Barton *et al*, 2009) and Cishan (Lu *et al*, 2009), the vast majority of carbonised grains are *Setaria italica*, suggesting this was the primary crop (Underhill, 1997:117). *Setaria italica* is well adapted for arid conditions (An, 1989:645) so would be an ideal dry farming crop, while *Panicum miliaceum* may have been insurance crop in case of drought (Lee, 2007:1091).

The evidence overall suggests widespread and large-scale millet cultivation of both *Setaria italica* and *Panicum miliaceum,* by 6000-5000 BC. Recent finds raise the possibility that *Panicum miliaceum* alone was the first cereal, brought into cultivation on the drier northern periphery, with *Setaria italica* added to, and becoming dominant later. The report of storage pits full of *Panicum miliaceum* spikelet (identified through phytoliths) at Cishan dates back to 8500-8000 BC (Lu *et al,* 2009). This new work at Cishan has clarified earlier problematic reports of millet phytoliths (Huang 1982, Tong 1984 in Hunt *et al,* 2007: table 1) by indicating the dominance of *Panicum* and the presence of some *Setaria italica* only in later samples from ca. 6000 BC. At Dadiwan in the northwest only *Panicum miliaceum* is associated with the Dadiwan Neolithic (6000-5500 BC) occupation in which millet-fed dogs (with a C-4 isotope signature) suggest cultivation (Barton *et al*, 2009). There is no evidence for pigs consuming millets at this time indicating that these were not a domesticated species (Barton *et al,* 2009:3). Flotation at Xinglongguo (Chifeng, Inner Mongolia) has produced abundant charred remains including more than 1500 charred millet grains from the middle Xinglonggua contexts (8000-7500 BP.). While the vast majority of these (>90%) are *Panicum miliaceum*, *Setaria italica* grains are also present (Zhao, 2006:2). At Peligang (Xinzheng, Henan) quantities of charred *Panicum miliaceum* and *Setaria italica* have been identified alongside a large number of agricultural tools, from 6000-5000 BC (Underhill, 1997:120, Yan, 1992:114). *Setaria italica* appeared in the Yilou valley along with early inhabitants during the Late Peligang (Lee and Bestel, 2007:5).

From 5000 BC, with the start of the Yangshao tradition (Banpo Phase), evidence for integrated dryland agriculture in north China becomes widespread. Sites normally produce evidence for both *Panicum miliaceum* and *Setaria italica* (e.g. Lee *et al*, 2007; Barton *et al*, 2009). In addition some sites have produced evidence of *Brassica* sp. seeds, such as Banpo and Dadiwan (Underhill, 1997: 120, Yan, 1992:114). It was into this cultural context, of the regionally expansive Yangshao tradition that rice was adopted alongside millet cultivation in the later 5th or early 4th Millennium BC.

Whether there was any rice cultivation during the Peiligang period is problematic. As reviewed above, Yellow River sites provide evidence for millet cultivation, but a few sites to the South or East have also produced finds of rice from this time period, including Jiahu in the Huai River Valley (ca. 7000-6000BC) (Higham and Lu 1998, Henan Institute, 1999, Liu *et al,* 2007), and two Houli culture sites of the lower Yellow River: Yuezhuang (equivalent Peligang Dates) (Crawford *et al,* 2006, Liu *et al,* 2007) and Xihe (Jin 2008, in Fuller and Madella, 2009). As already noted, however, the status of these early finds is unclear. Fuller *et al,* (2007; 2008) argue that these are too small to be domesticated, while Liu *et al,* (2007) maintain that they are domesticated. It is also probable that these sites are within the former range of the wild progenitor. Also, husk imprints have been found at Lijiacun and Hejiawan (Xiiang, Southern Shaanxi) (proto Yangshao culture) dating to the 6th millennium BC but the contexts are uncertain and systematic archaeobotanical study is lacking; also, it is not clear whether the rice is domesticated, cultivated locally or imported from the south (Ahn, 1993:211 in Glover and Higham, 1996:431). On the other hand, by sometime in the Early Yangshao period (4000-5000 BC) finds of rice in Henan, well beyond the potential range of the wild progenitor seem to be clear, and are confirmed by direct AMS dates from Early Yangshao finds at Nanjiaokou which date from at least 3900-3800 cal. BC (Ling Qin, personal communication).

4.10 Agriculture in the Yangshao Period

Cereal crops

According to the evidence from macro remains, the predominant staple crops in Yellow River area during the Yangshao period were millets, *Setaria italica* and *Panicum miliaceum* (Yan, 1992:114, Ho 1977, Smith, 1998, Lu, 1999, Fuller and Zhang, 2007:945). In the Ying valley *Setaria italica* dominates although *Panicum miliaceum* and infrequent *Oryza sativa* are both present. Panicoid weeds including *Panicum* sp. and *Setaria* spp. (as well as other grasses like *Digitaria*) have been identified in the macro remains (Fuller and Zhang, 2007:946). *Euphorbia* cf. *helioscopia*, a dryland crop weed with large seeds is only recorded in the Ying Valley during the Yangshao (Fuller and Zhang, 2007).

Setaria italica has been found at many Yangshao sites including, Banpo, (Xian, Shanxi) Beishouling, (Baoji, Shaanxi), (Yan, 1992:114), Huizui, (Yanshi, Henan) (Lee *et al,* 2007). At Linjia (Dongxiang, Gansu, quantities of *Setaria italica*, *Panicum miliaceum* and *Cannabis sativa* have been found in pits and ceramic pots on house floors (Teachers College, 1984 in Underhill, 1997:125, Yan, 1992:114). Evidence of unthreshed ears of *Panicum miliaceum* tied into bundles, dried in the sun and stored directly in the pits provided insight on storage practices (Yan, 1992:114). At Liuwan, Ledu, Qinhai, *Setaria italica* was found in guan-pots in burials.

Rice is present at many sites during the Yangshao period, but with increasing presence from the Middle and Late Yangshao periods. This is clear from systematic regional studies of macro-remains such as in the Yilou (Lee *et al,* 2007) and the Ying (Fuller and Zhang 2007). In the Late

Yangshao millets remain the predominant cereal crops in the Ying and Yilou Valleys. As already noted the earliest secure date (direct AMS) on rice grains from systematic sampling (flotation) comes from Nanjiaokou at 3900-3800 BC. Underhill (1997: table IV) notes incidences of *Oryza sativa* during the early Yangshao at Xiawanggang and Xiaji (Xiachuan, Henan), but dating evidence and lack of systematic sampling on these sites means that questions remain. Yan (1992:114) mentions paddy rice at Xiawanggang (Xichuan, Henan) and Hejiawan, Shaanxi, but does not describe how it was identified as such. Rice macro remains have been reported also from Yangshao sites Dahecun in Zhengzhou, Henan and Quanhucun in Huaxian in the Yellow River Valley but so far they are neither securely identified nor dated (Liu *et al,* 2005:85, Yan, 1982 in Liu *et al,* 2004:85). Andersson reported rice at Yangshao (now dated to Miaodigou II/ Late Yangshao 2600-2200 BC) (Glover and Higham, 1996: 431). There is phytolith evidence of rice at Late Yangshao Yulinzhuang (Lee *et al,* 2007: 1091, Liu *et al,* 2002), and charred rice grain in sediments from Zhaocheng (Lee *et al* 2000:1091). Carbon 13 analysis at Taosi (Xiangfen, Shanxi) points to 58 % of the Yangshao diet being made up of plants with a C4 photosynthetic pathway, which points to millets as the major staple (Cai, 1984 in Yan, 1992:115). This percentage would also include meat from millet-fed pigs. This relegates rice, pulses, fruits, nuts, vegetables and wild game like deer to comprising the other 42 % of vegetable diet.

OTHER CROPS

At the early Yangshao site of Banpo, Xian, Shaanxi, there is evidence of chestnuts, (*Castanea* sp.), hazelnuts (*Corylus* sp.), pine nuts (*Pinus* sp.) and Chinese hackberry, (*Celtis* sp.) indicating woodland gathering played a part in subsistence although agriculture is clearly characteristic of Yangshao economy (Chang, 1986:112). Remains of *Setaria pumila,* yellow foxtail millet, were also reported from Ching-ts'un (Bishop, 1933: in Chang, 1986:113), although it is unclear if this was gathered as a wild food. Later finds from the Middle and Late Yangshao, such as in the Ying Valley, indicate the gathering of several fruits (apricots, peaches, grapes, jujubes) as well as probably wild soybeans (Fuller and Zhang, 2007). The overall quantities of wild foods declined in the subsequent Longshan period suggesting the gathering was more important in the Yangshao era than later.

AGRICULTURAL TOOLS

Throughout the Yangshao agricultural tool technology advanced and a wide variety of stone tools were developed including many types of spade for soil turning, trapezoid, U shaped, heart shaped, rectangular with shoulders (Yan, 1992:115). At some sites there is evidence of long narrow incisions cut into the sides of pits possibly by a tool referred to in ancient texts as a '*lei*' (Yan, 1992:115). Stone and ceramic harvesting knives ('*zhi'*) were used for cutting millet panicles. The developments in sophistication of stone tool technology mirror progress in agricultural technology (Yan, 1992:115).

During the Yangshao stone spades become more uniform, trapezoid and rectangular with shoulders. Stone, but not ceramic, knives continued to be used as harvesting tools along with stone sickles and this continued into the Shang-Zhou period (Yan, 1992:115).

AGRICULTURAL SYSTEMS

There are questions regarding the favoured dry land agricultural system practiced during this period (Chang, 1986, Underhill, 1997:119, 126). Ho (1969, 1977, 1984) in Underwood, 1997:127) considers the fertile Loess soils as ideal for short fallow, while Chang (1986) believes large numbers of possible woodworking tools point a forested environment and so to swidden agriculture. Liu (1994 in Underhill, 1997:148) also suggests early Yangshao farmers practiced swidden farming. However, there is no evidence for shifting cultivation. This method of farming does not preclude settlement for fairly long periods, depending on the length of fallow. In current dry rice growing systems in Laos this can range from as little as 2 to around 40 years. (Saito *et al,* 2006, 176). Indigenous millet farmers in Taiwan use the same fields for up to 20 years (Fogg, 1983:100). Given the stability of known settlement sites for extended periods, however, Yangshao systems are likely to have been dependent on local environment and soil conditions (Underhill, 1997:127).

4.11 AGRICULTURE IN THE LONGSHAN PERIOD

CEREAL CROPS

Longshan sites are widely distributed across diverse warm and cool temperate zones in north central China. During the early Longshan foxtail millet (*Setaria italica*) continued to be the predominant staple grain crop in the Ying and Yilou Valleys of central Henan, Densities of broomcorn (*Panicum miliaceum*) also increased along with high proportions of Paniceae and Panicoideae weedy grasses (Fuller and Zhang, 2007: 946, Lee *et al,* 2007:1088). Carbon isotope analysis at Longshan Taosi, (Xianfen, Shanxi) suggests 70% of the diet was made up of C4 crops a considerable increase from the Yangshao (58%) (Cao, 1984 in Yan, 1992:115).

Rice (*Oryza sativa*) grains AMS dated to the Late Longshan are present in the charred remains at Huizui, and rice phytoliths have been identified at Nanshi and Luokou NE (Lee *et al,* 2007:1091, Liu *et al,* 2002). In these Yilou sites, there appear to be increased densities of manna grass (*Glyceria*) a rice crop weed suggesting greater use of wetland habitats (Lee *et al,* 2007:1091).

Wetland cultivation is also suggested by the increase in sedges and some grasses that were confined to the Longshan period samples in the Ying valley (Fuller and Zhang 2007): *Alopecurus,* (40-50cm), *Eleusine indica* (usually low growing sometimes 90cm), cf. *Pennisetum, Papsalam* spp. (Fuller and Zhang, 2007).

In contrast to sites in northern Henan, rice was the most common crop at Baligang in the Nanyang basin, southern Henan, (Fuller unpublished data, cf. Fuller and Zhang, 2007) and appears to have played a significant role in the agricultural economy in Shandong (Crawford, 2005:313). In general, there are grain impressions in ceramics and macro botanical evidence for rice from sites throughout much of northern China, suggesting that rice became more common from the Late Yangshao times, after ca. 2800 BC (Underhill, 1997:129). In Gansu, the evidence from Xishanping, which includes a direct AMS date, indicates that rice had diffused this far by 2600 ca. BC (Li *et al,* 2007).

Archaeobotany at Liangchengzhen in Shandong, points to a mixed agricultural economy with rice and *Setaria italica* as the key crops (Crawford *et al,* 2005:314-315). Residue analysis of ceramic containers suggests some rice was being used to produce alcoholic beverages (Chinese American Rizhao Liancheng Region Collaborative Archaeological Team n.d. in Crawford *et al,* 2005:315, see also McGovern *et al,* 2004)

Southwest Asian cereals, wheat (*Triticum* cf. *aestivum*) appear at Donghuishan (Li, 2002 in Fuller *et al,* 2007: 13, Crawford *et al,* 2005:) and wheat and Barley (*Hordeum*), at Xishanping (Li *et al,* 2007) by ca. 2500 BC. Elsewhere Longshan finds of wheat are rare (e.g. Crawford *et al,* 2005), and barley is absent, with both cereals, but especially wheat, becoming more frequent from the Bronze Age (e.g. Fuller and Zhang 2007; Lee *et al,* 2007). "Barley" reported from late Longshan Taosi, (e.g. Zhao, 2006 in Fuller *et al,* 2007:13), is now recognized as being a mis-identification of the inner kernel of *Xanthium* (Asteraceae) (Dorian Fuller, personal communication).

Li *et al,* (2007) have recorded a suite of eight rain fed main crops at Xishanping from 4600-4300 cal BP that include millets, rice and southwest Asian cereals (including possible oats) as well as buckwheat (from pollen) and soybeans indicating expansion of agricultural crops and techniques and /or farmers in northwest China at this time. As these soybeans are likely beyond the range of the wild progenitor, they provide the earliest unambiguous evidence for the cultivation of this species (Dorian Fuller, personal communication). Unfortunately the evidence from Xishanping was not quantified so it is not possible to judge the relative importance of different species recovered.

Edible wild plants were also used. This is indicated by fruit finds in the Ying (Fuller and Zhang 2007), the Yilou (Lee *et al,* 2007), and at Liangchengzhen (Crawford *et al,* 2005). Sometimes it is not clear whether species were wild food plants or weeds, such as *Chenopodium* and knotweed (*Polygonum*) (cf. Crawford *et al,* 2005; Lee *et al,* 2007). In addition finds of beans, adzuki (*Vigna* cf. *angularis*), soybeans (*Glycine* sp.), are ambiguous as to whether these were indeed cultivated by Longshan times or might have been gathered from wild populations. (Appendix 2).

INTENSIFICATION

The processes of agricultural intensification probably took place during the Late Neolithic Longshan period in north central China but the mechanisms have not yet been understood. Marked cultural changes took place during the 3rd millennium BC (see Chapter 2). Intensification would have been a primary factor enabling these changes to take place (Chang, 1986:250, Crawford, 2006:89). In the east at Liangchengzhen in Shandong rice increased in importance, wheat was added to crop complex (Crawford *et al,* 2001 in Crawford, 2006:89).

AGRICULTURAL SYSTEMS

Underhill (1997:132) suggests wet cultivation systems may have been developed. It seems likely the larger walled sites of the Longshan relied on short fallow systems to support higher yields (Morrison, 1994: 115, Underhill, 1997:139) but there is no direct evidence, such as new tools, to support this so far (Underhill, 1997:148). Before 2500BC most ground stone tool assemblages contained 30% to 40% axes, the rest being adzes, shovels and hoes but a change can be seen in Late Longshan assemblages and into the Erlitou period when harvesting tools, such as sickles and knives came to the fore perhaps reflecting changes in agricultural production (Wang, 2007 in Jing and Campbell, 2007:103).

4.12 AGRICULTURE IN THE ERLITOU PERIOD

CEREAL CROPS

In general there seems less evidence of what was happening to agricultural economies during the Erlitou period. In the Yilou valley the plant remains follow similar patterns to the Longshan period. There are increased rice remains, although only at larger sites, such as Shaochai (Lee *et al,* 2007:1091). Dry crops predominate, primarily *Setaria italica,* and some rice (Lee, 2004:186). At Huizui *Setaria italica* is still the major crop, 80% of all crop remains, followed by *Panicum miliaceum,* 8%, then minimal rice and wheat (Lee and Bestel, 2007:51). Rosen (2007) suggests this may point to social rather than economic factors. The increasing centralisation and stratification during the Erlitou period may have resulted in the use of rice as an elite prestige crop. Another interpretation could be an increased reliance on rice as an insurance or buffer crop with reliable productivity even during periods of climatic uncertainty because of heavy investment in the construction and maintenance of paddies (David Taylor, personal communication).

Wheat has been reported from the Erlitou period site of Zaojiashu (Louyang, Henan) but it is undated. Photographs of possible barley from the same site are inconclusive (Ye, 2000 in Crawford, 2005: 311, Louyang Relics Team, 2002 in Lee and Bestel, 2007:51), and are most likely misidentified inner kernels of *Xanthium sibiricum* (Dorian Fuller, personal communication).

The larger Erlitou period sites in the Yilou Valley demonstrate an increase in weedy grasses (Lee, 2007:1091), this is illustrated at Huizui by high proportions of panicoid grasses, including barnyard grass (*Echinochloa crus-galli*), foxtail grass (*Setaria italica* spp. *viridis*), and panic grass (*Panicum* cf. *bisulcatum*). All of these grasses thrive in agricultural fields so are likely to represent crop weeds. Some panicoid weeds at Huizui are very similar to domesticated millet (Lee and Bestel, 2007:54), which may point to established multiple cropping practices that can encourage mimicry. Fuller and Zhang (2007) report wild *Setaria* spp. and wild *Panicum* sp., but also many immature grains of *Setaria* and *Panicum,* presumably from the crop species. The quantities of immature grains may be a product of agricultural practices (with poor grain filling leading to higher quantities), or crop-processing (with winnowing concentrating immature grains/spikelets because these are much lighter than fully matured spikelets). Higher proportions immature millet grains tended to correlate with higher proportions of weeds, suggesting the presence of winnowing waste in some Ying valley macro botanical samples (Fuller and Zhang 2007). Legumes, chenopods and infrequent Cyperaceae are also present at Huizui (Lee and Bestel, 2007:54). There are no Cyperaceae recorded in the few Bronze Age samples from the Ying Valley (Fuller and Zhang, 2007). According to Lee and Bestel (2007: 57) both Luokou NE and Shaochai, contemporary regional centres, in the Yilou Valley, have far higher densities of crops and weeds than Huizui. Two small sites, Huilongwan East and Xinhougou-Yaochang East, have even greater densities of crops and weeds. However, due to the small number of samples taken it is not clear whether this is representative of the region as a whole (Lee and Bestel, 2007: table 5). There is limited evidence for nuts and fleshy fruits such as blackberry (*Rubus* sp.) and plum (*Prunus* sp.) (Lee and Bestel, 2007: 54, Lee *et al,* 2007:1091).

Pigs are still dominant but cattle, sheep and goats are becoming more common, and consequently the weed seed densities may be affected by foddering and dung-burning. Shifts in organization of labour for crop-processing, as discussed by Fuller and Zhang (2007), would also be expected to affect weed seed densities. Further analysis on this issue is needed.

AGRICULTURAL TOOLS

Although bronze technology existed it does not seem to have been used to cast agrarian tools (Lee, 2004:186). Finds at Huizui point to its role as a stone spade production centre during this period (Ford, 2004, 2001). Agricultural tools have been excavated from Erlitou (Lee, 2004:186) suggesting some cultivation was taking place on or close to site but the vast majority of food production almost certainly occurred in the hinterland and beyond.

4.13 SHANG

The cultural period following the Erlitou appears to have less archaeobotanical evidence of agricultural practices but by this time there is writing so some evidence is available from oracle bones and later poetry (Chang, 1980) which can give an indication of what followed the Erlitou, although the dangers of relying on ancient text for evidence of climate and vegetation should be considered.

According to various interpretations of the Shi Jing (Book of Songs) the environment consisted of a natural loess landscape of semi arid steppes (Ho Ping-ti, 1975 in Chang, 1980:144) or contrastingly, trees, forests, bamboo groves and water (Chang, 1980:145). Oracle bones contain symbols for trees that have been interpreted as apricot, medlar, willow and cypress. Symbols for economic cereals such as *Panicum miliaceum, Setaria italica*, rice, possibly soybean, wheat and wild rice have also been identified (Chang, 1980:148).

In the Yilou valley site sizes fall during this period, this is probably due to population decrease and the movement of the primary centre from Erlitou to Yanshi and then Zhengzhou (60 km away) (Liu and Chen, 2003, Lee *et al,* 2007:1091).The most significant development is the introduction of wheat. While *Setaria italica* remains the main crop, wheat is now second in abundance. Rice although rare is still present at Shangzhuan and Tianposhuiku (Lee *et al,* 2007: 1091).

Land cleared for farming is recorded in the oracle records. Trees were commonly cleared by ring barking. There is little evidence of metal agricultural implements during the Shang period, tools were still largely made from wood, stone, bone and shell, for example a large number of slate harvesting knives have been recovered. The most important agricultural tool during this period is the *lei*; a two pronged digging stick used to turn the earth over and possibly also as hand plough (Chang, 1980:224). References to collective cultivation suggest large production units of slaves or farmers depending on how the text is understood (Chang, 1980:226). What is clear is the large numbers needed to supply enough agricultural labour to feed the large urban populations of this period.

CHAPTER 5

SITES AND SAMPLING

5.1 SITES

INTRODUCTION TO SITES

The basis of this research is formed by the analysis of material collected from four key sites and seven peripheral settlements in Henan, north central China. The larger sites are Xipo, Baligang, and Huizui and Erlitou, (Figure 1) while the smaller villages are Gong Jia Yao, Mazhai, North East Gaoya, North East Zhaiwan, South East Zhaiwan, Peichun, and Yuangou.

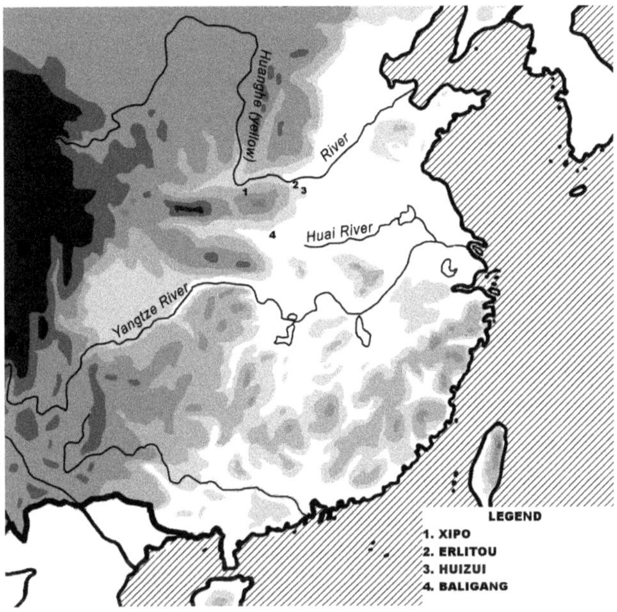

Figure 5.1 Map showing key sites

5.2 XIPO

XIPO, BACKGROUND TO THE SITE

The village of Xipo (34° 92'54, 80"N, and 110°40'03.33"E) is located in Lingbao county, western Henan, on the south-eastern Loess Plateau. Lingbao lies between 34° 7" and 34° 44" North latitude and 110° 21" and 110° 11" East longitude (Jia *et al.* 1992:104 in Ma, 2005). It borders Shaanxi province to the east and the Yellow River (Huanghe) flows from west to east along Henan's northern border with Shanxi. The Qinling and Xiaoshan mountain highlands in the south descend from the highest point at 2413.8 m. through central loess tablelands to the Yellow River terraces and sands to the lowest point of 208m (Jia *et al*, 1992:129 in Ma, 2005: 11). Seven tributaries flow from the mountains into the Yellow River dividing the central loess region into six flat tablelands of 450-600m above sea level.

Phase	Site	Date
Erlitou	Erlitou,	BRONZE AGE
Phases II-IV	Huizui West, Yuangou,	3850 - 3100BP
	SE Zhaiwan,	
Late Longshan	Huizui East, Baligang,	Terminal NEOLITHIC
Shijiahe	Mazhai, Jueshan, NE Gaoya,	4500 –3850BP
	NE Zhaiwan, Peichun	
Early Longshan	Huizui East, Baligang	Late NEOLITHIC
Qujialing		5000 – 4500BP
Late Yangshao	Huizui East Baligang,	Middle
Early Qujialing		NEOLITHIC
		5500 – 5000BP
Mid Yangshao	Xipo, Baligang,	Middle
Miaodigou	Gong Jia Yao	NEOLITHIC
		6000 – 4500BP
Early Yangshao	Baligang	Middle NEOLITHIC
		6900 – 6000BP

Table 5.1 Chronology – the widest date ranges are described

Due to erosion of the northern border of these tablelands and the Yellow River terraces are nearly vertical 30- 50m high cliffs, while deep gullies and riverbeds mark the eastern and western edges. The predominant soil type in Lingbao is loess (Ma, 2005:11).

Xipo is 3 km north of the Qinling Mountains, south of the Zhudingyuan, the Upper Sha River Valley. Its elevation is 460 – 472m above sea level sloping from southwest to northeast. The site sits between the Fufu and Linghu Rivers, two branches of the Sha River. It is situated in the northwest of the village and covers an area of approximately 40 ha. A modern road running from northwest to southeast divides the northern and southern parts of the site. There is also disturbance from modern agriculture and graves. The cut for the road reveals abundant archaeological remains such as, house floors, foundations and ash pits. Many similar features can be observed along the village pathways, terraces and gullies around the site (Ma, 2005:29, Li Xinwei, 2006, Personal comment, Personal observation) (Figures 5.6.)

Figure 5.2 Section through Xipo with large ash pit

The present climate of Lingbao County is temperate, arid and monsoonal with an average rainfall of 619.5mm. From June to July the average precipitation is 275.1mm contrasting with a more arid 25mm between November and February. There were 37 droughts between 1484 and 1985 and 37 floods between 1588 and 1985 (Jia *et al*, 1992:120 in Ma, 2005:12).

Over the past 40 years the average temperature has been 13.8°C, ranging from -17°C in January to 42.7°C in July. At present there are an average 215 frost-free days per year. This makes two crops per year possible in the lowlands and tablelands and one crop per year in the mountains (Jia *et al*, 1992:188 in Ma, 2005:12).
Pollen analysis from the neighbouring Wei Valley suggests a similar semi arid monsoonal climate during the Neolithic (Wang, 1988 and Zhou, 1963 in Ma, 2005:13) although there would have been variation throughout the period.
Today more than 300 species of coniferous and deciduous trees grow in this region. The foothills and gullies are covered by grassland that consists of a wide variety of Gramineae species. The loess plateau is covered by open forest, brush land and grassland (Ma, 2005:13)

FIELDWORK AT XIPO

There have been six archaeological surveys of Lingbao made between 1956 (Jia *et al* 1992: 775) and 2005 (Ma, 2005:13). Local archaeologists have located more than 100 Neolithic sites (ca. 6500 to 2000BC) and several Palaeolithic and early Bronze Age sites (Yang, 1991:168-169 in Ma, 2005: 13). Xipo has been under excavation by IACASS (Institute of Archaeology, Chinese Academy of Social Sciences) and Henan Provincial Institute of Cultural Relics since 2000 (Ma, 2005:29).

XIPO SAMPLES

I collected 122 samples from Xipo during the field season in April 2006. Of these 39 were processed and analysed for this project (Table 5.2).

Lab no	Sample	Homogenous ash layer/midden	Hearth	Laminate pit fill	Floor /foundation
X01	HLX06 X005				X
X02	HLX06 X006	X			
X03	HLX06 X007	X			
X04	HLX06 X011	X			
X05	HLX06 X060	X			
X06	HLX06 X064A	X			
X07	HLX06 X079	X			
X08	HLX06 X103	X			
X09	HLX06 X105	X			
X10	HLX06 X107	X			
X11	HLX06 X074A			X	
X12	HLX06 X052	X			
X14	HLX06 X053	X			
X15	HLX06 X015	X			
X16	HLX06 X041	X			
X20	HLX06 X020	X			
X21	HLX06 X099	X			
X22	HLX06 X104	X			
X23	HLX06 X081 F4		X		
X24	HLX06 X088				X
X25	HLX06 X096	X			
X26	HLX06 X024	X			
X27	HLX06 X008			X	
X28	HLX06 X004 F4			X	
X29	HLX06 X002 F3			X	
X30	HLX06 X049			X	
X31	HLX06 X037	X			
X33	HLX06 X059	X			
X34	HLX06 X051			X	
X35	HLX06 X009	X			
X37	HLX06 X019	X			
X38	HLX06 X065A	X			
X39	HLX06 X035	X			
X40	HLX06 X096	X			
X41	HLX06 X109	X			

Table 5.2 Xipo samples

There are also paths running through the village and surrounding fields and orchards that offer similar deep cuts through the Neolithic site providing a cross section of pits and floors which were sampled for both phytoliths and flotation

XIPO DATING

All the samples from Xipo have been dated by pottery typology to the mid Yangshao period (Miaodigou I).

XIPO CONTEXTS

The majority are from homogenous ash middens, while two are from house floor/foundations. Four are from laminated pit fills. During the 2006 season when the phytolith samples were collected the excavations at Xipo were focussed on the cemetery so there was little opportunity to collect sediments from excavation of structures. However, the modern village which overlies the Neolithic settlement is cut through by a modern road (Figure 5.2) providing a clear section through the village which exposes many ash middens and several structure floors, some containing hearths (Ma Xaolin, Personal Comment).

XIPO, HOW PHYTOLITH ANALYSIS WILL CONTRIBUTE TO THE SITE

So far little archaeobotanical work has been completed at Xipo. Flotation samples were collected at the same time as the phytolith samples and are awaiting completion of analysis (Gyoung-Ah Lee, Personal comment). Phytolith analysis should provide information on the agricultural economy of Xipo during the Yangshao and may address issues of labour organisation, crop choices and their relationship with social change.

5.3 BALIGANG

BALIGANG, BACKGROUND TO THE SITE

Baligang is a late Neolithic site situated in the village of Baizhuang, Dengzhou City, Henan, between 32 ° 22 ' and 32 ° 33' North latitude and 111 ° 00'38 " and 120 ° 00'20" East longitude (Jiang and Zhang, 1998).

The southernmost of the sampled sites, Baligang is situated in the northern margins of the northern subtropical zone, in the northern catchment of the Yangtze River. Baligang is sited inside the saddle shaped Nanyang Basin, which is approximately 100 – 140m above sea level. The terrain is flat, surrounded by mountains on three sides and open to the Jianhang Plain to the south. The Central Plains Channel an important traditional north /south route for trade and exchange lies to the south (Jiang and Zhang, 1998: 66).

Baligang is around 100m south of the Tuan River flowing from west to east. The local river system flows south to the Yangtze (Jiang and Zhang, 1998: 66).

The present climate is warm, temperate with four distinct seasons. The average annual temperature is 15°C but the temperature ranges from 38°-43°C in August to –10°C in February. The annual precipitation is 500 – 700mm with around 70% of rain falling during the summer (Jiang and Zhang, 1998: 66). According to Jiang and Zhang (1998:66) the Palaeoclimate was mainly subtropical, warmer and wetter with some climatic fluctuations.

Soils are predominantly yellow brown and black. The present vegetation on the limestone and yellow soils consists of evergreen shrubs, elm and deciduous broad-leafed trees while on sandstone, granite and gneiss in the mountains there are a few evergreen trees, deciduous oak forest and pine forest. Analysis of pollen from the site suggests the palaeovegetation consisted primarily of evergreen oak (*Castanopsis*), oak (*Quercus*), willow (*Salix*), walnut (*Juglans*) and pine (*Pinus*) (Jiang and Zhang, 1998:66).

Baligang lies between the millet farming areas of the north and rice cultivation regions of southern China. Evidence for both has been found (Jiang and Zhang, 1998:66).

FIELDWORK AT BALIGANG

The site was discovered in 1957 and was surveyed during the late 1970's and early 1980's. Baligang has been under excavation by Peking University from 1991 in collaboration with Nanyang Institute of Cultural Relics. Nine big scaled excavations have taken place (1992/94/96/98/2000/2002/2004/2007), so far the excavation area is about 8000 square meters in total. There are 4- 5m of cultural deposits with evidence for both Qujialing, which is contemporaneous to Late Yangshao in the Yellow River Valley, (6000-4800 BP) and Longshan settlement (Shijiahe 4800-4000 BP) (Jiang and Zhang, 1998:66).

Because of the unearthed mid-late Yangshao settlement, this site is now classified as National Protection Heritage. Baligang's Neolithic cultural deposits range continuously from the Peligang to the Longshan period, helping to build the complete chronology of the Nanyang basin, which is also important as part of the general chronological comparison between the mid Yangtze and Central plains area.

BALIGANG SAMPLING

Dorian Fuller, Qin Ling and students from Peking University collected the samples from Baligang in 2004 and November 2007 (Table 5.3). 29 samples were processed and analysed.

The Baligang samples were collected from either homogenous layers within ash middens or a long section of laminated layers from a large pit (CT1307). This pit section was at the edge of the 2005 excavated area and the samples ranged from the early Qujialing Period (3500BC) to the Western Zhou Period (700-249BC). The results from the Eastern Zhou samples have not been used in the major part of the analysis but provide a hint of what was happing afterwards. Eight samples from Baligang date from the early Longshan or Longshan period (2500-2000BC). There are seventeen samples from the Qujialing Period, which is contemporaneous to the Yangshao in the Yellow River Valley. The samples range from early Yangshao, through mid to late Yangshao (see Table 5.3).

BALIGANG DATING

The "Origin of Chinese Civilizations" project helped to produce 42 new radiocarbon dates from this site early this year, which provide a clear chronology of this site. The dates have not yet been published formally but the rough framework is:
Pre-Yangshao: 6500-6300BC
Early Yangshao: 4300-4000BC
Mid Yangshao: 4000-3000BC

Qujialing (Late Yangshao in central plain): 3000-2500BC
Shijiahe: around 2500 BC (very short in this area)
The Longshan has still to be radiocarbon dated (Qin Ling, Personal comment).

BALIGANG CONTEXTS

Samples were collected from pit fills ("ash pits") or cultural layers, and from Trench CT1307 from a long section of laminated layers. This trench section was at the south-western edge of the site area.

BALIGANG, HOW PHYTOLITH ANALYSIS WILL CONTRIBUTE TO THE SITE

Phytolith analysis has been undertaken previously at Baligang (Jiang and Zhang, 1998), chiefly to identify and reconstruct *Oryza* cultivation patterns and investigate the spread of rice agriculture from the analysis of 15 samples. Further phytolith analysis of a larger dataset can build on previous archaeobotanical analysis and add to the macro archaeobotanical data already gathered, as well as address the research questions relating to this project stated in chapter 1. Many of the phytolith samples come from sediments that do not contain charred material so can afford information on unburned plant material that has not been provided by the charred macro remains. The ash samples may supply evidence of plant material that did not survive burning in recognisable charred forms.

5. 4 ERLITOU AND HUIZUI

The sites of Erlitou and Huizui are in the Yilou River basin, part of the Zhengzhou-Louyang region, an extensive lowland area along the Yellow River in western Henan (Liu and Chen, 2003:22). The Yi and Lou Rivers flow from southwest to northeast through the valley where they join to form the Yilou River, which drains into the Yellow River (Liu and Chen, 2003:35, Liu *et al*, 2004:76). The valley is a large fertile alluvial basin with plains and terracing on a loess tableland 10-70m above the riverbeds (Liu *et al*, 2004: 77, Lee, 2004: 173), with the Yellow River and Mangling Hills to the north and surrounded by mountain ranges on all the other sides, including the Xiaoshan, Xiong'er, Funiu and Songshan. These ranges run from west to east and range from altitudes of 500 to 2000 m (Gonxian County, 1991:43, 69-70 in Liu *et al*, 2004:77).

The modern climate is 'temperate, sub-humid and monsoonal' (Ren, 1985:151-185 in Liu *et al*, 2004:76).

5.5 ERLITOU

ERLITOU, BACKGROUND TO THE SITE

Situated between 34°41′10″ and 34°42′23″ North latitude, 112°40′16″ and 112°41′55″ East longitude, Erlitou is situated in the Yilou River Basin south of the present course of the Luo River, between the modern villages of Erlitou and Gedangtou in Yanshi County, Western Henan Province. The ancient river flowed to the south of the site marking the boundary (Liu and Xu, 2007: 888, 893). Discovered in 1959 (Xu, 1959 in Lee 2004:176), and excavated intensively ever since, the site covers 300ha (Erlitou 2001, in Liu and Chen, 2003:29, Liu and Hu, 2007:886) or 400ha (Kaogu Yanjiusuo Zhongguo Shehui Kexueyaun, 1999 in Lee, 2004:173) and has been described as 'the largest urban settlement of the earliest archaic state developed in northern China' (Liu and Xu, 2007:887:886).

The primary centre for the Erlitou culture (c.a. 1900-1500 BC), there is evidence for three small settlements during the late Yangshao (ca. 3500 -3000 BC) and one small early Longshan village (ca. 3000 – 2500 BC). After a ca. 500-year gap between the Longshan and Erlitou occupation groups arrived and settled at c.a.1900 BC. During the Erlitou period a large urban site developed in phase II, peaked in phase III and declined during phase IV (1800 - 1500BC) to a village during the Erligang period (c. 1600-1250 BC) and subsequent abandonment (Liu and Hu, 2007:886-7, Liu and Chen, 2003:57-8).

During Erlitou phase I the site was a minimum 100ha and seems to have been the largest centre in the Yilou region. Liu and Xu (2007:888) explain this rapid development by migration from the neighbouring vicinities. Elite items, for example white pottery, ivory, turquoise and bronze, have been excavated from this phase (Institute of Archaeology, 1999 in Liu and Hu, 2007:888). Deposits from this period have been disturbed by later occupations so the layout of the settlement is difficult to see (Liu and Hu, 2007: 888). The site grew to 300 ha, its greatest extent, in Erlitou Phase II. Four 20m wide intersecting roads outlined a 12 ha palace complex (fig 5.11) (Liu and Hu, 2007:888). Rammed earth foundations, spread both inside and outside the palatial complex and a wooden drainage system has been discovered. Elite burials containing high status grave goods, such as bronze, jade, lacquer, white pottery, proto porcelain, turquoise, shells and cowries, were found for the first time within a palatial complex (Erlitou working team, 2005 in Liu and Hu,(2007:889).

The earliest evidence for the use of a wagon in China has been suggested by the discovery of parallel wagon tracks in the southern road although it is not clear whether human or animal traction was used (Erlitou Working Team, 2004b in Liu and Hu, 2007:889).

Craft production workshops, and a bronze foundry belonging to this phase have also been unearthed. These discoveries highlight the prosperity of Erlitou during this period as well as demonstrating urban planning (Liu and Hu, 2007:889).

Intensive agricultural and land levelling have destroyed a portion of the archaeological remains and three modern villages cover much of the site (Liu and Hu, 2007: 887). During many years of excavation a number of large rammed earth foundations have been revealed and are protected so only restricted parts of the site can currently be excavated. Recent excavation and surveys had added

Lab no	Sample	Period	Homogenous (ash layer/midden)	Hearth	Laminate pit fill
B1	CT1307 3:S1	Eastern Zhou			X
B10	CT1307 12:S1	Yangshao (Early Qujialing)			X
B11	H1634 3:S1	Longshan	X		
B12	H1632 3: S3	Longshan			X
B13	H1656 2: S3	Longshan	X		
B14	H1608 2: S2	Longshan			X
B15	H1646 1: S2	Longshan	X		
B2	CT1307 4:S1	Eastern Zhou			X
B3	CT1307 5:S1	Eastern Zhou			X
B4	CT1307 6:S1	Longshan			X
B5	CT1307 7:S1	Longshan (early or just before) Shijiahe period			X
B6	CT1307 8:S1	Longshan (early or just before) Shijiahe period			X
B7	CT1307 9:S1	Longshan (early or just before) Shijiahe period			X
B8	CT1307 10 (b) S2	Yangshao (Qujialing)			X
B9	CT1307 11 :S1	Yangshao (Qujialing)			X
B16	H1985	pre Yangshao	X		
B17	H1977	Early Yangshao	X		
B18	H1959	Mid Yangshao (earlier)	X		
B19	H1959 -2	Mid Yangshao (earlier)	X		
B20	CT601 ZS:1	Mid Yangshao (later)			X
B21	CT601 ZS:2	Mid Yangshao (later)			X
B22	CT601 ZS:3	Mid Yangshao (later)			X
B23	CT701 ZS:1	Mid Yangshao (later)			X
B24	H1906 -4- A	Mid Yangshao (later)	X		
B25	H1906 -4- B	Mid Yangshao (later)	X		
B26	DT 506 -5	Mid Yangshao (later)			X
B27	DT 506 -3	Late Yangshao			X
B28	DT 506 -4	Late Yangshao			X
B29	DT 506 -4- C	Late Yangshao			X

Table 5.3 Baligang samples

to understanding of changes in the spatial arrangement of the site over time (Erlitou Working Team, 2004a and b, 2005a and b, Xu *et al* 2004 in Liu and Hu, 2007:887).

Throughout Phase III the urban settlement at Erlitou the general layout remained the same as Phase II. However, 2m wide rammed earth walls were built around the palatial complex creating a distinct palatial town. Several palace buildings were abandoned or rebuilt creating a 'regulated palatial pattern'. In contrast to the construction of substantial palatial buildings and enclosures, domestic features, such as wells and storage pits declined (Liu and Hu, 2007:889). Outside this area was a large turquoise workshop (c.a. 1000 m^2). Bronze ritual vessels, *jue* and *he*, were being produced in the foundry (Liu and Hu, 2007:891).

Phase IV has been considered to mark the onset of Erlitou's decline as a major urban centre. The settlement still covered 300 ha but the cultural deposits were concentrated in the centre of the site. However, the bronze and turquoise manufacturing workshops and the roads were in constant use, new palaces were built and a large, 9m wide, rammed earth foundation constructed (Zhao *et al*, 2006 in Liu and Hu, 2007: 892) suggesting Erlitou was still developing.

There are very few remains from the next period, the early Lower Erligang (c.a. 1600-1450BC), which overlaps, to some extent, with Phase IV (Lui 2001in Liu and Hu, 2007:892). Liu and Hu (2007:892) suggest either the site was abandoned for some time between Phase IV and the early Erligang or the population decreased dramatically but remained manufacturing and utilising Phase IV goods until the Upper Erligang. The production of elite goods ceased after Phase IV (Liu and Hu, 2007:892).

A village of around 30 ha. had replaced the urban centre by the Upper Erligang phase (c.a. 1450-1300BC). The cause of Erlitou's collapse is unknown (Liu and Hu, 2007:892).

A 7.5 ha. Palace complex stands at the centre of Erlitou (Figure 5.12). The phytolith samples from the Erlitou period phases II to IV were collected from layers from sections cut through the palace area (Lee, 2005, pers comm.) The majority of these samples come from layers containing ash.

Sample	Unit	T	Provenience	Layer	Description	Phase
E01	1	107	South wall (west section)		grey/green waterlogged ash layer	ELT II
E02	2		South wall (east section)	5B	grey/ green waterlogged layer	ELT II
E03	3	104	South wall (west section, upper layer)	1	reddish brown layer with ash-charcoal	ELT II
E04			South wall (east section)	3	lightly greyish yellow brown ash pit 398	ELT II
E05	4	102	South wall (west section, uppermost)	ash	lightly grey, dense charcoal	ELT IV
E06			South wall (west section, middle)	17	ash layer	ELT IV
E07			South wall (west section, lowest)		greyish green waterlogged layer	ELT II
E08	5	115	South wall (upper)	1	Layer C, greyish brown	ELT IV
E09			South wall (lower)	2	below Layer F	ELT IV
E10	6	116	West wall (upper)	3	greyish, green?	earlier than IV
E11	7	117	South wall (west corner)	a	greyish brown	
E12				c	greyish brown (layers a and c are not clearly distinguishable)	ELT III>IV
E13				f	greyish (thin charcoal layer is boundary b/t layers e and f)	
E14	8	114	West wall (south corner, uppermost)	1	ash layer with yellow clay	ELT IV
E15			West wall (south corner, middle)	2	ash layer with packed clay	ELT IV
E16			West wall (south corner, lowest)	3	ash layer with dense charcoal	ELT II
E17	9	113	West wall	ash	greyish black? Yellow?	
E18	10	117	West wall (south corner)	ash	dark greyish brown hardened layer	ELT II
E19	11	110	South wall (east corner)	1	dark grey layer below layers	ELT III?

Table 5.4 Erlitou samples

ERLITOU SAMPLES

Gyoung-Ah Lee collected nineteen samples from the Erlitou site in November 2005. All of these were processed and analysed (Appendix 3). These samples are from the edge of the palatial complex. Most have been dated to phases II, III and IV, others have not been dated so precisely but are from the Erlitou period *(Table 5.4)*.

Because the only samples available from Erlitou are from the edge of a palatial area in contrast to samples from the other sites in this project they cannot be expected to offer the same kind of information.

HOW PHYTOLITH ANALYSIS WILL CONTRIBUTE TO THE SITE.

Phytolith analysis may point to the presence of crop plants at Erlitou. Very little archaeobotanical data has been analysed from Erlitou and none published in English so all phytolith analysis undertaken should contribute to the general understanding of the site.

5.6 HUIZUI

HUIZUI, BACKGROUND TO THE SITE

Huizui is a multi-component site situated close to a modern village on the west bank of the Liujian River, a tributary of the Li River, in Yanshi County, western Henan, around 15km southeast of Erlitou and 5km north of the Songshan Mountains (Liu and Chen, 2003:66, Liu *et al* 2004: Fig2, Ford, 2004:71, Rosen, 2007:42). The Liujian River originates in the Songshan Mountains and flows north into to the Yilou River system (Lui *et al*, 2004:91).

The site sits around 20m above the Liujian River on a Late Pleistocene terrace of alluvial silts (Rosen, 2007:42). The landscape, of the Liujian Valley has changed throughout the middle and late Holocene. Rosen's (2007) geoarchaeological study of the Liujian River floodplain at Huizui demonstrates a marked change from a stable landscape in the early Neolithic to a strong erosive stream flow at the beginning of the Late Yangshao period, closely followed by a waterlogged marshy landscape intercut by small streams probably brought about by the increasing precipitation during the Mid-Holocene Climatic Optimum. The valleys filled with fertile clays and silts brought in by annual flooding from amplified stream activity providing more land for cultivation, in particular the expansion of rice agriculture. This came to an abrupt end around 2400-2000 cal. BC when climatic conditions became markedly cooler and drier. At Huizui the stream cut through the sediments deposited during the Yangshao and Longshan periods drastically reducing the land available for irrigation and paddy farming (Feng *et al*, 2004; Rosen 2007: 46, Shi *et al*, 1993)

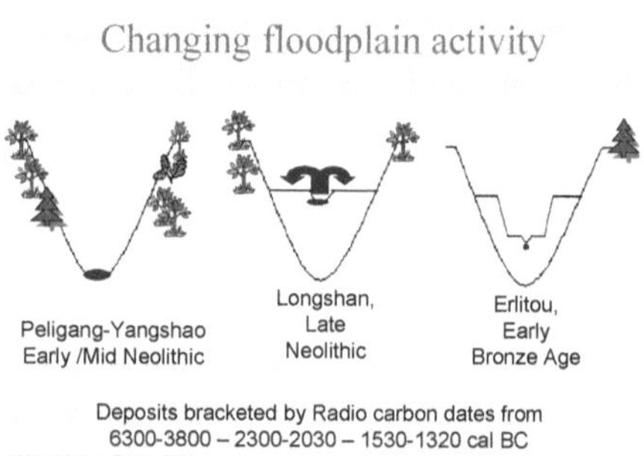

Figure 5.3 Changing floodplain activity

Covering an area of approximately 40 ha, Huizui's major occupation dates to the Erlitou period (3900-3500 BP). However, the Huizui area has been occupied at least since the Peligang (ca. 8000-7000 BP) (Rosen, 2007:39). Two areas have been excavated, Huizui West and Huizui East with a large modern gully dividing them (Ford, 2004:72, Lee and Bestel, 2007:49). The cultural remains of Late Yangshao Period and Longshan Period villages cover 10 ha of the eastern part of the site, while the remains of an Erlitou Period town have been uncovered in the western section of the site. Due to historical and modern disturbance few Erlitou period remains have been unearthed from Huizui East (Liu *et al*, 2004, Lee and Bestel, 2007:7). It has been suggested Huizui was a regional centre during the Neolithic Yangshao and Longshan cultures and developed into a secondary regional centre of the Erlitou Culture (Liu *et al*, 2004: 91, Liu and Chen, 2003:65-68). During this time the site occupied 25 ha. Evidence from survey and excavation at Huizui suggests a regional centre for stone tool production during Longshan and Erlitou periods (Ford, 2001, Chen *et al*, 2003, Henan 1st team 2003, 2005, in Chen, *in press*). Cultural deposits from the Yangshao to the Zhou periods are up to 4m thick (Liu *et al* 2004:89).

FIELDWORK AT HUIZUI

Huizui was excavated in 1959 (Henan Cultural Bureau, 1961, Henan Institute 1990) and has been surveyed and excavated by the Yilou Research Team since 1997 (Liu and Chen, 2007:37). So far an area of approximately 665m² has been excavated (Lee and Bestel, 2007:49) revealing ash pits, dwelling floors and burials dating to the Yangshao, Longshan and Erlitou periods.

HUIZUI SAMPLES

The Huizui samples are from three distinct areas, Huizui East, the Yangshao and Longshan settlements, Huizui West, the Erlitou site (*Figure 5.15*) and from a geological survey in the Liujian valley made in 2005 (Lee *et al*, 2007).

Gyoung-Ah Lee collected phytolith samples from Huizui West, Huizui East and the Huizui geological samples in November 2005. I collected 127 samples from Huizui East over a two-week period in November 2006. Of these samples 17 from 2005 and 39 from 2006 were processed and analysed.

HUIZUI EAST

Carbonised *Setaria italica* from Huizui East has a radiocarbon date of 5240-5210 cal BP (T4H7 Late Yangshao).

Samples were taken from homogenous ash layers (*Figure 5.12*). In addition, large pits with clearly laminated layers are typical of Huizui East. Floors with laminated layers of occupation debris are also common. This variation offers the opportunity to compare a general background to life at Huizui East with more precise events and activities.

Figure 5.4 Typical pit with ash layers at Huizui

Sample	T	Feature	Layers	Description	Period
HE1	2	H8	1	Bluish grey clay layer with dense charcoal, waterlogged?	Late Longshan
HE2			5	Hardened clay, waterlogged?	
HE3			6	Loose clay layer	
HE4			7	Yellowish clay loose layer	
HE5		F5	1	Packed clay layer (preparation for the floor above)	Late Longshan
HE6			2	Packed clay layer (preparation for the floor above)	Late Longshan

Table 5.5: Samples from Huizui East 2005 (Site Code 05HYH

The features the samples were taken from in November 2006 have been classified as 5 ash pits (H1, H4, H9, H12, H17) and one section cut through floors (F1). Some of these ash pits cut through house floors and laminated pits. One ash pit is a large bell shaped storage pit (H9).
H17 cuts through two Yangshao floors, a layer of living debris and a white layer containing fibres.
H12 also cuts through a floor and layers of fill below.
H4 is a Longshan pit with clearly laminated layers.
H9 is a large bell shaped storage pit dated to the Yangshao. There was a layer of charred seeds around 5 cm deep, including *Setaria italica* at the bottom. The pit was lined by burned earth with large grass leaf impressions lying diagonally. Samples were taken from distinct lenses. It would seem the pit had been lined with straw and seeds before burning. There is evidence from Iron Age Britain pointing to basket-lined pits for grain storage (Reynolds, 1979:74).
The samples from F1 are layers of Longshan floors and foundations. The floors consist of layers of fossiliferous tabular formed tufa slabs overlying a mud-plastered surface on a plant tempered adobe platform (McPhail and Crowther, 2007:103).

HUIZUI WEST

Huizui West was an Erlitou period settlement. Only 5 samples were taken from Huizui west so a very limited picture will be available. The samples come from ash

Sample	T	Feature	Layers	Description	Period
HE04	T1	H17	1	Foundation, rubble	Yangshao
HE05			2	Burnt reddish	
HE06			3	Grey layer	
HE08			4	Living residue, charcoal	
HE07			5	White layer with fibres	
HE97	T5	H1	2	Grey, some charcoal mixed	Yangshao
HE98			3a	Dark ash, mixture of clays	
HE99			3b	Brownish clay	
HE100			3ci	Brownish clay	
HE101			3cii	White ash	
HE102			3ciii	Grey/ charcoal	
HE103			3di	Burnt reddish soil, CBM	
HE105			3e	Mixed white lenses	
HE107			3ji	Ashy grey	
HE108			4a	Brownish clay, charcoal	
HE110			5	Burned sediment, CBM	
HE90	T7	H9		White patch	Yangshao
HE91				Less compacted	
HE92				White ashy	
HE93				Fibres from imprints, charcoal	
HE94				Yellow burned earth	
HE95				Charcoal	
HE96				Charcoal and seeds	
HE09	T4	H12	Floor	Compacted	Late Yangshao
HE12			Below floor	Packed layer	
HE17				Grey ash	
HE18				Grey and white ash	
HE19				Charcoal and white ashy	
HE52	T3	H4	North section layer 9	White fibrous	Longshan
HE53			8	Yellow	
HE54			6	Charcoal	
HE55			3	Pale brown clay	
HE56			2	Grey clay	
HE57			1	Hard white	
HE86		F1	6	Above rubble layer	Longshan
HE85			5	Rubble layer	
HE84			4	Compacted floor	
HE83			3	Rubble infill, foundation layer	
HE82			2	Compacted floor	
HE81			1	Below floor	

Table 5.6 : Samples from Huizui East 2006 (Site Code 06HYHE)

Sample	T	Feature	Layers	Description	Period	Depth cm
HW2	3	H2	2	Ash layer, west wall T3	Erlitou	54
HW1			3	Ash layer, west wall T3		107
HW3	3	H4	1	Ash layer, west wall T3		37
HW5			3	Ash layer, west wall T3	Erlitou	82
HW4			4	Ash layer, west wall T3		145

Table 5.7: Samples from Huizui West 2005 (Site Code 05HYHW)

Sample	T	Feature	Layers	Description	Period
HGS4	3	N1	6	Waterlogged layer with cultural debris	Yangshao
HGS5			8	Hardened waterlogged layer with dense deposit cultural debris; Yangshao potsherds, charcoal	Yangshao

Table 5.8: Huizui Geological Samples field season 2005 (Site Code 05HYH)

layers within two large ash middens. All samples analysed from here are from homogenous ash layers and represent mixed deposits, so they are likely to provide a view of the general background to the site

HUIZUI GEOLOGICAL SAMPLES

Radiocarbon dates for charcoal from geological sediment sections from about 1km downstream from Huizui on the Liujian River cal 8220-8130/8240-8030 BP and 8060-7980/8100-7890 cal BP. These might be the remains of Peligang exploitation of the floodplain (Rosen: Pers. Comm.) 2 samples from ash layers in geological sections in the Liujian River Valley were analysed. These are both dated to the Yangshao period using pottery sequences.

HOW PHYTOLITH ANALYSIS WILL CONTRIBUTE TO THE SITE.

Phytolith analysis will add to the already rich archaeological evidence from Huizui. The phytolith analysis already undertaken on the off site samples has suggested the earliest use of rice paddies in the central Yellow River Valley region (Rosen, *et al*, 2008). The analysis of the macro remains so far at Huizui has concentrated on the quantative relationship between features and vegetative remains (Lee and Bestel, 2007). The work using the phytolith dataset collected from Huizui for this project will focus on the uptake, expansion and decline of various crops and the relationship between these and social changes over three cultural phases.

5.7 SURVEY

SURVEY SAMPLES, BACKGROUND TO THE SITES

Gyoung-Ah Lee collected a further 11 samples from 7 sites in the Yilou River Valley in 2003 as part of the Erlitou Survey for Xu Hong. They are found outside the Yilou River Valley survey boundaries on the map (Figure 5.2). These samples are all from small sites and are dated from the Yangshao to the Late Longshan and Erlitou cultural periods. All of these samples come from ash pits. There was little further information available about the survey sites.

5.8 SAMPLING

COLLECTION

Samples were collected by first cleaning the surface of the sediment with a clean trowel which was re-cleaned after every sample. Next, a stainless steel measuring spoon was used to collect approx 100gms of sediment and cleaned after every sample. The sediment was double bagged and a label containing the date, site code, site name, sample number and initials of person who collected the sample was placed between the bags. In addition both bags were labelled on the outside with the same information. At Xipo two samples were collected from every context. Samples from laminated layers or

Sample no.	Site	River	Height above Sea Level	Coordinates	Period
S1	Gong Jia Yao		150~200m	34°36′02″	Yangshao
				112°38′20″	
S4	Mazhai	Shagou River	200~500m	34°32′48″	Late Longshan
				112°39′10″	
S6	North East Gaoya Lower	Shahe River	Below 150m	34°38′55″	Late Longshan
				112°42′18″	
S7	North East Gaoya Upper	Shagou River	Below 150m	34°38′55″	Late Longshan
				112°42′18″	
S9	North East Zhaiwan upper	Shagou River	200~500m	34°34′11″	Late Longshan
				112°42′53″	
S10	North East Zhaiwan	Shagou River	200~500m	34°34′11″	Late Longshan
				112°42′53″	
S14	South East Zhaiwan upper	Shagou River	200~500m	34°33′56″	Erlitou
				112°42′22″	
S15	South East Zhaiwan lower	Shagou River	200~500m	34°33′56″	Erlitou
				112°42′22″	
S8	Peichun	Liujianhe River	Below 150m	34°35′42″	Late Longshan
				112°44′56″	
S12	Yuangou A upper		150~200m	34°34′38″	Late Longshan
				112°36′10″	
S13	Yuangou A		150~200m	34°34′38″	
				112°36′10″	Erlitou

Table 5.9 Samples from the Erlitou survey field season 2003

thin lenses were collected using a knife to carefully slice the sample from the layer in order to avoid mixing.

5.9 CONTEXT TYPES

The contexts the samples were collected from can be grouped into 3 types; homogenised ash midden fills, laminated levels from pits, floor and floor foundations.

HOMOGENOUS ASH PITS

The homogenous ash pit fills were probably formed over time and are mixed fills that represent several dumping episodes from secondary and tertiary deposits, waste that has been created and disposed of elsewhere and moved to the ash pit for redeposition. Some may belong to a household, while others may be communal to the village. They may represent a solitary dumping event or several. The data from these deposits can answer general non-specific questions that give a broad picture of life in the village over a time. They are likely to be time averaged longer duration fills and represent the routine background noise to life.

The ash pits may contain:

A. Crop processing waste. This may have been used for fuel, fodder or building materials before being disposed of in the ash pit.

B. Remains from hearths, fuel and kindling.

C. General discarded waste, food and cooking residues, household waste, building materials.

A. Crop processing waste - plants, plant parts and phytolith types expected:

i. Rice/ *Oryza* – leaves, stems, husks – *Oryza* type keystones, double peaked glume cells

ii. *Panicum* – leaves, stems, and husks – trapezium shaped multicells, *Panicum* dendritic multicells.

iii. *Setaria* – leaves, stems, husks – *Setaria* type crosses, multicells with long strips of long sinuate cells with slightly scooped long bilobes, *Setaria* type dendritic multicells.

iv. Possibly *Triticum* and *Hordeum* and their associated crop weeds.

v. Dry crop weeds: *Bromus, Digitaria, Eleusine, Agrositis /Calamagrostis, Alopecurus, Lolium, Pennisetum, Urochioa/Eriochioa, Hordeae*, Paniceae, *Paspalum, Setaria, Stipa*, Asteraceae, Boraginaceae, *Chenopodium, Glyceria.*

vi. Wet crop weeds: *Coix lachryma-jobi, Echinochloa crus-galli, Eleusine indica*, Cyperaceae, *Phragmites, Zizania, Leersia, Ischaemum rugosum.* Some families, such as Cyperaceae, have species that can be found amongst wet and dry crops.

vii. Unidentified wild grass, especially panicoid grasses, would also be present and the silica skeletons from these probably make up a high proportion of the phytoliths in the assemblage.

The crop weeds present would be dependent firstly on the associated crop; for example, *Leersia* would be expected with *Oryza* rather than *Panicum*. However, the crop-processing residue from more than one crop could be present as the ash pits represent dumping over time.

A second consideration is the size of plant and the seed. Phytoliths from the husks of smaller light weed seeds, such as *Digitaria*, are likely to be found in earlier pre-storage stages of processing along with phytoliths from culms and leaves (Harvey and Fuller, 2005: fig 3), but as the deposits are mixed this is doubtful whether this would show in this type of sample.

There is likely to be difficulty distinguishing between crops and weeds from the same genera. This is why having several identification criteria is key.

While crop weeds belonging to Poaceae and Cyperaceae produce abundant identifiable phytoliths, dicotyledons such as Asteraceae, Boraginaceae and Chenopodiaceae do not. I have extracted identifiable phytoliths from modern plants from these genera but they are rare and not readily distinguishable to plant part.

LAMINATED LAYERS:

A. Pits

B. House floors and foundations

Laminated layers are apparently undisturbed layers of sediment, some only millimetres thick, which can represent *in-situ*, primary or secondary deposits. *In situ* deposits are often characteristically white sediments with high densities of phytoliths. Laminated layers should present a clearer picture of specific activities than homogenous ash pits because they are far less likely to contain such mixed deposits so should show short duration snapshots of activities taking place over a limited period. They have the potential to show social and inter annual patterning.

Similar information might be expected from floors and foundations. There is also one sample from a hearth, which might be expected to illustrate what was being used as fuel.

The laminated layers are likely to consist of:
A. Pit fills which can be *in-situ* primary deposits, as in storage pits and their linings, or contain secondary or tertiary waste disposal deposits.

i. Storage pits:
1. Pit lining – stems/leaves from large grasses, Panicum, *Phragmites*, Cyperaceae, high ratios of leaf/stem multicells and/or polyhedrons.
Sometimes the bottom of the storage pits contain a level of charred grain suggesting the pit was burned possibly as method of sterilisation (Lee: *Pers comm.*). In this case a lining layer containing a high density of husk phytoliths would show the grain was being stored in the husk.

2. Pit contents - the contents may be the remains of grain storage. If the grain has been stored hulled there will be few husk phytoliths present, as they are produced in the husk not the caryopsis itself. However, grain is more likely to be stored partly processed as this helps to prevent fungal infection. According to Reddy (1997:175) the closer the crop is to consumption the lower the probability of storage.

ii. One off or short term waste disposal/-dumping episodes.
These samples may contain all the components of the ash pits. If they are secondary deposits they are likely to be mixed and similar to the homogenous ash pit deposits.
If the samples consist of primary deposits they are likely to represent isolated dumping episodes and discrete activities or processing stages will be clearer and easier to interpret. These deposits could provide clearer pictures of specific activities.

B. House floors or foundations
The floors themselves usually consist of hardened rammed earth layers. The samples were taken from above the rammed earth and the foundations from the layers below.
a. Floor surfaces/layers, which are likely to contain *in-situ* and primary deposits. Plants, plant parts and phytoliths expected:

i. *In-situ* floor coverings, mats, bedding:
High densities of large grass leaf/culm, *Phragmites*, bulliforms, saddles, multicells from grass leaf/stems, Cyperaceae, cereal straw, crop processing by products, crop plant leaf/stem multicells, high proportions of long smooth cells and bilobes, very low proportions of platey and elongate cells from dicotyledons.

ii. Discrete events/ activities – food processing:
Food processing may leave crop and crop weed husks. Scalloped shaped morphotypes from Curcubitaceae rinds may also be visible.

iii. Crop processing:
It may be possible to distinguish different stages of crop processing in these samples as they are not necessarily mixed (see crop processing section below.

iv. Animal penning:
Phytoliths from dung or fodder (see above) might be deposited in animal pens. Corresponding faecal spherulites and high pH levels could support this.

Remains left on the floors by activities are likely to have been swept and trampled so will probably be sparse and spread. Macphail and Crowther (2007:105) found no evidence of occupation deposits between floors at Huizui and suggest the floors were swept and/or covered. Trampling may break multicell phytoliths so there could be fewer in these samples. However, the edges of floors close to walls may accumulate denser deposits.

b. Foundations are likely to contain primary deposits of packing materials. Plants, plant parts and phytoliths expected:
i. Mixed vegetation, so phytoliths of all types, such as straw and leaf multicells, single long and short grass cells and platey types. Crop husks are very unlikely.

None of the contexts appear to represent a latrine area but a selection of 46 samples were tested for pH. High acid levels could indicate either latrines or animal pens.

CHAPTER 6

METHODOLOGY, PHYTOLITHS FROM THE FIELD TO THE LABORATORY

In this chapter there will be an introduction to phytolith research, followed by a discussion of morphology and description of the reference collection created for this project, including identification criteria for the major plants, crops and crop weeds discussed. The archaeological and modern sample collection techniques will be described, as well as laboratory protocols and recording. Quantification and methods of analysis will be discussed in chapter 7.

6.1 Introduction to phytolith research

Phytolith, from the Greek meaning 'plant stone', has been used to describe both siliceous and calcareous mineralised secretions in higher plants (Piperno, 2006:5). Here it used only to describe silica bodies also known as plant opals or opaline silica. The majority of identifiable phytoliths are bio-mineralised particles formed within the intra and extra-cellular space of living cells in the culms, leaves, roots and inflorescences of higher plants. Phytoliths are created as ground water containing monosilicic acid is taken up from the soil. As the silica is transported in the transpiration stream it moves through the permeable plant membranes and is deposited as hydrated Opal A within the cell lumen and intercellular spaces, as well as forming external layers on the cell walls (Carnelli *et al*, 2001:425). Silicification occurs in Pteridophytes, Gymnosperms, and Angiosperms (Piperno and Pearsall, 1993:11, Mabberley, 1987:208, Kaufman *et al*, 1971, Jiang and Zhou, 1987, Sangster and Hodson, 1987, Madella, 1997:294, Pearsall, 2000:361, Piperno, 2006:5, Prychid *et al*, 2004:378). Biogenic silica is also found in unicellular organisms in lower plants, diatoms, and some animals, e.g. sponges. However, not all plants produce phytoliths. For those that do the same morphotypes are often produced across wide varieties of species and genera, so not all phytoliths are diagnostic in terms of taxonomic level.

History of research using phytoliths

Although phytoliths have been known for around 200 years the application of phytolith data to archaeological questions is relatively recent. Piperno (2006:2-3) defines four stages of phytolith research, discovery and exploratory (1835-1895), botanical (1895-1936), ecological (1955-75), archaeological and palaeoenvironmental (1978). Prychid *et al*, (2004: 378-80) summarise the discovery and early botanical history based on Netolitzky's 1924 early history of phytolith research. In 1814 Davey made early investigations of the epidermis of *Triticum*, *Avena*, *Arundo* and *Equisetum*. Struve (1835) discovered silica bodies in bamboo survived ashing but dissolved in caustic potash. Ehrenberg, the pioneering German microbiologist noted silica bodies in soil samples from a range of geographical areas, named them 'Phytolitharia' and developed the earliest classification system (Ehrenberg, 1841, 1854 in Piperno, 2006). Ehrenberg identified phytoliths in the red dust blown from the Sahara that Darwin gathered from the Beagle's sails as it anchored off the Cap Verde islands (Piperno, 2006:3). In 1857 Crüger located silica bodies in bark and parenchyma using Schulze's solution. He believed they were only found in dead cells. However, Von Mohl 1861, found silica filling whole cells in a variety of plants and concluded phytoliths could be found in living cells (Prychid *et al*, 2004:379). In 1867 Weisner distinguished short and long epidermal cells in *Saccharum* and *Zea*. Weiler (1893, 1897), and Grob (1896) discovered the cell lumina are filled with silica. Weiler (1893, 1897) demonstrated how the silicification fills the whole cell starting from the inside wall (Prychid, 2004:379).

From the mid 1890's studies of production, morphology and taxonomy were taking place in Europe, mostly centred in Germany (Piperno, 2006:3). The first incidences of phytolith analysis applied to archaeological questions began in the early decades of the 20th century, Netolitzky (1900, 1914) and Schellenberg, (1908) made identifications of wheat, barley and millet from ash pits and ceramics (Piperno, 2006:3). The majority of this early work was published in German and ended in the 1930's. From the middle of the 1950's to the 1970's phytoliths came to the fore in ecological research. Phytolith morphology, particularly of those produced in grasses, began to be investigated in detail by botanists such as Metcalfe (1960), Parry and Smithson, (1964, 1966), Twiss *et al* (1969 and Sangster (1970). Work took place on phytoliths in dicotyledonous broadleaf trees (Rovner, 1971, Geis, 1973, Bozarth, 1985, 1987, 1992) and Gymnosperms (Klien and Geis, 1978). Twiss *et al*, (1969) developed a classification scheme for the grass subfamilies, Pooideae, Panicoideae and Chloridoideae based on the morphology of short cells.

From 1978, Piperno's archaeological and palaeoenvironmental modern period, archaeobotanists have used phytoliths as independent proxies for past environments and to aid understanding human plant exploitation in the past (Piperno, 2006:3).

6.2 How and why phytoliths are produced in plants

Phytolith deposition

Phytolith development in plants is dependent on factors such as climate, environment, soil type, soil hydration, age and type of plant (Piperno, 2006:5). Piperno, (2006:12) summarises two basic processes of phytolith production in plants. The first originates in the plant's own genetic and physiological mechanisms and relates to phytolith production in designated cells and tissues. The second is associated with outside factors such as local environment and growing conditions.

Silica is an abundant element and a constituent of many mineral soils (Hodson and Evans, 1995, Prychid *et al*, 2004:380). Soluble silica is released into sediments and soils by the weathering of silicate minerals (Piperno, 1988, Prychid *et al*, 2004:380). Monosilicic acid (H_4SiO_4) is soluble in water and is absorbed into the plant with other minerals in the groundwater through the roots and carried in the xylem sap (Hodson and Evans, 1995, Prychid *et al*, 2004:380, Piperno, 2006:8). Piperno, (1988) suggests soluble silica can 'either be actively transported by metabolic processes or carried by passive non-selective flow in the transpiration stream'. Silicic acid becomes polymerised in plant tissues and is deposited within the growing plant as solid amorphous silicon dioxide (SiO_2) in the cell walls, lumina and in the spaces between the cells often taking on their form (Piperno, 2006:8).

Silica may be found in all plant parts including the roots but most of the silica is laid down in the aerial structures both vegetative and reproductive (Piperno, 2006:6, Prychid *et al*, 2004:380). In monocotyledons most phytoliths are commonly found in the epidermis (grasses and sedges) or in the sheath cells or vascular bundles (palms, bananas, orchids) (Prychid *et al*, 2004: 378). Silica bodies found in dicotyledons are often from the ray or axial parenchyma cells (Metcalf and Chalk, 1986, Welle, 1976: table 1).

SILICA AGGREGATE

Silica aggregate can signify bark, wood or deposits in the root epidermis (Albert and Weiner, 2001:256; Sangster and Hodson, 1992:241) Silica aggregates are soil minerals cemented biologically by an amorphous Si, Al, Fe and K rich mineral phase (Albert and Weiner, 2001:256). The soil minerals are taken up by trees with water, carried through the vascular system, then concentrated and cemented into an aggregate, mostly in bark, sometimes in wood (Albert and Weiner, 2001:256). Root and rhizome phytoliths have not yet been investigated thoroughly (Sangster and Hodson, 1992:241).

6.3 WHY PHYTOLITHS ARE PRODUCED IN PLANTS

Recent work has shown a number of reasons why plants produce phytoliths. For some plants, *Equisetum, Oryza, Beta vulgaris*, silica is believed to be essential for normal development, particularly for structural functions (Chen and Lewin, 1969, Takahashi and Miyake, 1977 in Piperno, 2006:12). Silica may also mitigate the toxic effects of heavy metals, aluminium and manganese, that are absorbed by the plant in the uptake of ground water (Hodson *et al*, 1997, Sangster *et al*, 2001 in Piperno, 2006:13). Another benefit can be increased resistance to fungi (Epstein, 1999, Marshner, 1995) and small herbivores, such as ants, caterpillars and beetles (Coley and Barone, 1996, in Piperno, 2006:13). Epidermal phytoliths are abrasive enough to offer mechanical resistance to large herbivores. The evolution of hypsodont (high crowned teeth) in horses has been linked to phytoliths in grasses (Jacobs *et al*, 1999, McNaughton and Tarrants, 1983, in Piperno, 2006:13, Stromberg, 2002, 2004). It is also thought that phytoliths prevent the collapse of cell walls during transpiration (Raven, 1983, in Piperno, 2006:14).

Patterns of phytolith production can be observed in some plant families, in particular where production is high (Piperno, 2006:7). In this study emphasis is placed on monocotyledons, in particular Poaceae, as the crop plants focussed on, *Setaria, Panicum* and *Oryza*, are all grasses and so are many of their weeds.

ORDER IN WHICH THEY ARE PRODUCED

Phytoliths are commonly deposited in, parts of the epidermis, sub epidermal tissue (for example in palms), and inflorescence parts of grasses (glumes, lemmas and paleas). But, other parts of the plant can also produce phytoliths, for example tracheids located in the xylem of vascular plants (Postek, 1981:125). Bulliform cells are commonly, but not exclusively, found at the base of the furrows in the adaxial surface of grass leaves. They can also form part of the abaxial epidermis and occasionally part of the subjacent mesophyll (Metcalfe, 1960: xxx, Evert and Esau, 2006:241, Figure 9.35). While there is some discussion as to the exact function of bulliform cells, they are known to provide water storage (Parry and Smithson, 1964) and movement of water inside the cells causes the leaves to fold preventing over transpiration (Delhon, 2009:178), so they are clearly are not designated for silica deposition as their primary function. Bulliform phytoliths commonly occur in grass species favouring watery habitats with a submerged root system and high transvaporation rate, such as *Phragmites* and *Oryza* (Sangster and Parry, 1968: 316, Parry and Smithson, 1964).

6.4 METHODOLOGICAL STRENGTHS AND WEAKNESSES

STRENGTHS

While phytoliths are subject to many of the same constraints as any archaeobotanical material, they have one great advantage over most macro botanical material, they preserve without needing to be charred, waterlogged, frozen, mineralised or desiccated so are often persevered where no other plant remains survive. Phytoliths are non-organic and survive well in most depositional environments, including acid soils. However, preservation can be differential according to the shape and size of the morphotype and the characteristics of the surrounding matrix, for example highly alkaline soils, above pH9, can often, but not always, dissolve phytoliths. Phytoliths from grass and sedges appear to be particularly resistant to dissolution (Piperno, 2006:19-22).

Phytolith data is useful for demonstrating *in situ* vegetation, as they tend to stay where they are deposited, although they can be carried by wind or water. These processes are likely to be visible on the surface of the silica bodies in the form of pitting and fractures (Rosen, 2000: 10; Madella, 2000:177).

Different plant parts produce characteristic phytoliths making distinctions between husk and leaf/stem possible, an advantage over macro remains which for the most part consist of charred seeds. The ability to see both leaves and husk is of particular advantage for this study as it provides an opportunity to investigate crop-processing stages (Harvey and Fuller, 2005).

Compared to other types of data used in archaeobotany, seeds, charcoal, pollen, phytolith research is still in a relative infancy. A major issue is taxonomy with many researchers using different descriptors for the same morphotype or the same descriptor for different morphotypes. For this reason an international nomenclature has been developed (Madella *et al*, 2005), which has been used here as far as possible. Identification can be challenging. The same shape can be present in plants from different taxonomic groups (Madella, 1997: 294). Although it is often possible to identify to genera and in some cases species in modern references, it is sometimes only possible to categorise to sub family in archaeological assemblages. This can be due to a variety of factors, for instance weathering or the small number cells preserved in a silica skeleton. For this reason conforming to several identification criteria rather than one or two is important (see case study below). The development of good reference slides is key. There are a growing number of online reference databases, but more are needed. A reference collection was created specifically for this project. In the case of this research there is a potential methodological discrepancy between rice and millets. Rice produces distinctive, morphologically specific phytoliths in both leaves and husk. In contrast, less reference work has been done on millets and while they produce characteristic phytoliths in the husks, the leaves are difficult to identify. The majority of phytoliths in all samples are non-diagnostic, for example hairs.

Different plant families produce phytoliths to varying degrees and in varying quantities, while some do not produce any. Grasses, which include many staple crops, (wheat, rice, millets) produce diagnostic phytoliths, but not all plants produce silica bodies and fewer create identifiable phytoliths. A specific problem is dicotyledons, which rarely produce diagnostic phytoliths. This means a phytolith dataset can produce a skewed picture and makes comparison between macro and micro datasets difficult.
Issues of quantification are dealt with in Chapter 7.

6.5 Taphonomy and formation processes

Questions concerning equifinality must be considered. There are many potential trajectories to arrive at an end state and consequently more than one possible interpretation of ratios produced by an assemblage. Identifying the types of questions that phytoliths can and cannot answer is key to using phytolith analysis successfully.

An understanding of the formation processes of each sample is essential to interpretation of the dataset. Possible routes of entry for mixed deposits need to be disentangled and considered (van de Veen, 2007: 968, 971). The origin of the deposit needs to be understood in order to interpret any patterns generated by the contents (Schiffer, 1983:675). Both the path phytoliths take to the assemblage, cultural and behavioural (C transforms) and natural (N transforms), and post depositional processes, need significant consideration before interpretation of an assemblage (Schiffer, 1972:158, Schiffer, 1976).

Phytoliths can be deposited into an assemblage in a number of ways. They have no physiological adaptation for mobility, so are not designed for movement over long distances, in contrast to pollen. They are frequently physically bound to both organic and inorganic components of soil, so are likely to be horizontally and vertically stable after deposition. Phytoliths characteristically stay in the same place, even in open environments. *In situ* deposition represents a localised environment (Pearsall, 2000:493; Madella and Powers-Jones, 1998:507; Piperno, 1985:247).

Phytoliths deposited in an archaeological context would most probably have arrived primarily through human agency (Pearsall, 2000, 493; Madella and Powers-Jones, 1998:507, Piperno, 1985:247, Hastorf, 1988:120). Concentrations of distinctive white layers with little soil or sediment are likely to point to *in situ* decay and deposition of vegetative material (Madella, 2001:177). These layers are characteristic of many of the laminated pit fills and linings sampled at Huizui and Erlitou.

Phytoliths can also be transported by aeolian means, for example the windborne dust referred to by Darwin (1846, 1860). However, usually only high velocity winds can carry them and generally only from dry, loosely bound matrices. Shapes with a greater surface area: volume ratio may be carried more easily. Phytoliths carried by wind may represent either local or regional environments, depending on vegetation cover, source and strength of wind (Madella and Powers-Jones, 1998: 503). This could have an effect on how the assemblage is interpreted and should be taken into consideration. However, wind-blown silica bodies are often recognisable by distinctive surface damage.

Phytoliths from the regional and local zone can be carried and dispersed by water percolating through soils, as well as carried by water flow (Bush *et al,* 1992, Madella and Powers Jones, 1988:503). In addition, they can be transported in the gut of an animal and redeposited with the faeces. One of the most interesting considerations for an archaeologist must be the proportion of phytoliths brought to the site as a result of human activity (Madella and Powers-Jones, 1998: 503).

Phytoliths are durable and stable in a wide PH range (3-9), they survive burning up to 1000°C so are commonly preserved in ash.

There are common features to many phytolith assemblages. Monocotyledons, particularly Poaceae and Cyperaceae, produce many more identifiable phytoliths than dicotyledons or gymnosperms. So while cereal crops are likely to be well represented other crops, such as fruit, oil seeds, nuts and green vegetables are difficult or impossible to identify. However, some Curcubitaceae do produce identifiable phytoliths in their rinds so these are potentially visible. It must also be remembered that the phytolith data can only represent a small proportion of the plants from a site and an even smaller part of all plants originally present and discarded (van der Veen, 2007:977).

Grasses form phytoliths in a specific order, for example, there are more likely to be hair phytoliths, which are often a natural defence against predators, than stomata (Parry and Smithson, 1964:177). This is due to physiology. Some cells such as stomata and bulliforms are not designed to take silica; in fact silicification will stop the cells functioning. If these become silicified, it is towards the end of the plant's life. Leaves and culms as a whole should produce many more phytoliths than husks because they make up a far greater proportion of the plant so higher percentages of leaf/culm phytoliths are to be expected. However, this refers to all morphotypes produced at varying rates in leaf/culms, for example, hairs, long cells and bulliforms. If phytoliths are separated into the component individual morphotypes it may be easier to see husks than leaves. For example, the distinctive double peaked glume cells from husks in *Oryza* are hairs so readily produced, but the diagnostic *Oryza* keystones from leaves are not so frequent. This is an area that would benefit from future exploration.

AGENTS OF DESTRUCTION

The two major agents of destruction are mechanical breakage by wind, bioturbation or compression, and diagenetic chemical weathering by alkaline elements in the matrix. Damaged silica bodies can sometimes be identified by pitting and breakage patterns (Madella, 2000:176). However, wind damage can be confused with chemical dissolution. Spines of dendritic cells and the central bar of bilobes are typically weak and easily damaged. If these are well preserved with smooth surfaces they can demonstrate lack of physical weathering and *in situ* deposition (Madella, 2001:177, Rosen, 2000:177).

SAMPLES

In this study, most samples represent mixed deposits formed from different dumping episodes and charring events over indeterminate periods. Having selected the samples, the varying sample types and their assemblages were compared between and within each site to establish uniformity or diversity, and to see variation through time within the site. A key question is whether the sites show consistency, is there one significant pattern or are there multiple patterns, and are differences context dependent or not?

The phytolith samples collected from these sites provide a robust archaeobotanical dataset. The varying contexts should supply information on general trends and routine daily life as well as more specific short-term events, thereby potentially affording clarification on social and inter-annual variation. A distinctive strength of phytolith analysis is the ability to see plant parts, broadening the scope of issues it is possible to address. However, it should not be forgotten that the sample types available limit the questions that can be answered. It may not always be possible to distinguish between alternative interpretations. For example, the presence of husk and absence of straw of a particular genus or species (e.g. *Oryza*) could be due to input of spikelets, i.e. a consumer site, (Thompson, 1996) or because threshing was carried out seasonally off- site (Fuller and Stevens, 2009).

6.6 REFERENCE COLLECTION

In order to better identify the phytoliths from archaeological contexts a modern reference collection of 291 samples was prepared to slide and photographed. A list of grasses and economic plants that are known to produce phytoliths was prepared using the Flora of Henan, (Zhang and Yi, 1981, Zhang, Yi and Yi, 1988, Zhang 1997,) Watson and Dallwitz (1992), the *Flora plantarum herbacearum Chinae Boreali-Orientalis* (Qin Zhongshi, nd), and information from Chinese macro remains (Fuller, Qin, Harvey, 2007, Lee *et al*, 2007, Crawford *et al*, 2005, DF unpublished data (Ying), Banpo Museum, 1992, Underhill, 1997, Liu *et al* 2005, Fuller unpublished data) .

The aim was to collect and process to slide as many plants relevant to the project as possible. Crops and their crop weeds were focussed on in particular, for example four populations of *Setaria italica* were processed as well as eight weedy *Setarias*.

Once processed to slide the phytoliths were photographed. The abundance was recorded into one of five categories: abundant, common, present, rare or none. Identification criteria were established for crops, *Oryza*, *Setaria*, *Panicum*, weedy *Setaria*s, common weeds, and Cyperaceae.

PLANT COLLECTION

The material for the reference collection came from a variety of sources: the Herbarium at the Institute of Archaeology, the Royal Botanic Gardens at Kew grass collection (thanks to Shelly Cleave), plants I collected in Henan, China in 2006, plants collected by Freea Itzstein Davey in the Yangtze Delta in 2006, plants collected by

Dorian Fuller for the Institute of Archaeology from 1996-2008. Yvonne Edwards collected grasses in northern China in 2006. I grew *Setaria italica, Panicum miliaceum* and *Brachiaria ramosa* in London in 2005/6 from seeds sourced from the United States Department of Agriculture.

PROCESSING REFERENCE MATERIAL

Two methods are commonly used to extract phytoliths from plant material, dry ashing or acid digestion. The samples for this collection were dry ashed. This technique was chosen due to its simplicity. No chemicals are necessary so it is cheap, safe and effective. Parr *et al* (2001: 884) made a comparative analysis of both methods and concluded any morphological differences, such as shrinkage, were insignificant.

Reference samples were washed in distilled or deionised water to remove dust and dried in a drying oven at <50°C. Once dry the plants were separated into leaf, culm, node, husk, inflorescence, floret, awn, rachis, glume, panicle, involucres, needles and bark where available. The dried plant parts were cut into several pieces of around 2mm and burned in a muffle furnace at 500°C for 3 hours to remove any organic material. Once ashed the reference samples were mounted in Entellen and dried. The samples were not rinsed with HCL to avoid the destruction of multicells that can occur while being centrifuged.

Some grasses (*Chenchrus, Saccharum*) melted to the crucible at this temperature. This may have been due to high starch or sugar content. They were reburned at 400°C, which provided references.

Due to the constrictions of time most references were not weighed during processing. Recent work has established ratios of the silica bodies to plant part (Tsartsidou *et al*, 2006, Carnelli *et al*, 2001) and this would be a useful area to explore further.

DATABASE

A database of the modern reference samples was prepared using Excel. Details of each sample processed were recorded. Photographs were taken with a Nikon Cool Pix camera through a Leica microscope at 100x, 200x and 400x. The majority were photographed at 200x as this provided the clearest images and features were still plain to see. Once processed the samples were photographed and scanned for abundance and identifiable characteristics (Appendix 6.)

6.7 IDENTIFICATION

MODERN REFERENCE COLLECTIONS

Identification of archaeological phytoliths was made by comparison to modern comparative reference collections, including one made specifically for this project and the collection of *Oryza* references compiled by Emma Harvey at the Institute of Archaeology, University College London. I also referred to published photographs and drawings of silica skeletons (Miller Rosen, 1992, figs 7.1-7.14; Miller Rosen, 1993:169: Greiss, 1957), dicotyledons (Bozarth, 1992:figs.101-10.8), grass silica bodies (Mulholland and Rapp, 1992: figs 4.1-4.15) Lu and Liu, 2000, Cyperaceae (Ollendorf, 1992: figs5.1-5.5), cereals (Ball *et al*, 2001:figs 1a-4b; Kaplan *et al*, 1992: figs 8.1- 8.37,) and food plants (Scott Cummings, 1992:Figs 9.1-9.17) as well as Wang and Lu's general photographic references (1992) and Kealhofer and Piperno, 1998, . In addition I used Dorian Fuller's unpublished SEM photographs of *Setaria* husks and Dominique de Moulin's SEM photographs of *Panicum* and *Setaria.* I also referred to the descriptions and drawings in Metcalfe's Anatomy of the Monocotyledons (1960). Krishnan *et al* (2000) have compiled a key of Indian grass phytoliths but only using single celled forms. The morphological descriptions are useful but in general this study is heavily reliant on size, which is significantly variable within and across species (Harvey, 2002:30). Lu and Liu (2003) have investigated representative lobate single celled phytoliths in grasses, looking at length of shank and shape of the lobes. This is useful as there are characteristic lobate shaped phytoliths, but limited as many grasses produce the same morphotypes across genera. I also consulted the web pages at www.missouri.edu/ ~phyto/index.shtml and www.bio.uu.nl/~palaeo/ Research/Namibia/dia-8l.jpg for reference photographs. Another useful work is Lu *et al* (2009) on identification of *Panicum miliaceum* and *Setaria italica.* Unfortunately this paper was published after the identifications for this project had been made. However, their results confirm many of the conclusions reached from my own identification criteria. In addition I had access to the reference collection created by Leftiris Zorzos at UCL, mainly of Greek plants, which I consulted for identification of morphotypes from dicotyledons, in particular polyhedrons.

IDENTIFICATION CRITERIA

An initial identification key was prepared for *Setaria italica, Panicum miliaceum* and *Oryza* as these were the key crop plants, and expanded to include *Setaria faberii, S. pumilla, S. palmiflora, S. verticillata, S. viridis, P. sumatrense.*

When looking for identification criteria for *Setaria* and *Panicum* specific questions about the morphology were raised. For the husk /seed coat/ inflorescence first I looked at the dendritic cells to identify if they were regular or irregular, the length and width of the lobes, the proportions of the space in the centre, the length of individual cells and whether they were long or square. Next, I looked specifically at the lobes to see if they ran along the sides or all the way round the cell and if so did they run all the way round the cell? Were they the same length all the way round or shorter at the ends?

A key was also developed for Cyperaceae. While it is possible to pick out varying criteria within the modern reference samples the archaeological samples proved very difficult to distinguish so were eventually classified as a single category, Cyperaceae. Although in the analysis all sedges were grouped together this is an area that could be developed. This was a common issue, many of the modern references offer the opportunity to develop identification keys but the archaeological samples are more difficult to categorise and often distinctions merge.

CASE STUDY SETARIA VERSUS PANICUM

Using several identification criteria as opposed to only one or two is important. An example can be seen in a simple case study using the average % per site. According to the archaeobotanical records of charred seed remains, the primary crop in north central China during the Yangshao was *Setaria italica* with a much smaller secondary crop of *Panicum miliaceum* (Lee *et al,* 2007, Chang, 1986) (*Figure 6.1*).

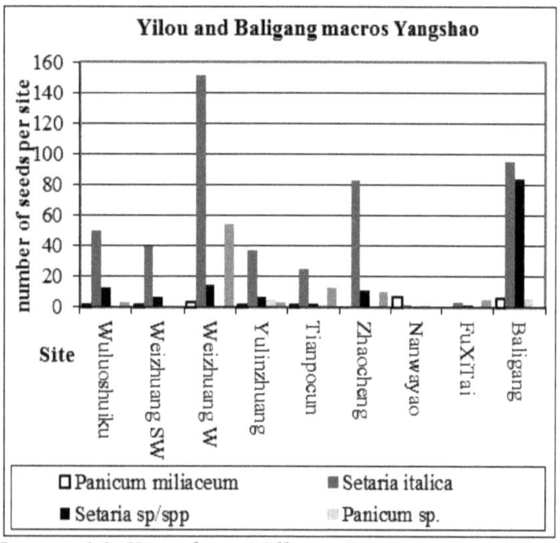

Figure 6.1 Yangshao Millet macro remains Yilou and Baligang (Lee et al, 2007, Fuller unpublished data)

To determine whether it is possible to differentiate between *Panicum miliaceum* and *Setaria italica* phytoliths reference slides from both plants were prepared and several distinguishing criteria identified. These were:

S. Italica
Inflorescence

1. Processes with a regular width
2. Width = height
3. Rounded flat topped papillae
4. Cells at least 6 processes long
5. *Setaria* type crosses – short fat square with dot
6. Long and narrow crosses

P. Miliaceum inflorescence

1. Wide space in the middle of long cells (wider than *Setaria*)
2. When long cells are alternate left and right shaped they are also short and broad
3. Processes are often fine and long
4. Processes are often irregular lengths and widths
5. Processes are at the ends of long cells as well as along the sides
6. Long elaborate cells are often shorter than *Setaria*
7. Short cells sometimes small cross or short thin bilobes – not distinctive
8. No dot in middle of short cells
9. Often (not always) 7 processes to the long side of long cells
10. Often 2-3 processes to short side long cells
11. Regular appearance to long dendritic cells
12. Processes often elaborate
13. No papillae

When the slides were counted silica skeletons with 5 or more criteria were identified to *Panicum* or *Setaria* husk. If they exhibited general characteristics but fewer than 5 of the criteria they were classified as Millet 1 (like *Setaria*) and Millet 2 (like *Panicum*). In the first chart below (*Figure 6.2*) all the husks that look like *Setaria* and *Panicum* but do not me*et al*l the criteria are included with those that meet the identification requirements, total millet 1, (cf. *Setaria*) v total millet 2, (cf *Panicum*).

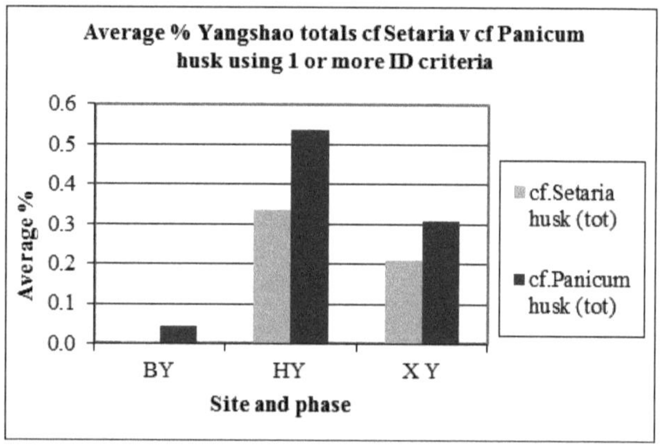

Figure 6.2 Average % Yangshao total cf Setaria v cf Panicum husk using one or more identification criteria

The proportions of cf. *Setaria* and cf. *Panicum* husk phytoliths are the opposite of the macro remains with *Panicum* as the dominant crop. However, if I use only the *Panicum* and *Setaria* husk multicells that fit five or more of the identification criteria the result is quite different (*Figure 6.3*). *Setaria* is the dominant crop at Huizui and Xipo.

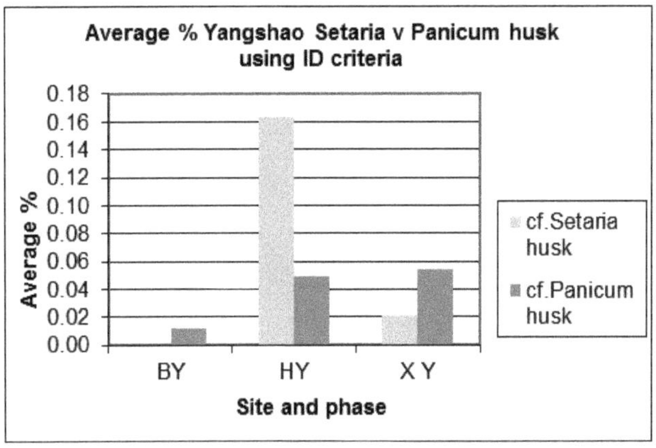

Figure 6.3 Average % Yangshao Setaria v Panicum husk using five or more identification criteria

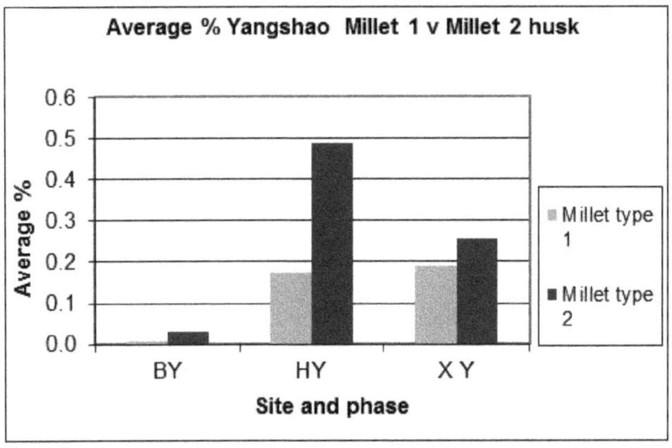

Figure 6.4 Average % Millet 1 v Millet 2 husk

There are differences between the Panicoid crop weeds, represented by millets 1 and 2, and *Panicum miliaceum*, the crop. However, they are small and difficult to see, especially in archaeological samples, so it is essential that a number of criteria for identification are established and adhered to. These groups include any crop weeds, such as *Setaria viridis*, that fall into each group (Figure 6.4).

This is a clear example of how adherence to strict identification criteria from good reference material is essential for meaningful phytolith analysis.
However, it may be noted there is still marked divergence between the Panicum miliaceum charred seed remains and the Panicum miliaceum phytolith husks, which are present in far higher proportions. There are a number of possible explanations. The phytoliths do not show whether they are from whole or partial or disintegrated seed, but the macros rely on a seed being whole or at least undamaged enough to identify, which may create a bias. The phytolith identification for Panicum miliaceum may need to be more stringent. This is something that can be developed in the future.

6.8 ARCHAEOLOGICAL SAMPLES

COLLECTION

When the archaeological samples were collected the section or floor was cleaned first and where possible samples of around 10 grams were taken. The samples collected by Gyoung-Ah Lee (Huizui 2005, Erlitou and the survey) and students from PKU (Baligang) ranged in size but were collected in the same way (Appendix 5).

pH of soils/ samples
The pH was measured in a selection of samples from all sites, periods and contexts in order to establish whether this affected the content and preservation of the samples and if there was an effect on how this impacted on the results as a whole. High levels of alkalinity, over 9, can be destructive (Piperno, 1988 in Piperno, 2006:22). However, the effect was minimal in the case of this study (see Results Chapter 8).

LABORATORY PROCESSING PROTOCOL

There are various procedures for removing phytoliths from archaeological sediments (Piperno, 2006:90, Coil *et al*, 2003, Rosen, 2001, Lentfer and Boyd, 1998, Zhao and Pearsall, 1998, Madella *et al*, 1998, Powers and Padmore, 1993). All the above methods involve removing the constituents of the sediments, minerals such as calcites, clays, organic matter, from the phytoliths leaving them clear for mounting.

I used Rosen's protocol (3/3/99, 2001) to process 147 sediment samples. This method was chosen because it is the standard protocol in the phytolith laboratory at the Institute of Archaeology UCL so all the equipment and chemicals were readily available. It is also a non-hazardous method as the use of corrosive chemicals is limited to diluted Hydrochloric acid. There are several stages to this procedure, which may result in the loss of some phytolith material, but the end result is clean so facilitates identification and counting. If the same procedure is used for all the samples any loss should be standard.

Every sample was dried at < 50°in a drying oven for a minimum of 24 hours. For each sample around 800 mg of dried sediment were weighed using an analytical balance, if there was less all the sample was processed. Next, the sediment was sieved through a 0.25mm mesh, and placed in clean PVA 30ml centrifuge tubes The samples were treated with 15ml 10% HCL to remove any pedogenic carbonates and washed in distilled water before centrifuging for 5 minutes at 2000rpm. The suspense was poured off. This was repeated twice more to remove all the acid. Next, any clays were removed by adding a 15 – 20 ml dispersant (5% Sodium hexametaphosphate), pouring the sample into a tall beaker and mixing thoroughly with distilled water. The silt and fine sand fractions were settled through an 8cm column of water, according to Stokes law. The samples sat for an hour and the suspension containing the clays poured off. This process was repeated until the suspension was clear. Then the samples were pipetted into ceramic crucibles and left to dry overnight in a drying oven at < 50°C. Once completely dry the samples were burned in a muffle furnace for 2½ hours at 500°C to remove any organic matter. Once cooled the samples were placed in 15ml PVA centrifuge tubes containing 3ml Sodium polytungstate solution (SPT), calibrated to 2.3 sp gravity for heavy liquid separation, and centrifuged at 800rpm for 10 minutes. Opal phytolith specific gravity ranges from 1.4 to 2.3 (Jones and Beavers, 1963, Prychid *et al*, 2004:380) while quartz is 2.7 so the phytoliths float. The suspension, containing the phytoliths was removed and washed twice with distilled water at 2000rpm for 5 minutes. The phytoliths were removed from the tubes by pipette and dried and weighed. Up to 2.5mg from each sample were mounted on slides in Entellen and dried.

The samples were analysed using a Leica microscope fitted with a polarising filter at 400x magnification.

RECORDING

Recording sheets were developed where possible following the identification criteria set down by Madella *et al* (2005) in the International Code for Phytolith Nomenclature. Occasionally there were no precise descriptors for the morphotypes present or the phytoliths were being identified to sub-family or genus, in these cases I used my own descriptions. One issue was the large number of different morphotypes present in the samples. The decision was made to group some into larger categories appropriate to the questions being asked. For example, many single long grass cells can be categorised separately according to the ICPN protocol but are common across genera. These were grouped together as either long smooth or long sinuate as finer definition would have had no effect on the analysis in the case of this research project. Hairs were also grouped together. Due to the limitations of time I decided to focus on morphotypes that might answer my questions, for example, *Oryza* keystones and husks. This meant a number of distinctive multi cells were categorised as indeterminate leaf/culm or indeterminate husk. I suspect many of these morphotypes represent crop weeds and with the development of a more comprehensive reference collection should be identifiable in the future.

General indeterminate categories for single and multicells were included. These were mainly degraded or broken phytoliths that could not be placed in other categories. Multicells were labelled indeterminate when it was impossible to distinguish which part of the plant they were from. I also recorded silica aggregate clusters. Fragments of silica aggregate were noted but not recorded, as they were too small to count. Large proportions of silica aggregate fragments in a sample obviously affect the weight percentage. This was one of the reasons I decided to use morphotype percentage per sample (see chapter 7) as varying proportions of silica aggregate across the body of samples inevitably affects the numbers per gram.

CHAPTER 7

ANALYTICAL APPROACHES: FROM PHYTOLITH ASSEMBLAGES TO ARCHAEOBOTANICAL ISSUES

7.1 INTRODUCTION

The dataset for this project is comparatively large (158 samples with between 400-600 phytoliths counted per sample) so there are issues that need consideration regarding the methodological approaches for dealing a dataset of this size. In this chapter counting and quantification methods will be discussed. I shall also consider context related variation, what effect the type of context the phytolith is from may have on the assemblage and its interpretation. The crop-processing model used to interpret stages in processing will be described and adaptations for phytolith analysis discussed, as will the methods of analysis needed to discuss agricultural expansion and whether it is possible to use phytoliths from archaeological contexts to discuss local environment. The methodology used for preparing and analysing comparative datasets will be described, in particular issues of comparing macro data with phytolith data.

7.2 QUANTIFICATION

How a dataset is counted and quantified inevitably affects the results and subsequently interpretation. The questions asked of a dataset have an influence on the kinds of analytical approaches taken.

COUNTING METHODS

How the phytoliths are counted is an important part of the data analysis. There are a number of counting and quantification strategies that can be used, qualitative, raw counts, and quantitative, absolute and relative counts. In contrast to macro remains morphotypes or classes of morphotypes are counted rather than taxa. Plant taxonomic differentiation based on phytolith morphotype counts often use quantative comparisons of ratios or relative frequencies. The aim is to reach a count that results in a reliable statistically robust dataset, (Stromberg, 2009:124). Insufficient counts can result in the loss of less abundant morphotypes (Albert and Weiner, 2001), but counting is time consuming, there can be many hundreds of thousands of phytoliths on a slide, so the loss of information has to be balanced against the practicality of counting a large number of morphotypes per slide. See Stromberg (2009) for a more detailed discussion on this point.

ABSOLUTE COUNTS

Following Madella *et al* (2002:706), for this project absolute counts were calculated first using the method devised by Albert *et al* (1999) and Albert and Weiner (2001).

Once removed from the sediment, between 2 and 2.5 mg of silica bodies were mounted on each slide under a 24 mm cover slip. This provides 48 fields per column and row, so each slide contains 2304 fields. Using a standard scanning procedure (Pearsall, 2000:304, Rull, 1987), solitary single-celled morphotypes from individual plant cells and multi- celled forms, or silica skeletons, consisting of attached adjacent cells from a section of silicified tissue were counted using a transmitted light microscope at 400x magnification. All phytoliths in each field were counted until the required number was reached. The number of each phytolith morphotype per gram of sediment in every sample was calculated in order to standardise the phytolith count. Where possible I counted and classified approximately 300-400 single cell phytoliths and 100-200 multicells on each slide, recording the number of fields. The counts were extrapolated up to the number per slide, (count / fields counted x total fields on the slide) and from that to number per gram, (number per slide / number mg mounted x total mg phytoliths / total mg sediment x1000) (Albert *et al,* 1999, Albert and Weiner, 2001:258). In this way I established 'absolute counts' of phytoliths for each sediment unit. This meant I was able to make comparisons of densities of phytoliths per gram between samples across the sites. Although useful, this method is not ideal because some slides contain higher levels of uncountable non-phytolith material than others, for example, if a sample is from a particularly clay filled context it may be impossible to remove all the clay which then adds to the phytolith weight. Some samples also have higher levels of minerals and others contain fragmented silica aggregates, which are difficult to quantify.

Lentfer and Boyd (1998:1163) use a system that involves giving each sample a clarity rating and also a visual estimation of the percentages of biogenic silica, non-biogenic mineral particles and clay and organic aggregates. While recording these particles may be useful this method is rather subjective. After the required number was reached the rest of the slide was scanned for the presence or absence of morphotypes that may have fallen outside the counts. Finding unrecorded morphotypes in this way was extremely rare. They were recorded as present but not included in the analysis.

PERCENTAGE PER SAMPLE, RELATIVE COUNTS

Next I calculated the percentage of each morphotype in relation to the others per sample. This had several advantages. The first, of standardising the data, which meant I was able to make comparisons between samples and also with macro data. This method has an upper limit of 100 making the charts far easier to read. I also needed to use relative counts for the correspondence analysis, as the Canoco programme could not deal in variation from zero to several million, which appeared in some absolute density charts. There are some disadvantages to this method; the emphasis tends towards the morphotypes with the highest values so the patterns provided are inevitably slightly different to those produced by absolute counts.

7.3 METHODS OF ANALYSIS

PRESENCE / ABSENCE

A useful method is to look at scales of abundance. A simple presence or absence analysis can be useful as it considers the absolute count as too heavily influenced by preservation so ignores it and instead looks at the number of samples in which a given morphotype appears within a group of samples (Popper, 1988:60).

UBIQUITY

Ubiquity is a quantative method that can be used to assess the relative importance of different morphotypes, and plants and plant parts. The ubiquity is shown as the percentage of samples in which a morphotype is present within a group of samples, such as, a site or phase (Popper, 1988:61). Ubiquity analysis can help strengthen the presence/ absence information and can provide added insight on broad patterns and trends within the data, for example changes in significance of a particular morphotype or plant. The score of one morphotype does not effect that of another so different morphotypes can be assessed independently (Popper 1988:60-1). However, it is important to remember that ubiquity measures frequency of recurrence not abundance, so is comparative rather than absolute. Frequencies can be used to compare within morphotypes but not between. Single frequency scores are only significant with others from the same morphotype (Hubbard, 1980:53 in Popper, 1988:61) and it should be remembered that they could also conceal patterns if the abundance changes but the frequency does not (Scarry, 1986:193 in Popper 1988:64) so the outcome of ubiquity analysis needs to be compared with an absolute method.

CORRESPONDENCE ANALYSIS

Multivariate analysis can prove useful for large datasets that include several sites and many variables such as the one generated for this project (Colledge, 2001:67). Correspondence analysis was selected as the multivariate technique most appropriate for this data because it graphically displays the relationship between the samples and variables (in this case phytolith morphotypes) illustrating patterns in the data such as combinations of clustering and seriation (Hoilund Neilson, 1988:30 in Colledge, 2001:67). Correspondence analysis uses a pattern-searching approach. It displays the rows and columns of a data matrix as points representing the relationships within and between the set of samples and the set of morphotypes, so morphotypes will be closer to samples in which they are more abundant (Powers *et al*, 1989:321). Here it has been used to describe sample-to-sample, morphotype to morphotype, and the coincidence of morphotype to sample relationships in order to interpret why these distributions occur.

CANOCO software, devised by Ter Braak (1988), was the programme used to carry out the correspondence analysis on the data. CANODRAW, developed by Smilauer (1992), was used to generate ordination diagrams showing correspondence between samples and phytolith morphotypes from an initial dataset of 145 samples and 64 variables (phytolith morphotypes). Relative percentage values, percentage of morphotypes per sample, rather than absolute counts (number per gram) were used because the absolute counts ranged from zero to millions, which could not be clearly expressed using this method.

The dataset gathered for this project is large. To clarify the data some non-diagnostic morphotypes were combined, such as all hair types to create a total hairs category. The dataset covers three cultural periods, four sites and a survey, and three main context types. Correspondence analysis is useful for large datasets and broad inter-site analysis so is a good starting point in tackling the broader issues raised by the large data set for this project.

Correspondence analysis has been used by archaeobotanists to explore abundance and presence/absence to compare contexts, sites and archaeobotanical samples with the products of activities, such as crop processing (Jones, 1987) and compare ecological weed groups (Colledge, 1998, 2001, Bogaard *et al*, 1999, Bogaard, 2002, Charles and Jones, 1997, Charles *et al*, 2003,) Powers *et al*, (1989) used correspondence analysis on a phytolith assemblage from the Outer Hebrides to recognise anthropogenic and non-anthropogenic depositional levels and interpret whether variation pointed to grazing or pasture in coastal sand dunes systems. They used absolute counts and the GENCORR programme. Absolute counts have the advantage of truly representing the presence of morphotypes within each sample. However, the nature of correspondence analysis means relative abundance of the morphotypes is the natural focus (Powers *et al*, 1989: 32), In contrast to Powers *et al*, (1989), I used % per sample in this analysis firstly because the variations in the numbers per gram of sediment were too great for Canaco to deal with. I also wanted to compare the relative proportions of morphotypes and groups of morphotypes within the samples temporally, spatially and according to depositional context.

7.4 METHODOLOGICAL QUESTIONS

There are also specific methodological questions that need to be addressed. The dataset for this project is comparatively large (158 samples with between 400-600 phytoliths counted per sample) so there are issues that need consideration regarding the methodological approaches for dealing with a large dataset, especially the extent to which taphonomic or diachronic influences dominate the composition of the assemblages and how these can be teased apart.

CONTEXT RELATED VARIATION

One of the major issues to be addressed is context related variation. Until we can identify to what extent differences in the composition of samples are related to context or whether they can inform larger issues, phytolith analysis is of little use.

Correspondence analysis should provide information on whether samples from the varying context types show similarities or differences according to type. Once this is established the question of how different compositions of phytolith assemblage relate to taphonomy, outside processes or post depositional change can be tackled.

WEIGHT PERCENT

The total weight percentage of phytoliths per sample was calculated to establish whether there was a relationship between phytolith density and context, phase or site.

PH LEVELS

The pH of forty-four samples, selected from Xipo, Baligang, Huizui and Erlitou, was measured. Eight from floor foundations, twenty-two from ash pits and fourteen from laminated pit fills were tested to see if pH was context, site or phase related.

DOMINANT MORPHOTYPES

Some morphotypes may be dominant. It is important to understand why. Some, for example hairs, may be produced in abundance, others, such as long smooth cells may be common across many genera. The question of dominant morphotypes must be considered before making analyses and they will inevitably affect any outcome.

ISSUES OF EQUIFINALITY

There are many potential trajectories to arrive at an end state and consequently more than one possible interpretation of data produced by an assemblage. This should be remembered while making analyses and all potential starting points should be considered.

7. 5 CROP PROCESSING

INTRODUCTION

How people organise their food supply is fundamental to everyday life. Crop processing would have been a basic recurring activity during the Neolithic and Bronze Age. Understanding crop processing stages and where and how they take place can offer insight to how a society organises production and distribution of crops and their by products, including aspects such as, labour organisation. This is particularly relevant to this project as changes in such a fundamental activity may suggest changes in organisation of the society as a whole.

USING ARCHAEOBOTANICAL MATERIALS TO UNDERSTAND CROP-PROCESSING STAGES

According to Dennell (1972, 1974, 1976) food production can be modelled from macro archaeobotanical remains, (grains, chaff and arable weeds), because they are more likely to be preserved during processing than consumption activities. Dennell's model of processing stages and likely archaeobotanical composition of each stage can be used with archaeological data to identify archaeobotanical samples from specific contexts. Hillman (1973, 1984) developed predictive models focusing on wheat and barley through ethnographic studies of traditional non-mechanised crop processing in Turkey. He contends there are a limited number of efficient methods of processing a cereal crop and each of these has a distinct effect on the make up of the crop products and by-products. A crop will experience stages of processing before consumption and each leaves a distinct and measurable effect on the composition of crop and by-product in an assemblage. Unlike Dennell, Hillman focuses on content of an assemblage rather than context. Archaeobotanical assemblages regularly exhibit distinctive characteristics in composition that can be related to the composition of modern crop products so it is possible to create an analogue (Hillman, 1984:39). This work was extended by Jones in Greece, concentrating on statistical analysis of characteristics of weed seeds from different stages of crop processing (1984, 1987). This 'cause and effect' model relies on the uniformitarian assumption that plants will only break apart into specific components due to their morphology (Harvey and Fuller, 2005:740). Crop processing models have been applied to a wide range of crops, including barley and rye in Fennoscandia (Englemark, 1989, in Thompson, 1996:119) and quinoa in the Andes (Sikkink, 1988, in Thompson, 1996:119). Thompson (1992, 1996) used the Hillman model as a basis for her work on rice processing at Khok Phanom Di in Thailand, while Reddy, (1992, 1997) applied it to her work on the archaeobotanical remains of millet processing in India. Both the rice and millet processing models Thompson and Reddy use may be applicable with phytoliths as well as macro remains (Harvey and Fuller, 2005).

The composition of each assemblage is distinct at each stage so furnishing a possibility of identifying distinctions between producer and consumer sites (Hillman, 1984: 4; Reddy, 1997:168, Thompson, 1996:121). As assemblages from many contexts are made up of more than one deposit and these are often secondary and tertiary waste it is the broad patterns created by regular routine activities, mainly after storage, for example, dehusking for consumption. Absent stages can be considered non-routine. However, the classic consumer model assumes absence of early stages is indicative of a consumer site. But these early stages may have been carried out annually off site making them archaeologically invisible (Stevens 2003). Crops may be traded unprocessed, for example, in Tamil Nadu specialist metalworkers are often paid in sheaves of rice, immediately after harvest for work done in the previous season (Srinivas, 1976, in Harvey and Fuller, 2005:242), and crops may be fully processed off site by some producers.

Evidence of early crop processing stages has been taken to point to local cultivation. Distinguishing cultivation as opposed to consumption of crop plants is key to understanding more complex human economic behaviour. Cultivation, on or close to site, would suggest food was being produced rather than purely procured through trade and exchange (Reddy, 1997:162). Late or final processing stages are more likely to occur in domestic contexts so may afford information on daily routine or repeated patterns of activity over time (Fuller and Zhang, 2007:943). As crop processing involves systematic repetitive activities (De Garine, 1994:236), which may leave visible patterns in the archaeological record (Fuller and Zhang, 2007:955), the results of these routine actions should stand out in the data.

CROP PROCESSING STAGES

(Figure 7.1) Distinct stages of processing may be evident, depending on the crop and eventual use, harvesting, threshing, winnowing, sieving, pounding, and grinding. Each processing stage eliminates different plant parts and so produces assemblages of product, selected for succeeding stages, and by-product or residue which is processed out (Hillman, 1984: 4, Thompson 1992:250, Reddy, 1997:168). Some residues have economic importance, such as fodder (van der Veen, 1999:211), or as payment for labour (Srinivas 1976:102 in Harvey and Fuller, 2005:741).

Hillman (1984:7) argues for regional variation in processing methods and different processing pathways for different crops. Variation in processing method is likely to be dependent on seed morphology, especially winnowing, as smaller seeds, such as *Panicum* would be more likely to be lost during winnowing by wind (Reddy, 1997; 171). Whether the seed is hulled or naked has an affect on which processing stages are required (Hillman, 1984:7; Reddy, 1997:172). Both native Asian millets, *Panicum miliaceum* and *Setaria italica* require dehusking (Harvey and Fuller, 2005:743).

Archaeobotanical representation of each stage depends largely on where the processing takes place. If these activities occur in or close to the fields rather than on site they will be invisible archaeologically. However, if processing takes place on site the discarded residues may be visible as patterns in secondary and tertiary refuse.

7.6. MILLET PROCESSING

Panicum miliaceum and *Setaria italica*, the two millet crops found in the samples from Henan, have different sized seeds so are unlikely to have been processed together even if cultivated around the same sites. However, they are both small panicled millets (Fuller, 2006:3) and fall into Reddy's small-headed millet category (Figure 7.2). Some adaptations need to be made for use with phytoliths. Although panicles should contain high levels of husk, there should also be low proportions of phytoliths from leaf/culms. Likewise, chaff will contain some leaf/culm phytoliths.

HARVESTING

Ethnographic work on non-mechanized millet processing in India by Reddy (1994, 1997) has demonstrated grains can be reaped or gathered by uprooting and cutting the inflorescence from the rest of the plant. This is more common with fodder crops (Reddy, 1997:169). More usually, the seed heads are cut from the plant. There are variations in this method according to crop, for example in India, isolated seed heads of larger panicle millets, *Sorghum bicolor* and *Pennisetum typhoides* (= P. glaucum) are selected and cut individually either at the head or base, selecting against most weeds and culms. Smaller varieties, such as, *Panicum miliare* (=P. *sumatrens*) are harvested by gathering a number of stems and cutting at the base with an iron sickle (Reddy, 1997:169). This method inevitably incorporates culms and crop weeds so can be distinguished in the crop processing waste. Both *Panicum miliaceum* and *Setaria italica* are small panicled millets so may have been harvested from the base.

After harvesting crops can be stored and exchanged as panicles or sheaths although they are usually processed. The early stages are typically carried out in fields, or on threshing floors at the edge of villages in which case they are likely to be invisible in the archaeological record (Reddy, 1994).

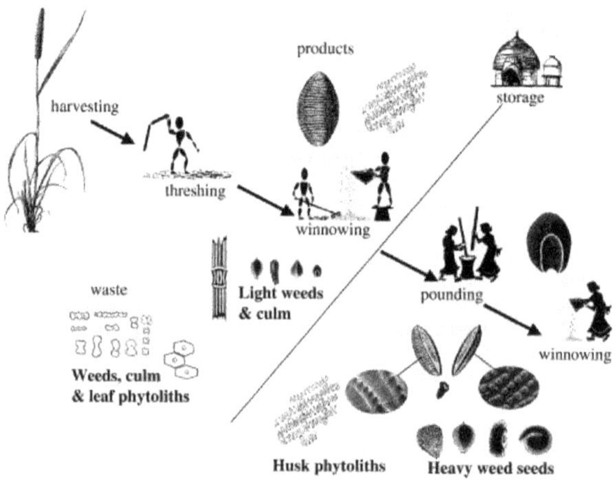

Figure 7.1. Millet processing stages (Harvey and Fuller, 2005, Figure 5)

MILLET HARVESTING

SMALL HEADED HULLED MILLETS

Cut at base Panicles , leaves/stem, weeds	**Cut at top** Panicles
Phytoliths Crop husk, weed husk, crop and weed leaf culm	**Phytoliths** Crop husk, some crop leaf/ culm

THRESHING AND WINNOWING
Beating + shaking

Spikelets/long straw + weeds Same sized weeds /heavy chaff	Spikelets/heavy chaff Grain
Phytoliths Crop husk, weed husk, crop and weed leaf culm	**Phytoliths** Crop husk, some crop leaf/culm

STORAGE

Spikelets and same sized weeds (only hulled crops)

Phytoliths
Crop husk, weed husk

Temper Fuel Fodder

PARCHING

MILLING and POUNDING

Grain/Chaff+ small weeds	Grain/Chaff
Phytoliths Crop husk, weed husk	**Phytoliths** Crop husk, weed husk

Chaff store + use

Figure 7.2 Model of millet crop processing and expected phytolith assemblages (adapted from Harvey and *Fuller: 2005: fig. 4 based on Reddy, 1994)*

THRESHING

The next stage is threshing to release the grain. Ethnographic observation shows there are three common techniques; beating with a stick, using feet to loosen the seeds or animal trampling (usually cattle). The method is dependent on the maturity and quantity of the crop (Reddy, 1997:169) and is selected for the specific crop type to optimise grain retrieval and time invested, for example there would be little point in threshing a small seeded grain such as *Panicum miliare* with cattle as there would be considerable wastage (Reddy, 1997:170). By-products from threshing should contain higher proportions of hulled or immature millets (Fuller and Zhang, 2007:955, Lundstrom-Bandais *et al*, 2002).

WINNOWING

Winnowing, or separating the grain from the chaff, straw, spikelets and weeds by weight, can by done by wind or by shaking in a wedge shaped, winnowing basket. Winnowing can be time consuming and labour intensive as it is a process that needs to be repeated many times until there are no contaminants left with the grain. The lighter by-products of winnowing, light weeds, leaves and straw (Harvey and Fuller 2005: fig 3), are often collected as fodder (Reddy, 1997:170).

These two processes can be distinguished in the macro archaeobotanical record by the presence of charred culm nodes and light weed seeds (Harvey and Fuller, 2005:745). While it is not yet possible to clearly identify phytoliths from specific millet culms the phytoliths from the weed husks should be visible. A relatively high proportion of culm phytoliths than in later processing stage by products are also expected. After winnowing, the heavy product is raked or swept with a broom to separate incompletely threshed panicles for return to the threshing floor (Harvey and Fuller, 2005:743).

SIEVING

Sieving is a stage where components larger than the grain are separated by size. While sieving is an important stage in *Triticum* and *Hordeum* processing, (Hillman, 1981; Hillman, 1984 in Harvey and Fuller 2005:743, Stevens, 2003) winnowing is more commonly used to separate according to size in millet processing (Reddy, 1997:172; Harvey and Fuller, 2005:743).

The by-products from these stages may be used as fodder, fuel or building materials (Reddy, 1997: fig 6; Harvey and Fuller, 2005:745), or returned to the fields as compost. Spikelets in their protective husks can be stored for later processing after winnowing or sieving (Reddy, 1997: fig. 6). Processing these spikelets to obtain clean grain for consumption requires a more concentrated labour investment in people and time so without organised communal labour pools it is unlikely that processing will be carried out during harvest time when labour is in high demand (Harvey and Fuller, 2005:745, Fuller and Stevens, 2009:9).

Well-ripened grain is often dried in the sun but if only partially ripe it may be roasted or parched over a fire to ease separation from the husks, (Reddy, 1997:170) although this stage is unlikely to be identifiable through phytolith analysis.

POUNDING

Once the grains such as *Panicum miliaceum* and *Setaria italica* are clean they are pounded repeatedly in a mortar with a heavy wooden staff to remove the husks (Reddy, 1997:170, Lundstrom-Bandais *et al*, 2002). Inevitably some spikelets and grains will be lost during this stage. Once the husks are broken or separated the crop is winnowed again to remove lemmas, paleas and weed seeds (Harvey and Fuller, 2004; 745). Finally, the crops are picked over by hand to remove any remaining visible contaminants. Lemmas, paleas, weed seeds and some smaller grains, producing mixed husk phytoliths, should stand out in waste assemblages from this stage (Fuller, 2002, 2003). As the chaff from small millets is diminutive it is very likely to be destroyed by charring and only visible through phytolith analysis.

GRINDING

Grinding is the process which transforms the grain to flour. There is no by product although a considerable amount of flour is spilled at this stage (Reddy, 1997:171). While there is little likelihood of macro remains from this activity, it may be visible in the phytoliths but only if any husk is remaining, as phytoliths are not produced in the caryopses. *Panicum miliaceum* and *Setaria italica* husk phytoliths were used to interpret a find of possible ancient noodles in Longshan China (Lu *et al* 2005).

STORAGE

Crops can be stored at different stages of processing. The closer a crop product is to consumption the lower the probability it has of being stored for later processing. In contrast, crops that are ready for consumption are very likely to be stored. These stages can be distinguished by the presence or absence of non-crop grain components. Storage for subsequent processing depends on the processing stage, what is being stored and the farmer's other activities. Factors such as rodent and small animal infestation need to be considered and crops are particularly vulnerable to bacterial or fungal infection if they are stored after pounding but before winnowing (Reddy, 1997:175). Some dehulled grains do not store well as their lack of husk and damaged pericarp makes them vulnerable to insects or fungus, in contrast to hulled cereals, such as *Setaria italica* (Reddy, 1997:175). Once stored, crops are likely to be processed routinely as needed in and around domestic spaces, usually by women either from an individual household or the combined labour from neighbouring households (Thompson, 1996:121, Peletz, 1988:48 in Harvey and Fuller, 2005: 745).

WEEDS

Despite being common rice weeds, some sedges (Cyperaceae), especially *Cyperus*, can occur with dry crops (Fuller and Zhang, 2007:952). Many grasses can be found in millet crops, including smaller seeded *Panicum* and *Setaria* species (Behrendt and Hanf, 1979:7)

HOW MILLET PROCESSING MIGHT BE SEEN IN PHYTOLITH ASSEMBLAGES

Direct evidence of harvesting is unlikely to be found in the samples collected for this project as it takes place off site. However, methods and tools may be inferred in the phytolith data by examining relative proportions of phytoliths from crop and weed husks. As stated above using finger knives to harvest just the panicles selects against weed husks and straw so sites or phases with relatively high husk to weed husk and straw may be harvesting high on the stem. The inclusion of high levels of weed husks and leaves suggests cutting lower on the stem, maybe with sickles, or uprooting. In theory it should be possible to distinguish uprooting from sickle harvested threshing waste by the presence of low growing weed seeds that would be missed by cutting higher up the plant. However, this remains to be tested with phytolith data and context needs to be taken into consideration before any interpretation is made.

Assemblages with higher proportions of crop husks may represent storage or dehusking waste. High percentages of weed husks with low crop husks may point to winnowing waste. Both weed and crop husks in high proportions might point to various cultivation practices, for example un-weeded fields, or mixed by products from threshing and dehusking.

7.7 Rice processing

Thompson's ethnographical work in lowland Thailand suggests crop-processing practices had not changed to a great extent over generations until recently (1992:122). Thompson identifies five stages of traditional crop processing; harvest, threshing and winnowing, dehusking, storage and cooking and consumption, (Thompson, 1996:121).

Harvest

As with millets there is choice in harvesting height and tools, which can have a crucial effect on resulting assemblages. Finger knives used to cut single panicles select against the incorporation of crop weeds. Sickles are used in a similar way to cut handfuls of panicles at a time. Most panicles are cut with 60-70cm stems, but some is cut close to the base leaving a long stem for hand threshing. The rice is bundled into sheaves and left to dry before threshing (Thompson, 1996:124). Straw is considered a valuable commodity and sometimes collected in a second round of reaping and at times is used as part payment for labour (Sharp, *et al*, 1953 in Thompson, 1996:124,) Parallels can be made with wheat (van der Veen: 1999) and payment of iron smiths with whole stems in South India (Srinivas in Harvey and Fuller, 2005: 741)

Methods of cultivation also influence the presence of crop weeds, broadcast seeds, more often the choice in dry farming, are more likely to grow in accompanying stands of weeds such as *Eleusine*, whereas the rice in rain fed lowland fields and waterlogged paddies is usually transplanted which seems to remove some weeds (Thompson, 1992:252, 1996:146). Control of water level during early rice growth may also help to control many weeds by drowning them.

Threshing and winnowing

Threshing often takes place on a specially prepared threshing floor. The separation of spikelets from straw has various methods. People or buffalo can tread over the pile (mainland South East Asia), sometimes a stone roller is used (Turkey) (Hillman, 1984:123), flailing the crop using sticks (Indonesia) or pulling the panicle and straw through a wooden comb (Japan and Korea). The straw is turned and the straw stems are raked away (Thompson, 1996:138-9). Some grain is lost at this stage but very little.

Once threshed the rice is sometimes dried before winnowing, increasing the weight difference between the filled and empty immature spikelets. The spikelets are tossed into the air to separate the grain-filled spikelets from the chaff, including unfilled or immature spikelets. The waste is used as animal food and the semi-cleaned spikelets are dried before storage (Thompson, 1996:125-129).

Storage

The rice is stored dehusked as 'rough rice' or husked as 'semi-clean spikelets', sometimes for up to 10 years, although some may spoil in which case it is usually burned so it may enter the archaeological record as charred spikelets (Thompson, 1996:140).

Dehusking

Up until this stage the rice is treated in bulk and requires organised labour. However, once stored, the crop can be used when it is needed so is likely to be processed at a domestic level in small proportions on a daily basis. Milling often takes place using wooden mill and sometimes a second winnowing takes place before manual pounding, hand cleaning and consumption (Thompson, 1996:129-134, 140).

Alternative methods used with wild-gathered rice are to soak the husks in water and then rub them apart by hand (Isaacs, 1987:115, Crib and Crib, 1974:101-2 in Thompson, 1992:249) or dry in bundles in the sun and burn to separate the grains from the straw (Specht, 1958: 484 in Thompson, 1992:249). Either of these methods would leave a different type of phytolith assemblage. The first husk only, the second mixed husk and leaf/culm.

Cultivation and crop weeds

How the rice is cultivated and the environment where it is grown will have a bearing on the crop processing assemblages. Rice can be grown in different environments, dry, wet and deep water (Moody, 1989) Each environment has a range of accompanying crop weeds, some overlap, for example Cyperaceae, others are particular to that environment, for instance, *Chloris barbarta* can be found with upland rice but is unlikely to thrive in a more humid environment (Thompson, 1996: Table 32) (Table 7.1). The way deep water paddies are created can select against weeds of any kind, as the paddy base is cleared and tamped, then waterlogged, drowning many potential weeds.

Weed seeds can play an important role in distinguishing which crop processing stages may be represented in an assemblage (Jones 1984B, 1987A, Englemark, 1989 in Thompson, 1996:145). However, when compared to other cereals, such as, wheat and barley, or dense small panicled millets, rice processing rarely produces large quantities of weeds. Thompson found no weed seeds in the rice processing by products from lowland rain fed systems in Thailand, and neither did Yen, (1982) in upland harvest samples. This study is of rice temper, mainly from dehusking waste, in pottery so identifiable crop weeds may not be expected. This paucity may be due to a number of reasons such as local environmental conditions and cultivation practices as many non-aquatic

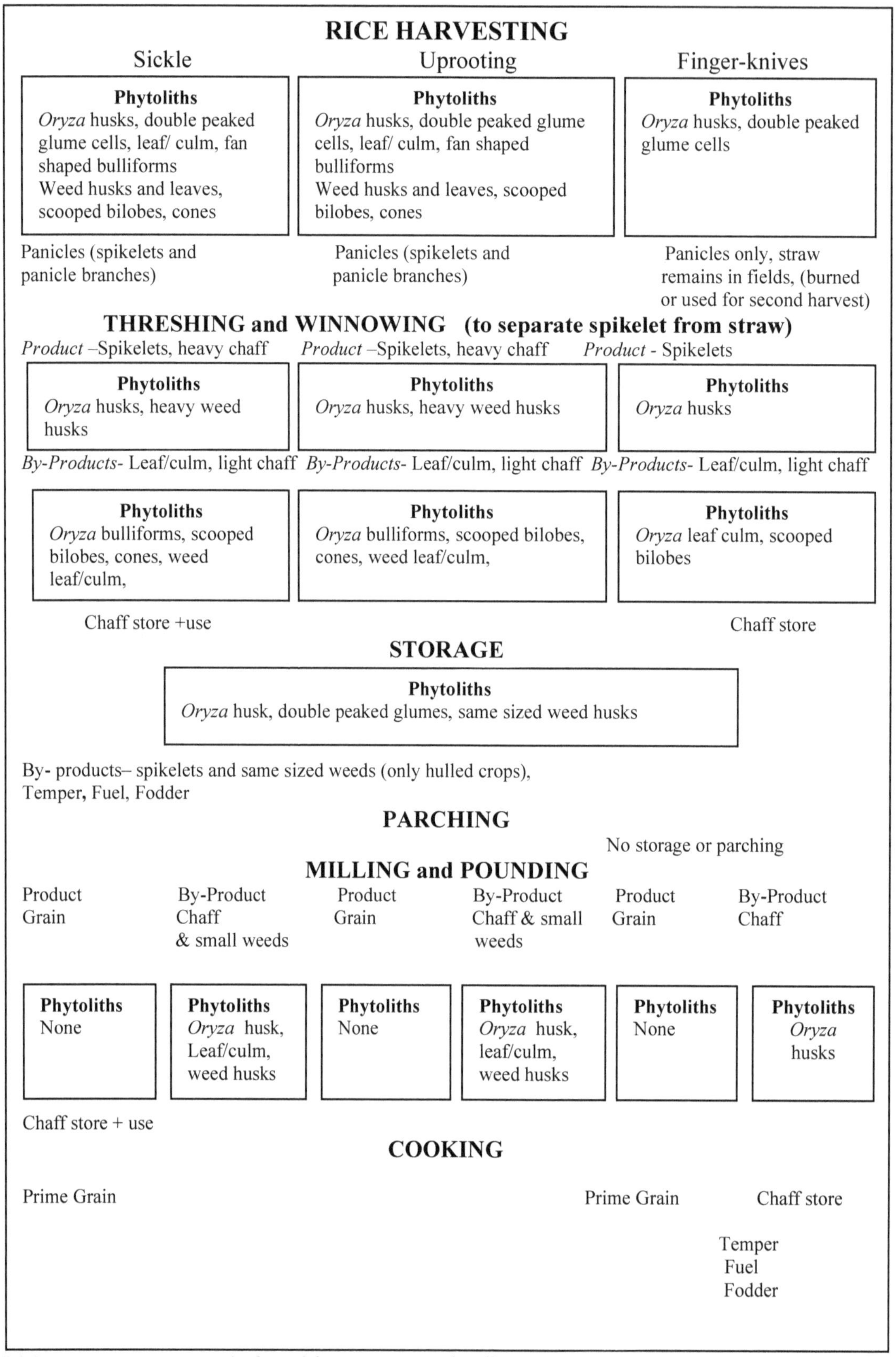

Figure 7.3 Rice processing stages(adapted from Harvey and Fuller, 2005 after Thompson, 1992, 1996)

segetals are drowned when flooded (Fukui, 1978 in Thompson, 1992:145). Nevertheless, semi-aquatic grasses and sedges (Cyperaceae) grow happily in inundated conditions and compete with rice crops.

Upland rice, where seed is broadcast into dry soil rather than transplanted, has the greatest weed infestation and diversity (Moody, 1989). Transplanted seeds have the advantage of lack of competition as seedlings and subsequent replanting in standing water, which prevents weed regrowth.

Weeds of rain fed rice	Weeds of deep water rice	Weeds of upland rice
Cyperus difformis	*Echinochloa colona*	*Dactyloctenium aegyptium*
Cyperus pulcherrimus	*Panicum cambogiense*	*Digitaria adscendans*
Echinochloa crus-gali	*Leptochloa chinensis*	*Eleusine indica*
Fimbristylis littoralis	*Echinochloa crus-gali*	*Eragrostis temella*
Leptochloa chinensis	*Paspalum scrobiculatum*	*Panicum repans*
Scirpus juncoides	*Ischaemum barbatum*	*Cynodon dactylon*
	Cyperus pulcherrimus	*Chloris barbata*
	Fimbristylis littoralis	*Brachiaria reptans*
	Cyperus rotundus	*Cyperus rotundus*

Table 7.1. Rice weeds that produce phytoliths (after Thompson, 1996: Table 32 Sources Kittipong, 1983, Vongsaroj, n.d.)

Swamp fields seem to produce the fewest weeds, probably due to plot clearance destroying the majority of weeds (Syarifudden *et al*, 1983 in Thompson, 1996:145).

Harvesting methods will affect the number and diversity of weeds in an archaeobotanical assemblage, for example using a finger knife will select against many weeds (Thompson, 1996:145), as observed by Reddy (1994) in big dense panicled millets (*Pennisetum* and *Sorghum*)

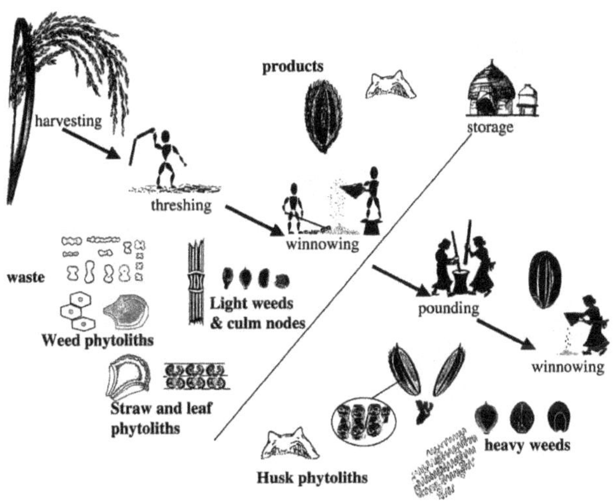

Figure 7.4 Rice processing stages (After Harvey & Fuller 2005, Fig 3

HOW THESE MODELS CAN BE ADAPTED FOR USE WITH PHYTOLITH DATA

Fuller and Zhang (2007:955) suggest evidence of variation in crop processing may be discerned by comparing grain to weed ratios, following the principles of Stevens' analysis of wheat and barley processing (2003). This should be possible using phytolith data as both millet and rice husks are clearly identifiable. Varying cultivation and harvesting practices may also influence the proportions of weed seeds. In macro remains sample size may impact diversity of weed seeds (Fuller and Zhang, 2007:945), this has not yet been tested for phytoliths.

There is the question of whether immature grain will have produced phytoliths and if so are they distinguishable from mature grain. It seems unlikely this would be visible in archaeological phytoliths given the difficulty in identifying to species compared to modern reference samples.

In theory phytolith husks and leaves can be treated in the same way as seeds and chaff, in practice the plant parts preserved are somewhat different. The vast majority of phytoliths preserved in this project are from leaves and stems and are not distinguishable as chaff.

Although both rice and millets are found in all sites they will be dealt with separately as their processing should leave different distinguishable residues.

HOW RICE CROP PROCESSING STAGES MIGHT SHOW IN THE PHYTOLITH ASSEMBLAGES

Identifying rice harvesting practices raises the same issues as with millet crops. There are various harvesting methods and tools, which can produce different assemblages. Finger knives may be used to cut the panicles only, sickles or up rooting would produce leaf and culm phytoliths in addition to the crop husks. Rice has an advantage over millet because it produces readily identifiable phytoliths in its leaves, *Oryza* bulliforms. It also produces scooped bilobes, which although they may be found in other Oryzoideae, are likely to be *Oryza* or *Oryza* crop weed leaves. Looking at the proportions of *Oryza* husk to *Oryza* bulliform, *Oryza* leaf multicells and scooped bilobes should give an indication of how the crop was harvested.

Threshing bi-products should consist of crop and weed leaves and stems. Thompson's ethnographic observations note few crop weeds are found with rice (Thompson, 1996). This may not have always been the case, especially with dry farming and is certainly an area that could be usefully investigated in the future.

Rice early winnowing residue should consist of high proportions of leaf waste and light weed husks; for example, some sedges (Cyperaceae), like *Cyperus Fimbristylis,* produce very small seeds so are likely to be removed in the early stages of crop processing (Fuller and Zhang, 2007:952), although others such as *Scirpus* spp. are likely to be large, heavy weed seeds (Fuller pers. com). These produce identifiable phytoliths and should be visible, although identification to genus level remains problematical (see Chapter 6.). Cyperaceae grow in all types of rice cultivation regime (Table 7.1). With more extensive references it may even be possible to identify agricultural practices by the presence/absence of the rice crop weed husk phytoliths.

Samples containing evidence of high proportions of *Oryza* husk with few weed husks and leaves might suggest storage, but they could also point to dehusking. This could be better clarified by the presence of husks from large weed seeds, such as the distinctive *Coix lachryma - jobi*, although these can also be used as a crop, that are heavy enough to fall with the rice grains during winnowing and spikelet residue.

7.8 FODDER AND FUEL

Crop processing by products in ethnographic studies were used as animal feed so may be traced in dung used as fuel (Charles, 1988, Miller, 1984, Hillman *et al*, 1997, Reddy, 1997, Murray, 2009). However, as pigs are the predominant domestic animal at the sites sampled in this project, it is doubtful whether their dung, due to its consistency, would be retained for fuel. It would be more likely spread back onto the fields as manure along with other waste. While it was not possible to do ethnographic work as part of this project, animal and human waste was recycled along with vegetable waste at Xipo and spread back onto the fields as fertiliser and compost. This is certainly an area of investigation that could prove invaluable in the future.

7. 9 METHODS OF DATA ANALYSIS USED TO SHOW CROP-PROCESSING STAGES

There are several methods of analysing data to show crop-processing stages. In this study I used correspondence analysis first to see if there were obvious broad patterns within the composition of the samples related to crop processing. Once this had been established I generated tri-plots using three axes to compare ratios of crop husk: weed husk: leaf/stem and where possible, for example with Oryza morphotypes, crop husk: weed husk: Oryza leaf/stem. From there bar charts were used to clarify the picture, in relation to particular plant parts or ratios that were of interest.

7.10 GRASS SUBFAMILIES AND HABITAT

Phytoliths from non-crop plants may provide insight on the input of vegetation other than crops. In turn this could afford further understanding of the local environment and vegetation around each site and any spatial and diachronic variation.
It is likely that much of the non-crop vegetation found within the sites is from crop weeds. Some may be from building materials, thatching or temper for mud bricks. Unpicking the human from natural influences on the local vegetation may prove problematic. However, it is worth examining variation in the levels of different grass sub families and also comparing proportions of grasses with C3 v C4 photosynthetic pathways. Different Gramineae subfamilies produce morphologically varied phytoliths (Twiss *et al* 1969). Broadly speaking, Panicoideae, many of which grow in warm humid conditions, tend to make bilobes and cross-shaped phytoliths. Pooids thrive in cooler damper conditions and produce rondel shaped short cells. Chloridoideae, which indicate a warm dry habitat, produce short saddles. Bambusoideae produce collapsed saddles and indicate a sub-tropical – tropical environment (Lu 2002:382, Twiss, *et al*, 1969, Piperno and Pearsall, 1998, Piperno, 2006). Plants belonging to the Oryzoideae sub-family can produce scooped bilobes and are often distributed in seasonally inundated wetlands so are a good indicator of local environments. However, as *Oryza* is an important economic crop throughout this region it could be difficult to disentangle the phytoliths produced by the crop plants from those found in plants from the local natural habitat, especially as the samples analysed in this project are from on rather than off sites.

The ratio of rondels to bilobes/cross to saddle can provide an indication of cool/arid to warm/moist to warm/arid (Twiss, 1992:113). However, this is only in the broadest terms and this method should be approached with caution. While Pooids do not generate bilobes or saddles Panicoids do occasionally produce rondels (Watson and Dallowitz, 1992,). Some Panicoids thrive in cooler drier habitats. Arundinoids, such as the ubiquitous water loving *Phragmites*, produce saddle shaped phytoliths that can correspond morphologically to saddles from Chloridoids, for example *Eragrostis minor*. Slight differences can be discerned in the modern reference collection material, but since deposited material is likely to have been subject to taphonomic processes and weathering, there is a strong likelihood that single cell saddles are indistinguishable to subfamily in the archaeological record. One solution might be to refer only to multicells/ silica skeletons present in the sample, as these might contain features to assist in identification.

A similar method is to distinguish plants that have C3 or a C4 photosynthetic pathway (Iriate, 2006, Parker *et al*, 2004, Fredlund and Tieszen, 1997, Twiss, 1992,) and group them accordingly as this can give an indication of general habitat. The C4 photosynthesis pathway is the result of anatomical and biochemical modifications within the plant, increasing photosynthetic efficiency where conditions cause high levels of photorespiration, such as rocky, arid or saline environments (Sage, 2004:341-2, Schulze *et al* 1996 in Sage, 2004:342, Ehleringer and Monson, 1993:415, Collins and Jones, 1985:127). The majority of C4 plants are grasses or sedges found from warm temperate zones to the tropics and are the most common grasses found in arid landscapes (Archibald, 1995, Sage *et al* 1999 in Sage, 2004:341). Changing climatic conditions should produce alterations in the ratios of C3 to C4 grasses (Parker *et al*,

2005:666, Nelson *et al*, 2006:567, Collins and Jones, 1985:122). Grasses have a short life span and adapt to environmental change fairly rapidly. Changes in levels of rainfall and temperature are likely to alter the C3:C4 proportions (Ficken *et al*, 2002 in Parker *et al*, 2005:666, Pyankov *et al*, 1999:15, Niu *et al* 2005: 2874). However, this approach needs to be considered carefully, while many C3 plants are Pooids and prefer a cool environment, *Oryza*, like many other Bambusoideae, grows in warm humid habitats but is also a C3 plant.

As stated earlier with regard to *Oryza*, agricultural crops will influence the results. *Panicum miliaceum*, which was one of the most important staple crops in North Central China throughout this entire time span, is a panicoid so is likely to influence the number of bilobes present in the samples. However, as *Panicum* obviously thrived in this region, it is very likely that this also has some reflection on the local environment.

7.11 Bulliforms as indicators of transpiration

As stated in the methodology chapter (chapter 6) bulliform shaped phytoliths can be linked with hydrophilic grasses and increased evapotranspiration (Delhon, 2009:179, Webb and Longstaffe, 2002, 2003). Comparing variation in proportions of Bulliform morphotypes across periods and sites may provide an indication of evapotranspiration. Delhon (2009:179) uses the ratio of bulliforms to long grass cells to establish a proxy indicator for pedosedimentary sequences. Application to the samples in this project is more problematical as they are from archaeological sites, and so culturally influenced. However, using the bulliforms in conjunction with other indicators such as diatoms and sponge spicules, or phytoliths from water loving plants, (for example, *Phragmites*), may provide insights into the local environment. Cultural impact on the samples must be considered carefully, however, before coming to any conclusions.

7. 12 Comparative dataset, phytoliths from Neolithic India

Comparative datasets afford the opportunity to explore another facet of the results. Comparison of the database created for this project and a phytolith database created by Emma Harvey (unpublished PhD thesis, UCL, 2006) from 7 sites in India, three in Uttar Pradesh and four in Orissa, offers the opportunity to explore methodological questions such as whether the same models can be tested on phytolith datasets compiled by different people. In the case of the Indian dataset here although both researchers worked in the same laboratory, used the same methodology and identification criteria the Chinese samples were examined under both polarised and non polarised light while being counted, whereas the Indian were analysed using only non polarised light. This may have an influence on the identification of phytoliths from dicotyledons because while they are invisible under polarised light and non silica inclusions are highlighted, they sometimes look very similar to non silica inclusions under non polarised light so could be ignored if polarised light is not used. Grass and sedge phytoliths have more specific morphologies so are possible to identify without checking with polarised light. The phytoliths from each dataset were recorded with slightly different categorisation appropriate to the different regions and environments sampled. To address this matter a data table was compiled which combined and grouped both sets to 36 morphotypes.

Comparison of these datasets may also highlight variation in crop selection and possibly cultivation or harvesting practices between Neolithic China and India as well as drawing attention to any broad differences in local environmental conditions. Differences between dry (India) versus wet (China) might be expected (Fuller and Qin, 2009, Rosen, 2008).

CHAPTER 8

GENERAL RESULTS

8.1 Introduction

The dataset gathered and analysed for this project was large and produced abundant results. These are presented over the following two chapters. The phytolith morphotypes were analysed using correspondence analysis, site averages, presence absence, morphotype ubiquities, relative frequencies and absolute densities to establish broad and then finer resolution relationships within the data. 147 samples were counted and analysed (Appendix 3). Four of these were dated to either preceding or subsequent to the periods examined so not included in the general analysis.

The results focus mostly on crop and economic plants to establish patterns that may provide insight to the questions outlined in Chapter 1. Phytoliths from non-crop plants are also considered.

In order to understand the effect of the varying pathways of the phytoliths into the samples and then what may have happened to them once deposited, the results of analysis for basic taphonomical issues are shown first. These are followed by the results generated by correspondence analysis to provide a broad general picture, followed by results for each period and site. Once this has been established, the following chapter presents results showing crop changes, beginning with a temporal and then spatial analysis, including any which demonstrate agricultural expansion. An analysis of samples for crop processing stages is next, followed by an investigation of the non-crop phytoliths and their relationship with the local environment. Finally, the results of the comparative data set will be presented.

8.2 Taphonomy

The phytolith weight percentages per sediment sample for each sample were calculated to establish the density of phytoliths preserved. The percentages were calculated by dividing the total phytolith weight extracted by the sample weight and multiplying by 100. The pH of a selection of forty-four samples was also examined and compared to the weight percent in order to ascertain the extent to which alkalinity levels in the sediments had affected preservation. Correspondence analysis was used to draw out the broader patterns within the dataset, including the effect of contextual variation on the constituents of the samples. These results are described in the correspondence analysis section (8.3).

Weight percentages

The weight percentage of phytoliths in each sediment sample was calculated to establish the densities of phytoliths preserved per sample (Figure 8.1). The majority of samples at Xipo, Baligang, Huizui East and Erlitou consisted of <3% phytoliths but there were exceptions, for example three samples at Xipo, Baligang and Erlitou consisted of >60% phytoliths. The survey samples and those from Huizui West also contained generally higher densities, the exception being the one survey sample from the Yangshao (S1). The two highest density samples (X3, B25) were both from ash middens. The lowest densities were from Huizui (HE 98) and Baligang (B10) < 0.02%, both lenses from laminated pit fills.

Figure 8.1 Weight % phytolith per sample

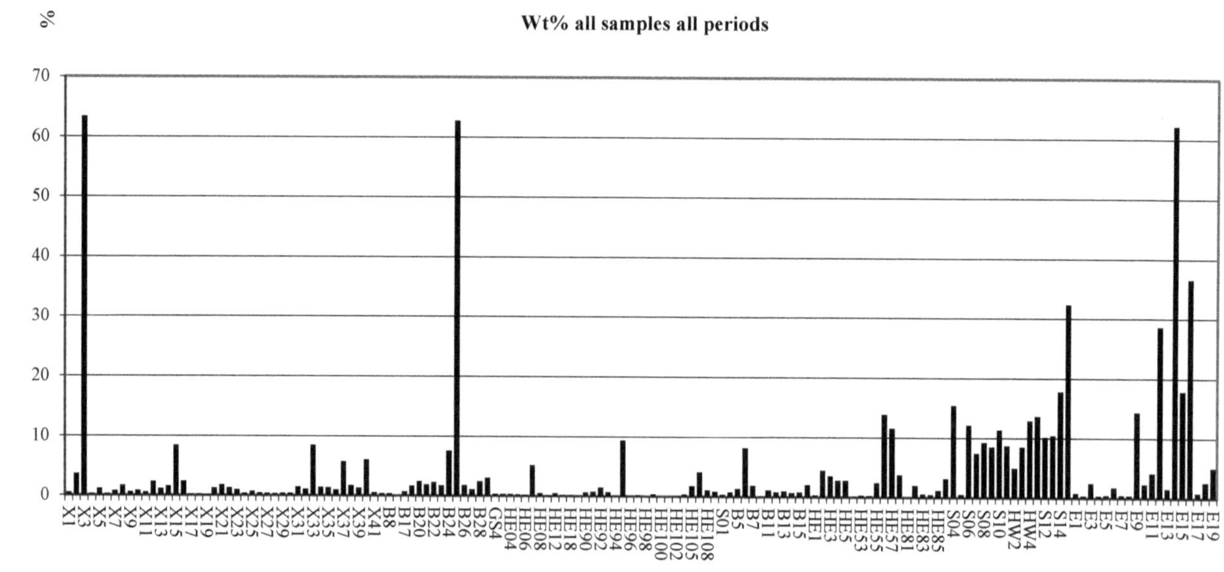

The Longshan and Erlitou period survey samples ranged from 7.2% to 32.29%. The Huizui West percentages, while falling within a slightly lower range (4.95 % - 14.34%), were still consistently higher than the majority of other samples. These were all from ash middens. However, there is a very high-density sample from an Erlitou laminated pit fill (62.2). Interestingly, it is from an ash layer, stressing how phytoliths accumulate and preserve well in ashy contexts, perhaps unsurprisingly as most ash generally consists of at least some vegetative material.

DENSITY RELATING TO CONTEXT TYPE

The range of phytolith weight percentage per sample from ash middens is from 0.12 % to 63.4%. Most are <2% but a third are > 6%.The densities from the laminated pit fills are all <18% apart from two ash layer samples from Erlitou (62.2, 36.6). The vast majority are <2% phytolith and half of those are <1%.The floor/ foundations present a similar picture with most samples containing <3% silica bodies. This highlights the importance of considering context when analysing the samples quantatively since it seems to have a direct bearing on the quantity preserved.

DENSITY RELATING TO PHASE

A second possible issue is how long the samples have been in the ground. The Yangshao ash middens, despite containing the two densest samples, are otherwise all < 10% silica bodies. There is an increase in the Longshan from the survey samples, and another general increase in the Erlitou but the majority of samples from the Erlitou site itself are very sparse. There seems to be a trend towards higher densities in the later periods. This could be related to human influence on the samples, for example, were a greater number of phytoliths being deposited due to more intensive crop processing taking place on site? Or it could be simply higher deposition of vegetative waste.

PH

The sediment samples from the ash middens varied from pH 7.14 to 8.36.

The pH of a selection samples from each site, context and period was measured and recorded using the standard procedure described in chapter 6 (Figures 8.2 – 8.5). This may provide insight into taphonomic processes and preservation as phytoliths survive well in acid and neutral sediments, whereas highly alkaline soils can dissolve or degrade silica bodies (Piperno, 2006:22).

The three different context types were examined first (Figures 8.2-8.4). The laminated pit fills had the lowest and highest measurements (6.87 and 9.19), reflecting the variation in the constituents of the lenses that make up this type of context (figure 8.3). Unfortunately there was no sediment remaining from Erlitou period laminated pit

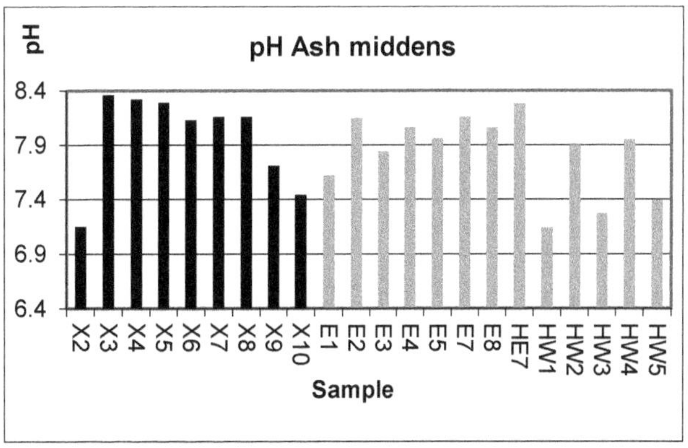

(X= Xipo, E=Erlitou, HW = Huizui West)

■ Yangshao, □ Erlitou

Figure 8.2 pH Ash middens

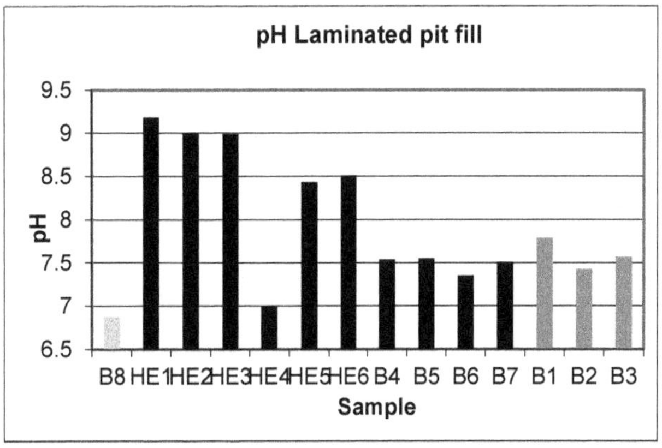

B= Baligang, HE = Huizui East

□ Yangshao, ■ Longshan, ■ Eastern Zhou

Figure 8.3 pH Laminated pit fills

fills after phytolith processing as the focus was on looking for a distinction over time rather than anything specific to the Erlitou period the later Eastern Zhou samples were used. Although the single Yangshao sample had the lowest pH, unsurprisingly the period does not have a bearing on the pH of the sediment in the laminated pit fills.

The floor/foundations all were above pH 8 apart from one and all within the neutral range (figure 8.4). There was no great difference between site and period, although the samples from Huizui East were generally slightly higher, all samples measured, except one (pH7), were above pH 8. The pH fell within similar ranges at Xipo and Erlitou, and again at Huizui West and Baligang. At Huizui some samples were at the higher end of the scale and there was a broader range in general, but overall the range across all the samples is slight. None of the sediment samples were acid so it is unlikely they came from either latrines or animal pens.

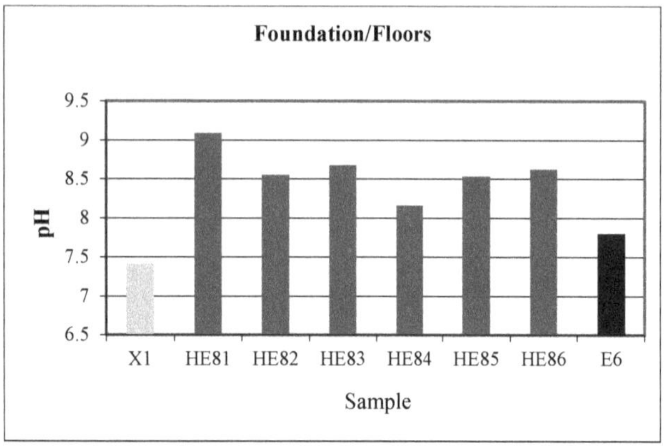

X = Xipo, HE=Huizui East, E =Erlitou

Yangshao, Longshan, Erlitou

Figure 8.4 pH Foundations/floors

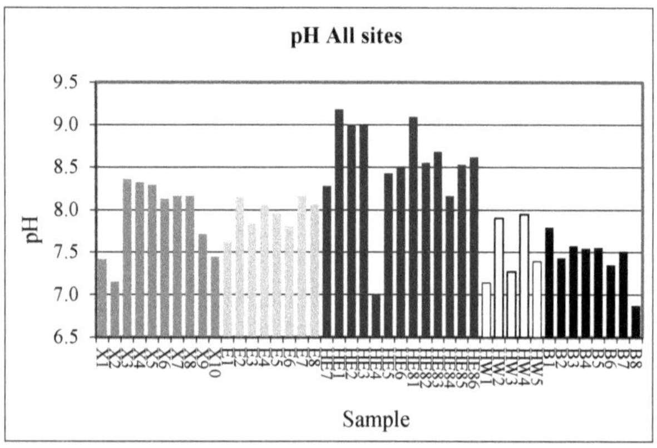

X = Xipo, E = Erlitou, H = Huizui East, HW = Huizui West, B = Baligang

Figure 8.5 pH all sites

WEIGHT AND PH

Weight % and pH were compared to see if there was a relationship between them that might highlight preservation issues. There seemed to be little connection in the case of these samples, although the floor/ foundation contexts did have generally lower weight percentages and very slightly higher pH values, (Figure 8.6. 8.7).

8.3 CORRESPONDENCE ANALYSIS

Correspondence Analysis was used to establish broad patterns in the data by highlighting patterns of correlation in the composition of samples. Relative frequencies were used with CANOCO because the very large numbers generated by the absolute densities meant there was too great a variation between some of the counts for the CANOCO programme to work efficiently. Correspondence analysis was chosen because it is a useful way of simplifying and visually presenting a large

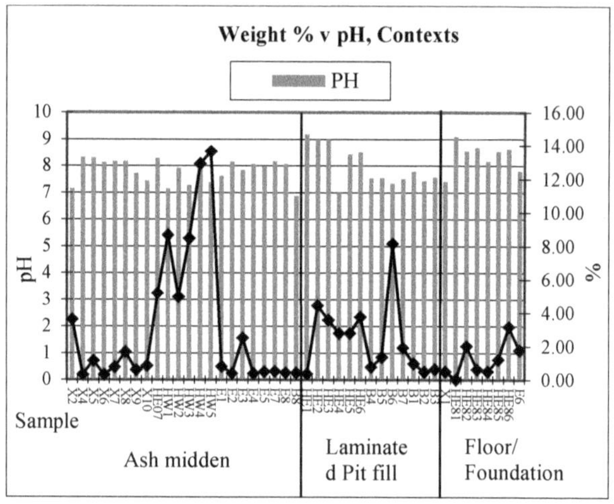

Figure 8.6 Weight % v pH Contexts: ash midden, laminated Pit fill, floor/foundation

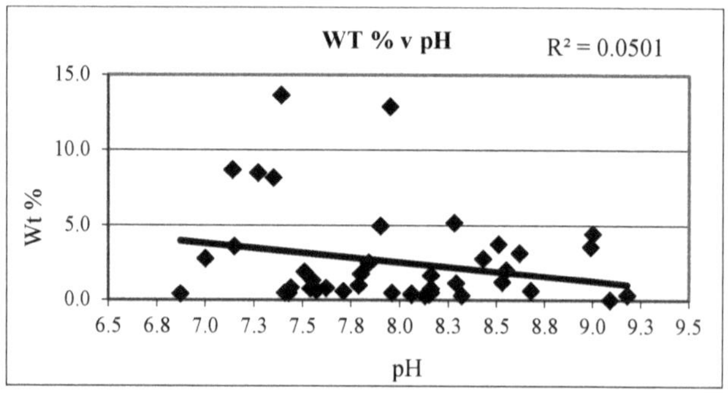

Figure 8.7 Scatter plot showing poor correlation between weight % and pH

dataset containing many samples and variables of morphotype without losing detailed information. It shows the relationships between samples in terms of similarity of content, so can demonstrate if particular groups of sample, for example from a site or cultural phase, are similar or different in terms of proportions of their constituents to other specific groups of samples. In this way it may be possible to discern relationships between specific groups of morphotype and sites or phases.

Using CANOCO plots representing samples and morphotypes (labelled as species) were produced. The samples are also represented as pie charts showing the composition of each sample according to selected morphotype groups. Biplots demonstrate the coincidence of species to samples. The ordination axes show degrees of variation in the data. The principle axis (axis 1, x), represents the greatest variation, followed by the second, (axis 2, y), and third on a decreasing scale (Colledge pers. comm.). Clusters mean the samples are composed of similar groups of morphotypes.

The greater the distance between samples from left to right along axis 1 the greater differentiation there is between the proportions of the morphotypes within the samples. The % of variance from the point of origin, 0, is shown below for axis 1 and to the left for axis 2. When discussing the charts I use the quadrants as a tool to describe the position of samples or variables on the chart. There is no special or uniform significance to the quadrants themselves.

Unknown, rare and ubiquitous forms: mesophyll, sheet, starch, stomata, leaf/culm stomata, indeterminate single and multi cells, total hairs, *Bromus* and large Graminae husk, were deleted from the full dataset leaving 148 samples and 53 variables. Next the four samples that fall out of the range of the three major cultural periods examined (B16, B1, B2, B3) were also deleted. The table containing the remaining data was used to generate the plots described below.

The morphotypes were categorised into 4 groups: millet crops and weeds, *Oryza* and weeds, water loving plants, woody and dicots, and grasses. These groups were selected to highlight any variation between crop types and morphotypes from broad plant groups.

CONTEXTUAL VARIATION

When the complete dataset was examined there was little evident differentiation between context types according to the makeup of the samples (Figure 8.8). I examined the results for the whole dataset for each site then for each context, ash middens, laminated pit fills and foundation/floors to see if there was discrimination on the basis of composition between context types. As the charts were generated I deleted outlying samples and variables (morphotypes) if necessary in order to provide a clearer general picture.

I repeated this with the phases, Yangshao, Longshan and Erlitou.

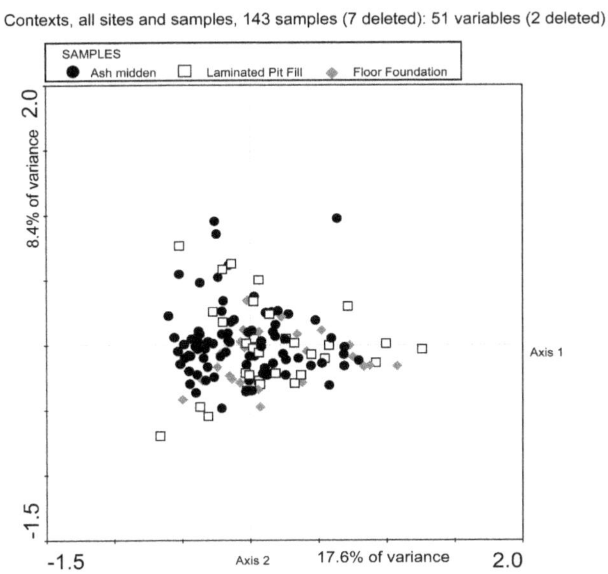

Figure 8.8 Contexts, all samples, 143 samples (4 deleted):51 variables

Samples from the ash middens, laminated pit fills and floor foundations are spread all over the chart demonstrating little differentiation related to context between the samples as a whole, although few floor foundations are present in the top left quadrant and those close to the centre of the chart. The species chart (Figure 8.9) shows millet husks in the upper part of the top left quadrant. This could suggest millet husks are more likely to be found in ash middens and laminated pit fills than on floors at the sites investigated.

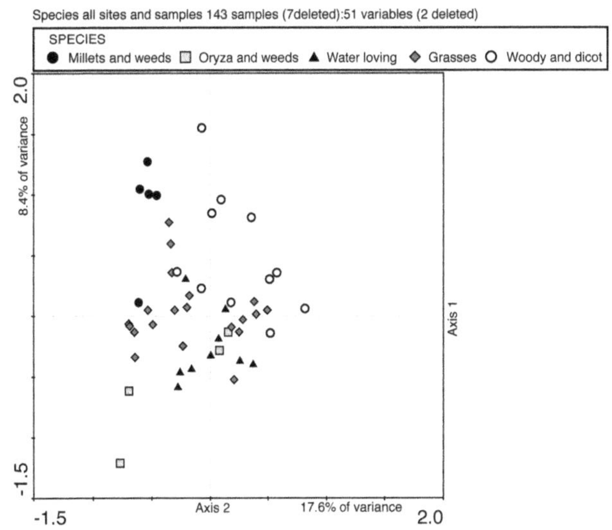

Figure 8.9 Species all sites and phases, 143 samples (7 deleted): 51 variables (2 deleted)

A similar pattern can be seen in the Yangshao, there seems little differentiation in composition between context types (Figure 8.10). The % of variance increases only slightly in both axes.

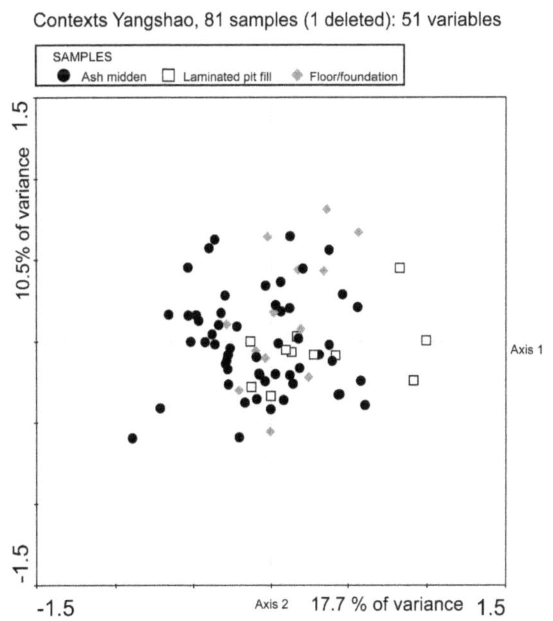

Figure 8.10 Yangshao contexts, 81 samples (1 deleted): 51 variables

However, in the Longshan a change can be noted (figure 8.11). The contexts are still mixed along axis 1, but less so than in the Yangshao plot. The Survey ash midden samples are all on the left of axis one and high on axis 2, while the lone Baligang ash midden sample is isolated from survey on the upper right side of the chart, but is close to other Baligang samples. The laminated pit fills tend to lie towards the lower part of the chart. There is differentiation between these and the ash midden samples along axis 2. The floor/foundations are mixed with the laminated pit fills but all towards the right. The position of the samples suggests the site may be more significant than the context.

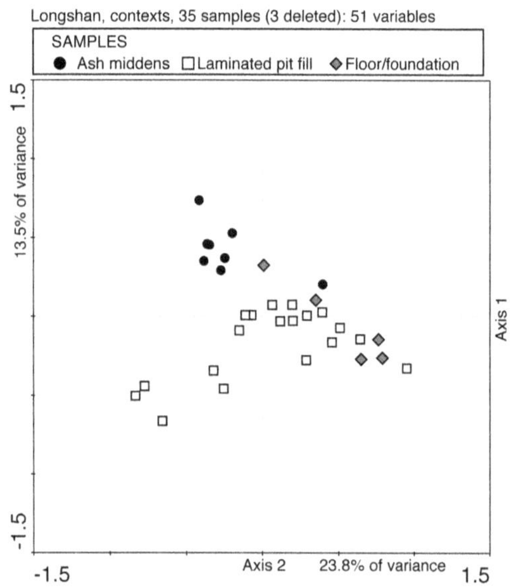

Figure 8.11 Longshan contexts, 35 samples (3 deleted) : 51 variables

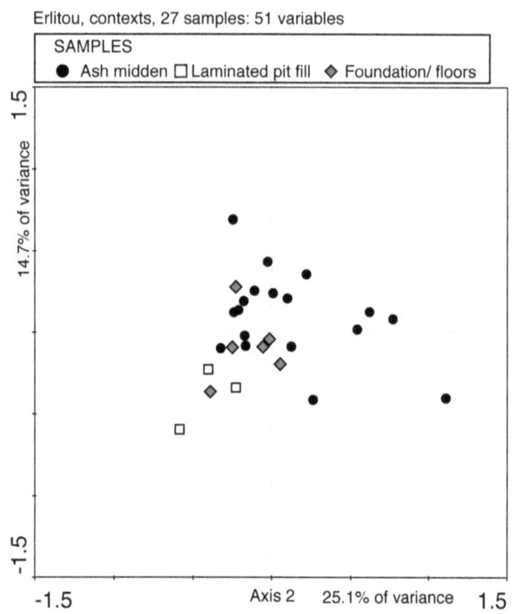

Figure 8.12 Erlitou contexts, 27 samples: 51 variables

The Erlitou period ash midden and floor /foundation samples are mixed and there seems little differentiation overall despite the higher percentages of variance, particularly along axis 1 (25.1%). Ash midden samples are present in all quadrants, whereas the floor foundations tend to be positioned towards the left. The laminated pit fills are all in the lower left part of the chart, separate from the ash middens on axis 2 but with some overlap on axis 1 (Figure 8.12).

The Huizui samples are all from ash middens and in the upper part of the chart along with the survey samples, which are confined to the upper left quadrant.

The results of the correspondence analysis on the constituents of the context types suggest despite some slight differences most samples share basic phytolith morphotype components. Despite the some clustering of floor / foundations and laminated pit fills in figure 8.10, the % of variance on both axes does not demonstrate a significant difference in the constituents of samples according to context. The differences seem to be more related to site than context.

This is supported by the results of the weight percentages and pH as discussed earlier.

8.4 CORRESPONDENCE ANALYSIS AND MAJOR PATTERNS GENERATED

ALL SITES AND PHASES

The full data set was examined to see if there was discrimination the basis of composition of samples between sites (Figure 8.13). Charts were generated from a full dataset with unknown, rare and ubiquitous forms deleted (see above for list of deleted morphotypes). Four samples were deleted: HE55, HE56, and HE57 due to high relative proportions of millet densities. These samples are all from a large storage pit, which is described separately below.

First the different cultural periods were highlighted (Figure 8.13). The % of variance along both axes is quite low (16% on axis 1, 8% on axis 2). Samples from the Yangshao and Longshan periods are present in all quadrants. The Erlitou period samples are absent from the lower right quadrant and mostly towards the left on axis 1. The four samples on the right are on the upper part of axis 2 but generally they are mixed with samples from the other two periods. There is no clear differentiation in the makeup of the samples.
The species plot shows differentiation between millets and weeds, and morphotypes from the water loving plants group on both axes (Figure 8.14). *Oryza* husks and keystones are in the centre of the plot, while scooped bilobes, which can be found both in *Oryza* and *Oryza* weed leaves, are to the left with *Oryza* leaf/culm. Most woody and dicot forms are towards the right of axis 1. Grasses are ubiquitous, although morphotypes associated with short cells from grass leaves, apart from rondels, are alongside the millet husks in the upper left quadrant which suggests they may be from millet crop weeds. Bulliforms (bulliform, keystone, *Oryza* keystone) are separate from the short cells, in the lower right hand quadrant, closer morphotypes associated with water.

Most woody and dicot forms are towards the right side of the chart.

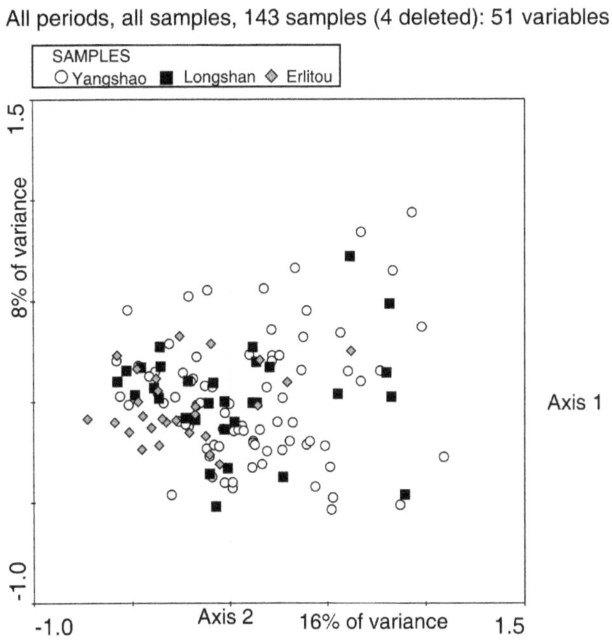

Figure 8.13 All periods, all samples, 143 samples (4 deleted): 51 variables

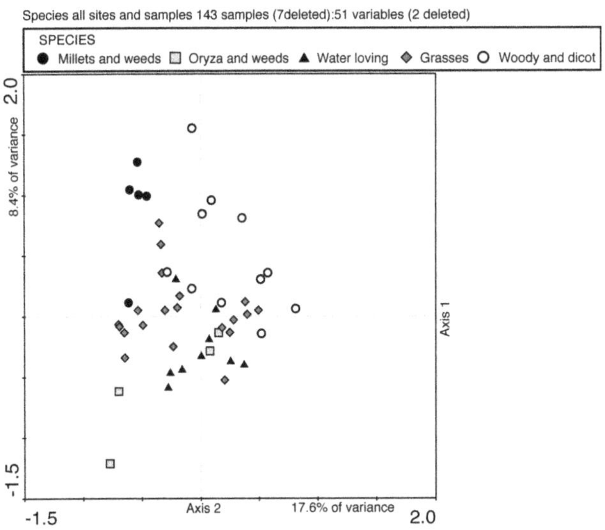

Figure 8.14 Species, all samples, 143 samples (4 deleted): 51 variables

.Next, the sites were brought to the fore (Figure 8.15). The samples are mixed. There is little differentiation as a whole, although there are some patterns. The survey sites are clustered together (apart from the lone Yangshao sample), as are those from Huizui West. Next, the sites were brought to the fore (Figure 8.15). The samples are mixed. There is little differentiation as a whole, although there are some patterns.

The survey sites are clustered together (apart from the lone Yangshao sample), as are those from Huizui West. Baligang samples do not appear in the upper left quadrant, this is probably due to the lack of millets in the Baligang samples.

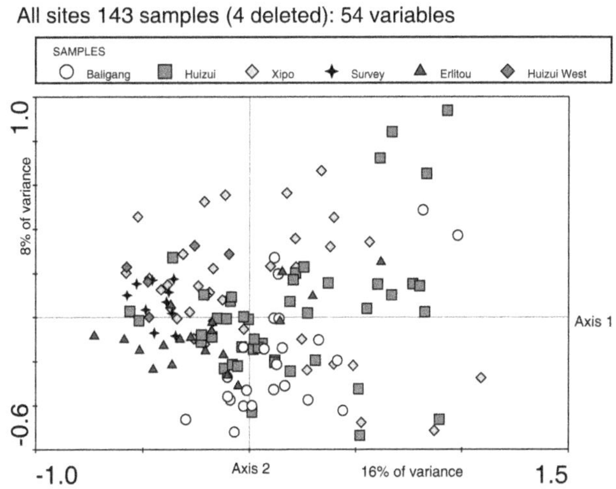

Figure 8.15 All sites, 143 samples (4 deleted):54 variables

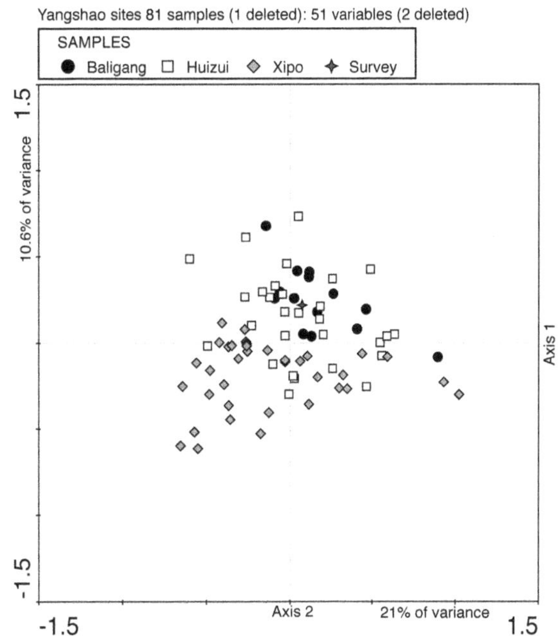

Figure 8.16 Yangshao sites 81 samples (1 deleted):51 variables (2 deleted)

YANGSHAO

The Yangshao sites chart has greater % of variance on both axes (Figure 8.16). Baligang samples are mostly in the upper right hand site of the chart. Huizui samples are present in all quadrants. Xipo samples are predominantly low on axis 2. The single Survey sample is in the upper right quadrant among the Huizui and Baligang samples. There is some differentiation between the Xipo and Baligang samples.

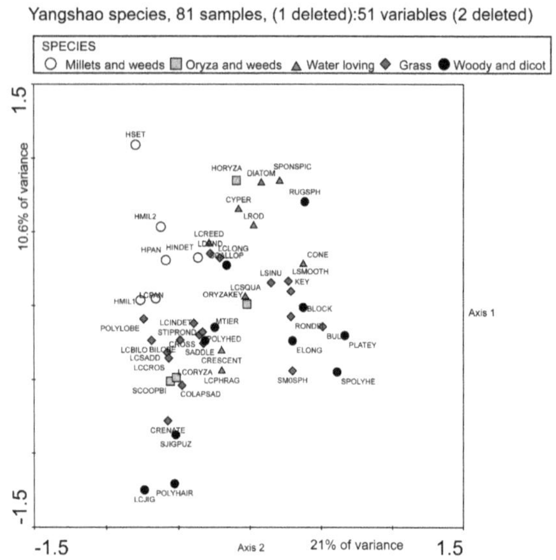

Figure 8.17 Yangshao Species, 81 samples (1 deleted): 51 variables (2 deleted)

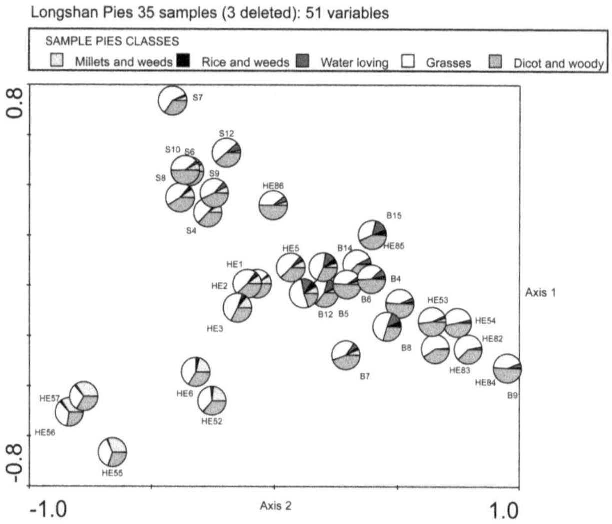

Figure 8.19 Longshan Pie chart, 35 samples (3 deleted): 51 variables. 23.8% of variance on Axis1, 13,5 % variance on Axis 2

The species plot shows clear differentiation between millets and weeds, and water loving and woody and dicot forms on axis 2 (Figure8.17). The millets are to the far left of axis 1 and in the upper part of axis 2, while the woody dicot samples mostly fall towards the right. Water loving morphotypes cross axis 1 but are below axis 2, apart from *Phragmites*. *Oryza* morphotypes are spread across the chart. There are some interesting groupings within the groups of morphotypes, especially the grasses. Initially the grasses appear mixed across the whole chart but short cells from panicoid, chloridoid and Bambuseae are all to the right side of axis 1 and towards the upper part of axis 2. Morphotypes from keystones, bulliforms, and *Oryza* keystones are clustered together in the lower right quadrant. This type of phytolith is found in leaves of grasses that enjoy a humid environment (Bowdery, 2007:139).

LONGSHAN

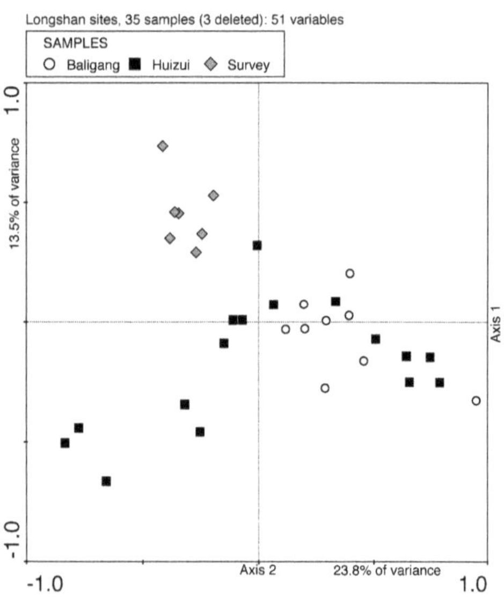

Figure 8.18 Longshan sites, 35 samples (3 deleted) :51 variables

In contrast to the Yangshao, the Longshan samples show marked differentiation (Figure 8.18). Baligang and the Survey samples are on opposite sides of the plot on both axes, while Huizui samples are present in all quadrants.

All samples have high proportions of grasses (Figure 8.19). The samples from Baligang tend to contain water loving and *Oryza* morphotypes, reflective of water based rice agriculture. The survey samples have high levels of non-crop grasses, some woody and dicots, less water loving and very little or no millet or *Oryza,* suggesting a different agricultural system in this area. The Huizui samples are in clusters, one in the lower left quadrant containing high levels of millets and weeds, a second in the lower right quadrant consisting mainly of grasses and woody and dicot morphotypes, and a third cluster in the centre of the chart that contain water loving and *Oryza* as well as grasses and woody dicots. This seems more related to the specific feature the samples were collected from (for example, samples HE81 - 86 are from layers of a floor and foundation, samples HE 1-4 are from layers of a laminated pit, samples HE 52- 57 are from another laminated pit) than the context type as a whole. The variation in the constituents of the samples from Huizui points to a mixed agricultural economy with both rice and millet farming supporting the evidence provided by the data analysis in chapter 9.

ERLITOU

The Erlitou site samples are present in all quadrants (Figure 8.20). The Survey and Huizui samples are in the upper part of axis 2 and the survey samples are clustered in the upper left quadrant.

There is clear differentiation between millets and weeds and *Oryza* forms on axis 2 (Figure 8.21). The only visible millets are in the upper left quadrant. In contrast, *Oryza* forms are only present the lower left and centre of the chart. This suggests rice and millet were treated separately. All samples are composed of high proportions

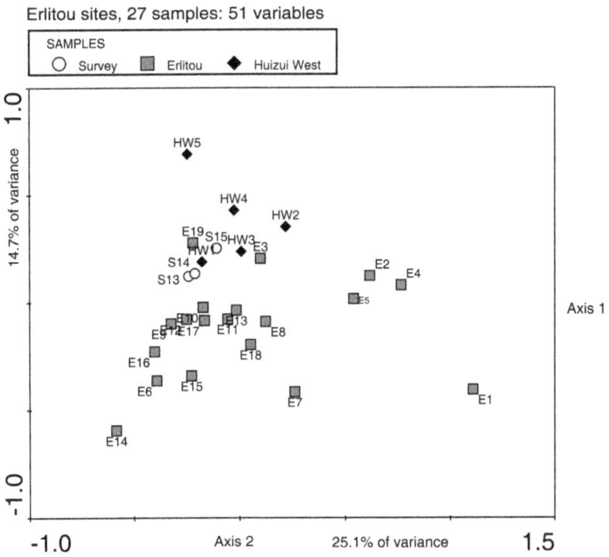

Figure 8.20 Erlitou Sites, 27 samples: 51 variables. 25.1% of variance on Axis 1, 14.7% of variance on Axis 2

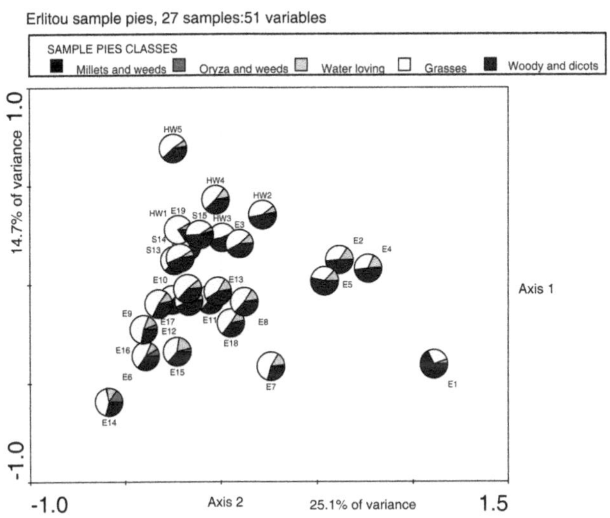

Figure 8.21 Erlitou pie chart, 27 samples:51 variables

of grasses. The samples in the lower half of the chart are all from the Erlitou site and those in the left quadrant have more *Oryza* and weeds and fewer woody and dicots. In conclusion, although when the samples are looked at all together there is little differentiation patterns emerge when the cultural periods are observed individually. There are changes from the Yangshao to Longshan and again to the Erlitou. These differences are clearly related to the geographical positions of the sites with millet and mixed farming in the Yellow River Valley contrasting with rice agriculture at Baligang in the south. The most clear cut differentiation is during the Late Neolithic Longshan perhaps reflecting new developments in arable farming systems.

8.5 Crops and Weeds

The next group of charts consider crops and weeds only, in order to gain an idea of general patterns of varying crop choices. The morphotypes representing crop weeds consist only of those identifiable to groups that are likely to be found specifically with either *Oryza* or *Setaria*/*Panicum*). There are morphotypes, such as bilobe or rondel, produced by weed grasses found with both types of crop. However, as these are ubiquitous they have not been used here.

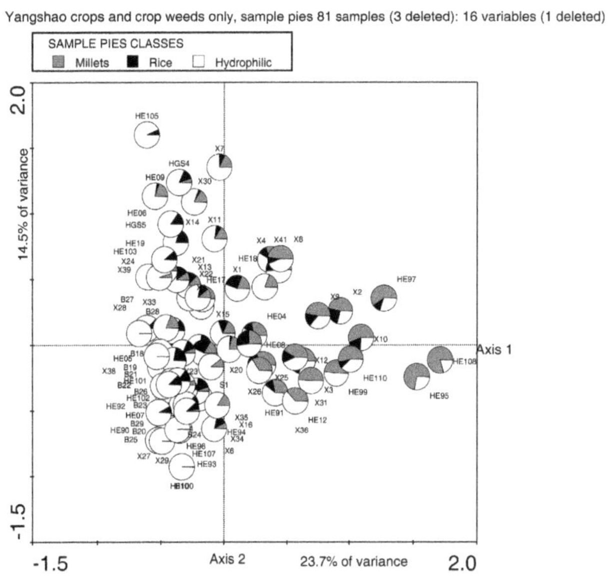

Figure 8.22 Yangshao crops and weeds only, 81 samples (3 deleted): 16 variables (1 deleted)

Yangshao sites crops and weeds only

The Xipo and Huizui samples appear in every quadrant and show no real differentiation. The Baligang samples are confined to the left side of axis 1 (Figure 8.22). Millets and panicoid weeds are pulling some Xipo and Huizui samples to the right in contrast to the Baligang samples. The *Oryza* and *Oryza* weeds are mostly confined to the left side of the chart. The high proportions of *Oryza* weeds and lack of millets are the reason why the Baligang samples are all to the left of axis 1. The outlying Huizui samples to the right below axis 2 contain high proportions of millet type 1 (panicoid grass) and *Setaria* husk and are both from contexts with abundant charred material.

Longshan sites crops and weeds only

The Longshan samples from Huizui and Baligang show a similar pattern to the Yangshao (Figure 8.23). Huizui samples are present in all parts of the chart. Baligang samples are confined to the left. The Survey samples are all in the upper central part of the chart. The samples in the cluster in the right lower quadrant from Huizui are all from the same feature and are from floor/foundations. There is clear differentiation between millets and weeds, and *Oryza* and *Oryza* weeds and this differentiation is the cause of the cluster of Huizui samples with high proportions of millet husks from the same feature, in the bottom right quadrant

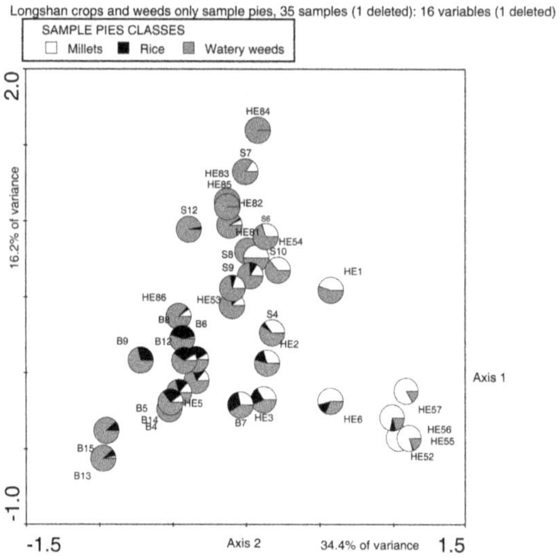

Figure 8.23 Longshan crops and weeds, 35 samples (1 deleted): 16 variables (1 deleted)

The survey samples all contain *Oryza* weeds; some also have millet but none *Oryza*. All samples contain *Oryza* weeds but the samples from Baligang have higher relative proportions of *Oryza* and fewer millets.

ERLITOU SITES CROPS AND WEEDS ONLY

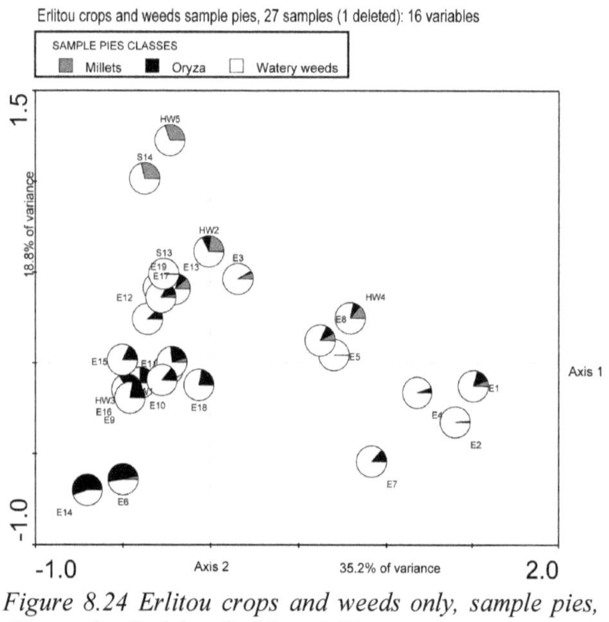

Figure 8.24 Erlitou crops and weeds only, sample pies, 27 samples (1 deleted): 16 variables

The Erlitou period is more difficult to read (Figure 8.24). The Erlitou site samples are present in every quadrant as are the Huizui samples, except on the lower right, while the survey samples are in the top right hand quadrant only. There seems little differentiation between sites, although there is no *Oryza* in the survey samples. There seems a clear division between samples containing either *Oryza* or millets but not sites.

ERLITOU CONCLUSIONS

In conclusion, the correspondence analysis using crops and weeds only highlights the differences between site relating to geographical position as well as rice and millet agriculture seen in the previous set of charts. As before the division between rice and millet farming is clear during the Yangshao and even more so in Longshan period before becoming more mixed in the Erlitou.

8.6 RESULTS FOR INDIVIDUAL SITES

Presence/ absence and ubiquities were examined for each site. The results were very similar so have been considered together (for example, Figure 8.25) All tables and charts are in Appendices 8.3 and 8.4

8.7 XIPO UBIQUITY AND PRESENCE/ABSENCE (39 SAMPLES)

All 39 samples analysed from Xipo are dated to the Mid-Neolithic. The majority were taken from ash middens (30). Six samples were from laminated pit fills. There were also three samples from floors and foundations and from a hearth).

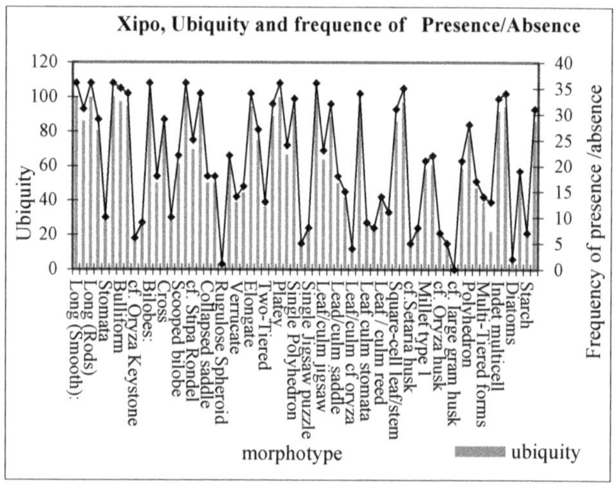

Figure 8.25 Xipo, ubiquity versus frequency of Presence Absence

When ubiquity and presence /absence are compared the results correspond very closely (Figure 8.25). For this reason the results have been amalgamated.

XIPO GRASSES AND CEREAL CROPS

Of the single cells from Graminae, all samples contained long smooth cells from leaf/culms as well as hairs, rondel shaped short cells, which are common to Gramineae but can also be present in Panicoids. Bilobes, which are indicative of Panicoid grasses, are present in every sample and bilobe multicells are in 32 of the 36 samples. Saddle shaped short cells, representative of Chloridoideae, are present in all but one sample and saddle multicells are preserved in 18. Long saddles are produced by *Phragmites* (Arundinoideae) but were indistinguishable in the archaeological samples. There is a possibility that some of the saddle morphotypes are from *Phragmites* as other *Phragmites* morphotypes are present in the Xipo samples.

Collapsed saddles suggesting Bambuseae are present in 18 samples. Crosses, which can be from panicoid grasses that enjoy marshy habitats such as *Coix lachryma jobi* or *Echinochloa,* occur frequently. Bulliform cells from grass leaves often associated with humid conditions (Bowdery, 2007:139) are represented 35 samples. Square celled leaf/culms, which consist of abundant bulliform cells, occur frequently. Long smooth multi cells occur in 34 samples. *Oryza* type keystones are present in 6 samples, while scooped bilobes that can be found in *Oryza* and other Oryzeae such as *Zizania* are present in more than half the samples (22). *Oryza* leaf/ culm multicells are rare (4 samples) while *Oryza* husk multi-cells are a little more common but still infrequent (7 samples). Distinctive *Setaria* type single cell bilobes appear in 11 samples. *Setaria* husk multi cells are infrequent (5). *Panicum* husk multi cells occur in 8 samples. Multi celled husk phytoliths that did not meet enough of the identification criteria for *Setaria* or *Panicum* yet have some of the characteristics were classified as Millet type1 (similar to *Setaria*) and Millet type 2 (similar to *Panicum*). These were found in 21 and 22 samples respectively.

XIPO CYPERACEAE AND OTHER MONOCOTYLEDONS

Single cell phytoliths from Cyperaceae occur frequently, rods are found in every sample, while cones appear in 11 samples. Over half the samples contain multi- cells from Cyperaceae (21). One sample contains possible Arecaceae.

XIPO DICOTYLEDONS

Of the morphotypes from dicotyledons, platey cells occur in every sample. Single polyhedrons and multi polyhedrons from dicotyledon arboreal leaves, such as *Quercus*, *Betula*, *Corylus* and herbaceous leaves, such as *Artemisia* (Bozarth, 1992: 194) are common (36 samples). Both single and multi-celled jigsaw puzzle shaped morphotypes also suggestive of dicotyledon leaves from deciduous trees and herbaceous plants, for instance *Boehmeria* and *Urtica* (Bozarth, 1992:200) are present. Scalloped shaped morphotypes often found in Curcubitaceae rind are present but rare (4 samples).

XIPO OTHER

Stomata cells are not designed to take silica so are among the last cells to silicify (Parry and Smithson, 1964) and indicative of a plant that is at the end of its life. These are common, especially in the multicells. This might also point to high transvaporation. Diatoms are present in 2 samples and sponge spicules are found in around half (19).Indeterminate leaf /culm cells and indeterminate husk multi cells occur in every sample and indeterminate phytoliths are also frequent.

XIPO CONCLUSIONS

The individual results from Xipo show rice was being farmed in the cooler more arid part of Henan by the Yangshao, although millets, in particular Panicum seem to be the main crop. The frequency of Panicoid and Chloridoid grasses and the presence of Arecaceae demonstrate warm conditions and the presence of dicotyledon leaves suggest possible deciduous tree cover. Cyperaceae, diatoms and sponge spicules point to watery plants being brought to the site.

8.8 HUIZUI UBIQUITY AND PRESENCE/ABSENCE (53 SAMPLES)

Fifty-three samples from Huizui were analysed, forty-six from Huizui East, five from Huizui West and two off site geological samples. The five from Huizui West have been dated to the Erlitou period. Two samples from Yangshao contexts are classified as geological because they were collected from ancient fields off the main site and part of the sediments were used for geoarchaeological analysis (Rosen, pers.com). Of the forty-six samples from Huizui East, twenty three are from the Yangshao, five of these from the Late Yangshao, twelve from the Longshan and six of these from the Late Longshan.

The contexts the samples were retrieved from range from ash middens: eighteen from the Yangshao period at Huizui East, five from the Erlitou at Huizui West, floor/foundations, six from the Longshan period at Huizui East, and laminated pit fills, ten dating to the Yangshao and twelve to the Longshan (see Chapter 5 for details).

HUIZUI GRASSES AND CEREAL CROPS

Phytoliths from grass leaves and stems are found in every sample. Rondels are the most frequent of the diagnostic short cells. There are also relatively high occurrences of *Stipa* type rondels suggesting quantities of Pooid grasses. Bilobes are the next most common, followed by saddles demonstrating the presence of Panicoid and Chloridoid grasses. Morphotypes from *Oryza* appear in few of the samples. *Phragmites* and reed multicells are relatively rare. *Setaria* bilobes and husk multicells occur in around a third of the samples, as do *Panicum* husks. Millets type 1 and 2 are more frequent.

HUIZUI CYPERACEAE AND OTHER MONOCOTYLEDONS

Single cells from Cyperaceae leaves are very common. Cones and Cyperaceae multicells are also fairly frequent, appearing in more than half of the samples. Rugulose spheroids appear in 6 of the 53 samples.

HUIZUI DICOTYLEDONS

Platey forms are found in every sample from Huizui. Silica aggregates are common. Jigsaw puzzles, that suggest dicot leaves, are present in close to half the samples. 9 of the 53 samples are scalloped forms possibly from Curcubitaceae rinds.

HUIZUI OTHER

29 samples contain sponge spicules and 18 diatoms.

HUIZUI CONCLUSIONS

Huizui has a wide range of morphotypes perhaps reflecting the temporal range of the samples. In contrast to Xipo the most common short cell is rondel suggesting a greater presence of Pooid grasses. The wide range of phytolith types could also point to mixed farming as Huizui has all three crop plants as well as frequent Cyperaceae, diatoms and sponge spicules suggestive of wet rice farming.

8.9 ERLITOU PRESENCE/ ABSENCE AND UBIQUITY VALUES OF MORPHOTYPES (19 SAMPLES)

19 samples from Erlitou were analysed, 7 from phase II, 6 from phase IV. The other samples are not securely dated beyond the Erlitou period. 1 is earlier than IV, 1 III-IV and 1 possible III (Figure 8.26).

ERLITOU GRASSES AND CEREAL CROPS

Long single cells are present in every sample, as are short grass cells: bilobes, rondels and saddles. Cross morphotypes are in 17 samples. Long indeterminate multicells from leaves and indeterminate husk multicells are also in every sample. There are frequent multicells containing bilobes, crosses and long smooth cells suggesting abundant varied grasses. Bulliforms are in all the samples; keystone forms in all but two and square leaf/stem multicells are in 18. Their presence points to grasses that prefer damp conditions. Either *Phragmites* or reed multicells appear in the majority of samples also pointing to a humid environment.

Oryza keystones are only present in fewer than half the samples although multicells from *Oryza* leaf/culm are found in close to half the samples (9). *Oryza* husk is rare, only appearing in 3 samples. Scooped bilobes, which can be found in other Oryzeae, are more common (15).

Phytoliths identifiable to *Setaria italica* are rare. *Setaria* type bilobes occur in three samples while husks are only found in one. *Panicum miliaceum* also only appears in one sample. Millets 1 and 2 are more frequent but still uncommon.

ERLITOU CYPERACEAE AND OTHER MONOCOTYLEDONS

Cyperaceae single cells appear frequently, rods in all samples, cones in sixteen and *Cyperaceae* multicells in seventeen. Rugulose spheroids are present in only 1 sample.

ERLITOU DICOTYLEDONS

Both elongate and platey cells from dicotyledons are present in all the samples. Silica aggregates are common. There were no scalloped shaped phytoliths.

ERLITOU OTHER

Diatoms and sponge spicules are quite frequent each appearing in more than half the samples.

ERLITOU CONCLUSIONS

Very few samples contain crop husks, only one each Panicum and Setaria and three rice suggesting crop processing was taking somewhere else. Most samples have wide range of morphotypes from a variety of grasses. There are also dicotyledons commonly found in tree bark in all the samples.

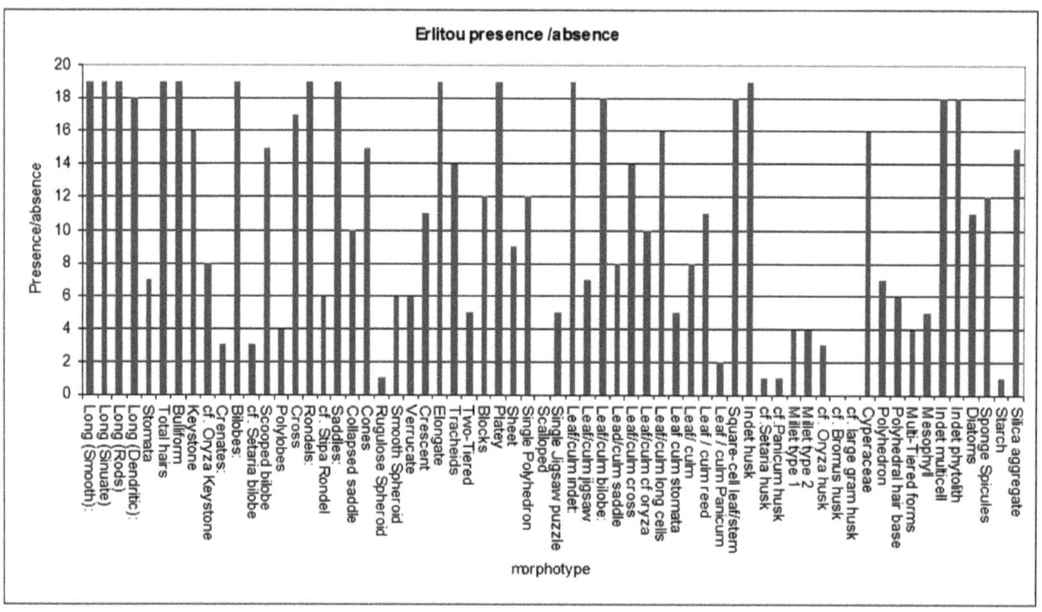

Figure 8.26 Erlitou presence/absence

8.10 SURVEY UBIQUITY/ AND PRESENCE /ABSENCE (11 SAMPLES)

Eleven samples were analysed from the seven survey sites, one from the late Yangshao, six from the late Longshan and four from the Erlitou period. All were from ash middens.

SURVEY GRASSES AND CEREAL CROPS

A wide variety of Graminae long and short cells appear in high frequencies across all the Survey sites. Bilobes, crosses, rondels and saddle single cells are recorded in all the samples along with bilobe multicells. Collapsed saddles in only two samples suggest Bambuseae was not common. *Oryza* leaves are not present and *Oryza* husk appears only in one sample, however scooped bilobes, which are sometimes used to indicate rice cultivation, are present in 4 of the samples. *Setaria* bilobes appear in seven samples and *Setaria italica* husk in five, while *Panicum miliaceum* husk is in three. Millets type 1 and 2 are more common and present in more than half the samples.

SURVEY CYPERACEAE AND OTHER MONOCOTYLEDONS

Rods appear in every sample and cones in all but one, while Cyperaceae multicells are present in 9 of the 11 samples. There are no rugulose spheroids.

SURVEY DICOTYLEDONS

Scalloped shaped phytoliths indicating cucurbit rind do not appear. Platey forms are found in all but one sample and silica aggregates are present in all. There are no jigsaw puzzles. However there are single and multi polyhedrons which increase in the Erlitou period samples.

SURVEY OTHERS

Diatoms and sponge spicules are very rare.

SURVEY CONCLUSIONS

Setaria and *Panicum* are infrequent but more common than rice, which is very rare in the survey sites. However rice weeds, such as Cyperaceae are common .The bulk of the morphotypes are from weedy grasses. This may be because crops, in particular rice, were being grown around these sites for export to larger centres, although the lack of diatoms and sponge spicules could point to dry farming. There single and multi polyhedrons along with platey and silica aggregates suggest available wood for burning.

8.11 BALIGANG UBIQUITIES AND PRESENCE /ABSENCE (29 SAMPLES)

Twenty-nine samples were analysed from Baligang. One is pre Yangshao and three are from the Eastern Zhou, which are immediately before and after the time periods covered by this project, one early Yangshao, ten mid Yangshao, five late Yangshao, and nine Longshan. Nine are from ash middens, sixteen are from laminated pit fills, and four are from cultural layers but have been classified as ash pits as they are also mixed deposits.

BALIGANG GRASSES AND CEREAL CROPS

Phytoliths from Graminae in general are widely distributed. There is a wide variety of short cells in many samples. *Oryza* keystones are common (18) and husk multicells even more so (22). Scooped bilobes appear in twelve samples and *Phragmites* and reeds are both present in nearly half the samples. *Panicum* husks only occur in two samples and while millets 1 and 2 are more common they occur in less than half. *Setaria* husk is only present in the Longshan samples. More crosses appear in the Longshan.

BALIGANG CYPERACEAE AND OTHER MONOCOTYLEDONS

Cyperaceae occur very frequently, multicells are present in 28 of the 29 samples. Cyperaceae are frequent throughout both phases but cones from the inflorescences are more widespread during the Yangshao. *Phragmites* leaf/culms increase in the Longshan, but reed leaf/culms, multicells that look like *Phragmites* but do not me*et al*l identification criteria, decline slightly. Both diatoms and sponge spicules occur more frequently in the Longshan. Rugulose spheroids, suggesting palms are present in 6 samples, more than at any other site.

BALIGANG DICOTYLEDONS

Platey forms are present in all samples for both periods. Jigsaw puzzles from dicotyledon leaves are common. Scalloped shaped phytoliths increase in frequency from the Yangshao to the Longshan. Polyhedral cells decrease, apart from the hair bases. Silica aggregates decrease in the Longshan but are still frequent.

BALIGANG OTHER

Diatoms occur in nearly a third of the samples (9) and sponge spicules are common (20).

BALIGANG CONCLUSIONS

Baligang stands out from the other sites. The frequency of rice and rice weeds along with diatoms and sponge spicules points to an emphasis on wet rice farming. The presence of palm phytoliths shows the climate was warm and there is evidence for other trees and herbaceous plants.

CHAPTER 9

RESULTS: SPECIFIC

9.1 CROP CHANGES

To see the broad general patterns of crop changes over the three cultural periods I looked first at the average percentages of morphotypes per period and then at the absolute densities.

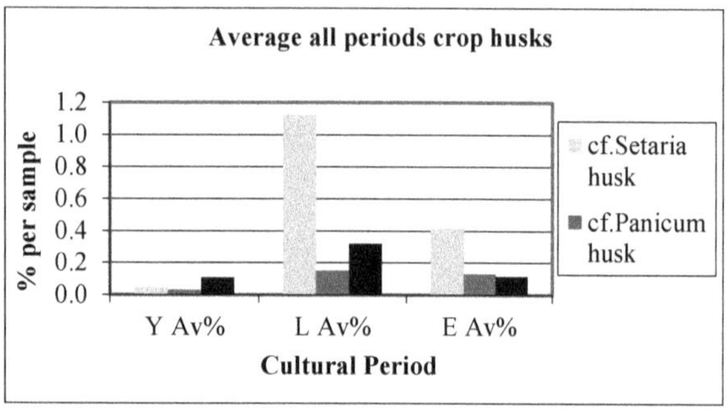

(Y=Yangshao, L= Longshan, E = Erlitou)

Figure 9.1 Average % per period crop husks, all cultural periods

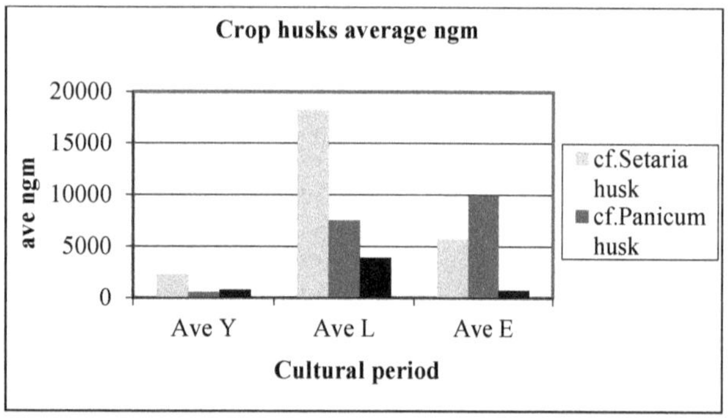

(Y=Yangshao, L= Longshan, E = Erlitou)

Figure 9.2 Average number per gram per period crop husks all cultural periods

During the Yangshao the crop husks make up very low proportions of the total phytoliths. The highest levels are *Oryza*, followed by *Setaria* and then *Panicum*.

There are higher proportions overall of all crop husks during the Longshan period, with particularly high levels of *Setaria italica*. The Erlitou period sees a drop in average percentages of all crop husks. *Setaria* continues with the highest proportions, *Panicum* and *Oryza* have similar average percentages per sample.

However, if the average absolute densities are looked at the results are different, there is more *Setaria* than *Oryza* during the Yangshao, in the Longshan there are higher levels of both types of millet crop than *Oryza* and the highest density in the Erlitou period is of *Panicum*, while *Oryza* levels fall very low. This highlights the different results that can occur depending on the method of analysis used, in this case the difference between using absolute versus relative counts. The relative count is more useful here because it highlights the relative importance of the crop within the sites. More samples were taken from the northern part of the region, Xipo and Huizui, than from the site in the South, Baligang. While some *Oryza* is present in these northern sites there is no *Setaria italica* and little *Panicum miliaceum* at Baligang in the Yangshao samples. Rice however, is present at Baligang and in relatively high proportions. The same is true for the Longshan; more samples were taken from the Yilou Valley region in the north of Henan, Huizui and the survey samples, than from Baligang in the south and again for the Erlitou period where all samples originate from the Yilou River Basin, Huizui, Erlitou and the survey samples.

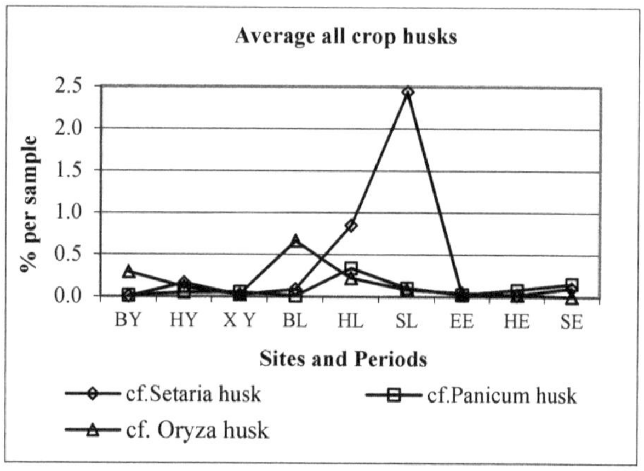

(B= Baligang, H= Huizui, X =Xipo, S= Survey, E = Erlitou, Y=Yangshao, L=Longshan, E= Erlitou)

Figure 9.3 Average % all crop husks, sites and periods

The high relative proportion of *Oryza* in the Yangshao is due to the results from Baligang (Figure 9.3) illustrating the geographical situation of the site has an important bearing on crop choice. As in the Yangshao period, during the Longshan the comparative results are site rather than period related. Both Huizui and the survey samples, from the Yilou river valley to the north of Henan produced relatively high levels of *Setaria*, while Baligang in the south had low *Setaria* and highest average proportions of *Oryza* husk.

The numbers per gram of each crop husk morphotype were averaged per site and period by totalling the number per gram and dividing by number of samples for each site and period.

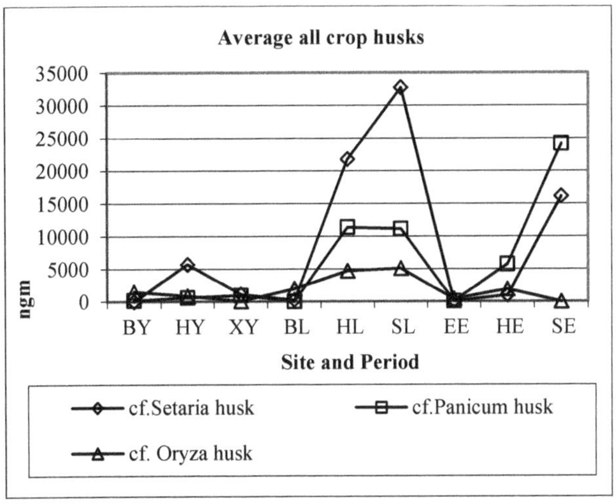

(B= Baligang, H= Huizui, X =Xipo, S= Survey, E = Erlitou, Y=Yangshao, L=Longshan, E= Erlitou)

Figure 9.4 Average number per gram all crop husks, sites and periods

For example the sum of the number per gram of *Oryza* husk multicells in the Yangshao samples at Baligang was divided by the total Baligang Yangshao samples.

The absolute densities reinforce the increase in levels of all crop husks in the Longshan in particular millet crops in the Central Yellow River valley sites (Figure 9.4). The most striking result in the average absolute counts during the Erlitou period is the increase in *Panicum*, especially in the survey samples. *Setaria* and *Oryza* husk numbers drop at Huizui and the survey sites, while the Erlitou site has scant crop husk phytoliths of any type.

The averages, both relative and absolute, demonstrate major crop choices are related to local environment as can be seen by Baligang which is in the south of the region, within the Yangtze catchment area and has the highest relative proportions of *Oryza* during the Yangshao and Longshan period. Although average densities at Huizui and in the survey samples are higher than at Baligang they are lower than both *Panicum* and *Setaria* at these sites. Unfortunately no Erlitou period samples were available from Baligang.

The sites from the north and west of the region selected millet crops as their main crops in all periods examined. However, there is some difference; Xipo to the west has more *Panicum* both absolutely and relatively, while Huizui has more *Setaria* in the Yangshao and Longshan, but levels drop drastically during the Erlitou period when there is an increase in *Panicum*. The Survey samples also show high *Setaria* in the Longshan and fall in the Erlitou with a corresponding increase in *Panicum* in the absolute counts.

The results of the average counts, absolute and relative, indicate crop choices are primarily connected to geographical position and local environment. Rice predominates at Baligang throughout the Yangshao and Longshan. *Panicum* is the crop of choice at Xipo during the Yangshao in the drier cooler northwest.
However, there are differences between periods as well.

There is a general increase in crop husks from the Yangshao to the Longshan and subsequent decrease into the Erlitou. It would seem that the main crop in the north central sites, Huizui and the survey sites, also changes from *Setaria* to *Panicum*.

In conclusion, the lower levels of crop phytoliths during the Yangshao period at all sites might suggest crops were being supplemented by gathering wild food, as demonstrated in the Ying Valley (Fuller and Zhang, 2007). The general increase in crop husk remains during the Longshan may be related to population growth, an increasingly heavy dependence on cultivated grain and could be a result of more successful agricultural practices, possibly related to more intensive and extensive farming. The fall in the Erlitou period may suggest crop processing was taking place in a different setting from previously, or that partially processes or dehusked grain was being imported into larger centres, such as Erlitou.

9. 2 CROP CHANGES SITES

As geography and environment clearly play a part in crop selection examination of individual sites is next. First Xipo, it is situated in the north west of the region and all samples are from the same cultural period, Miaodigou, the Mid Yangshao. Next, Baligang, and the changes from Yangshao to Longshan, general and then in closer detail as some samples have been identified to early, mid and late Yangshao. Then crop changes in the sites in the Yilou River Valley, Huizui, the survey sites and Erlitou, will be assessed. Relative counts were used here because of the variation in densities between some of the samples made the charts difficult to read. Also percentage per sample was used for the correspondence analysis and comparative datasets.

XIPO

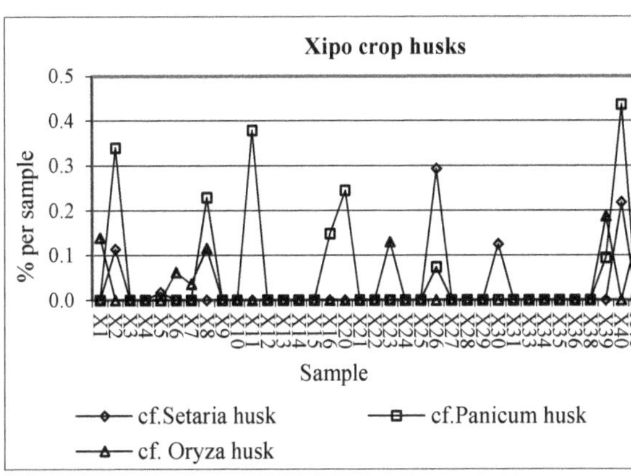

Figure 9.5 Xipo crop husks

The relative densities suggest *Panicum* as the most common crop at Xipo (Figure 9.5). *Panicum* husks appear in the most samples and usually in the highest proportions, apart from one sample where *Setaria* is higher (X26). This fits with the suggestion that *Panicum* was more commonly grown in the drier cooler climate in the north and west.

Oryza husks are present in low proportions in several samples. The presence of morphotypes from *Oryza* leaves (leaf/culm multi cells and *Oryza* keystones) (Figure 9.6) suggests local cultivation close to the site, although there is not enough evidence to suggest whether it was being dry or wet farmed or grown in paddies.

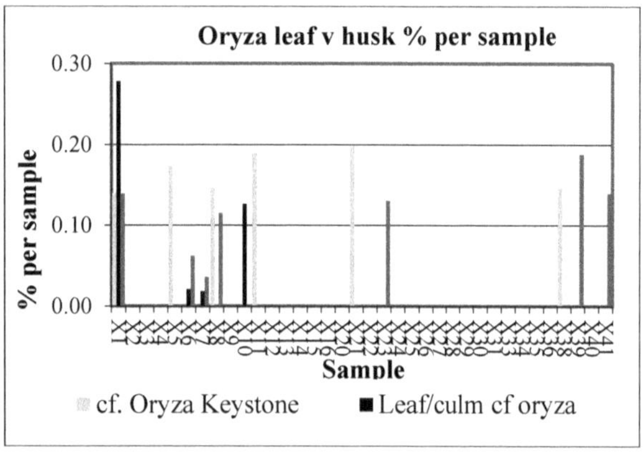

Figure 9.6 Xipo Oryza husks and leaves

BALIGANG

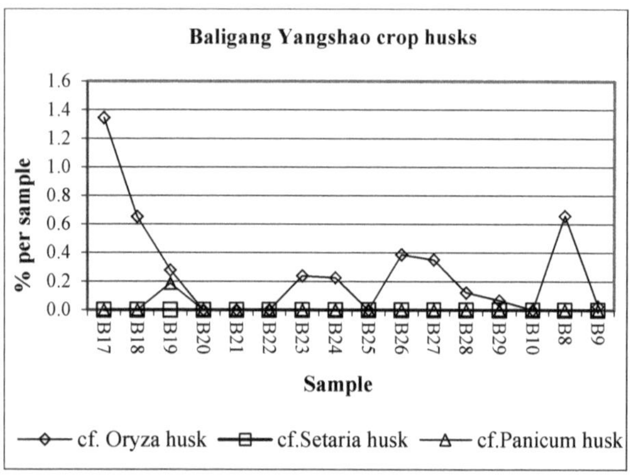

Figure 9.7 Baligang Yangshao crops

In contrast to Xipo, *Oryza* husk is predominant at Baligang in the Yangshao samples (Figure 9.7). Low percentages of *Panicum* appear in one sample. The paucity of *Panicum* suggests it may be a crop weed rather than crop. There is no *Setaria italica* in the Yangshao samples.

The crop choice pattern during the Longshan at Baligang shows a similar picture to the Yangshao, although the crop husk percentages per sample are higher (Figure 9.8). Rice predominates. There is one sample with low levels of *Panicum* and two containing *Setaria*. As suggested for the Yangshao the *Panicum* husks probably represent weeds rather than a *Panicum* crop, particularly given the warmer, more humid climate at Baligang, which is less suitable for *Panicum miliaceum*. Two samples contain *Setaria italica*, which may be a secondary crop.

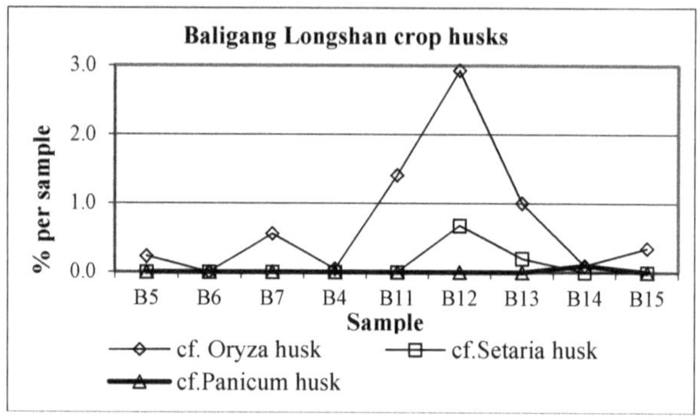

Figure 9.8 Baligang Longshan crop husks

However, the main crop, *Oryza*, remains the same at Baligang throughout the Yangshao and Longshan periods.

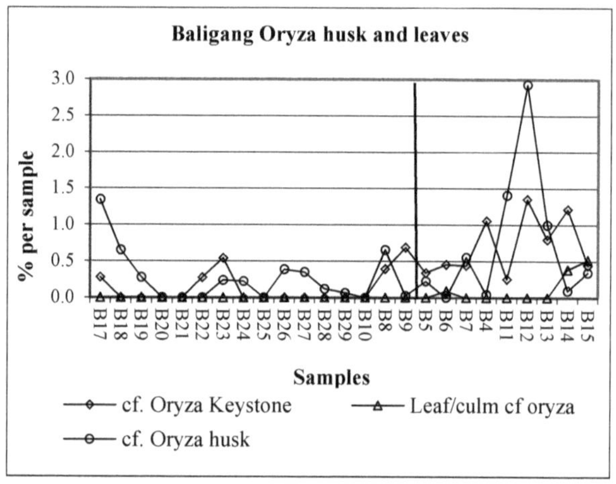

Figure 9.9 Baligang Oryza husk and leaves

The high proportions of morphotypes from *Oryza* leaves accompanying the husks at Baligang point to local cultivation and processing (Figure 9.9). This is supported by the presence of phytoliths from *Oryza* crop weeds such as Cyperaceae, as well as accompanying diatoms and sponge spicules, (Figure 9.10). A possible explanation for the far higher proportions of cone shaped Cyperaceae husk phytoliths during the Yangshao may be related to a change in harvesting practices (see crop processing section).

There are samples from Huizui from all three cultural periods. Two Yangshao samples, HGS4 and HGS5 are off site.

Oryza, Setaria and Panicum husk are present in the Yangshao samples from Huizui (Figure 9.11). One sample has a relatively high proportion of Oryza husks and they are present in lower proportions in three other samples. Setaria husk is the most common, appearing in 8 samples, while Panicum is present in 5. Where Oryza husks are found they are the only crop husks in all but one sample (HE97), suggesting Oryza was dealt with separately from the millets.

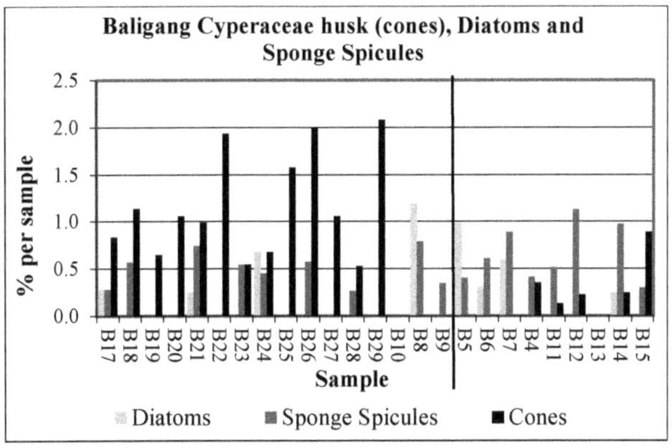

Figure 9.10 Baligang Yangshao and Longshan Cyperaceae husks (cones), diatoms and sponge spicules

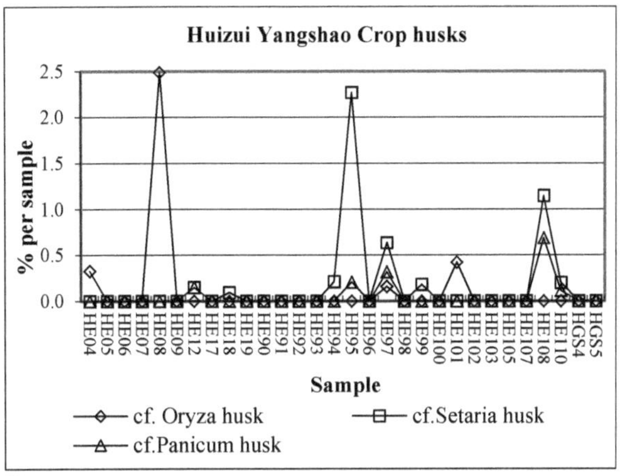

Figure 9.11 Huizui Yangshao Crop husks

HUIZUI

Panicum is always found with *Setaria* and usually in lower proportions. *Oryza* keystones from leaves are also present at Huizui although rarely in the same samples as the husks (Figure 9.12). The exception, HE95, is an ash midden so is likely to represent mixed deposits. The results suggest all three crops were being cultivated in the environs of Huizui during the Yangshao period. Although one sample contains exceptionally high levels of *Oryza* husk (HE08), *Setaria* seems generally the more predominant crop.

The two off site samples have relatively high levels of *Oryza* leaf and no crop husks of any kind (Figure 9.12).

During the Longshan period *Setaria Panicum* and *Oryza* husks are all present at Huizui and generally in higher proportions than in the Yangshao (Figure 9.13). A broad correlation between *Panicum* and *Setaria* continues, while *Oryza* again seems to be treated separately from the millets. *Oryza* is present in similar proportions to *Panicum* suggesting they are both secondary crops. There is a definite change in the Late Longshan when all crop levels per sample drop, in particular *Panicum*. *Setaria* continues as the most frequent crop and *Oryza* is common. *Oryza* keystones from the leaves correlate with the husks in samples from a laminated pit fill from the

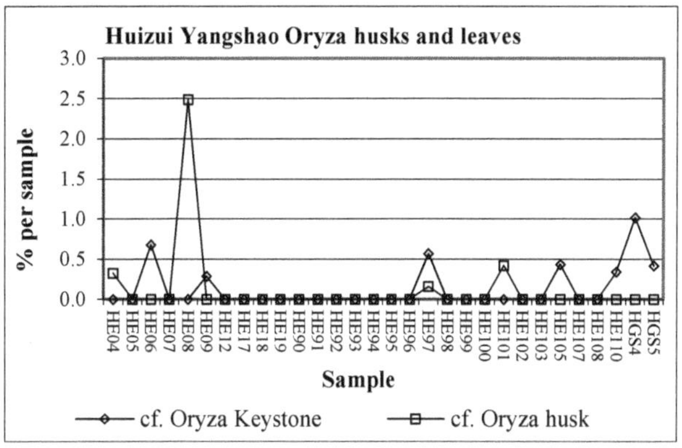

Figure 9.12 Huizui Yangshao Oryza husks and leaves

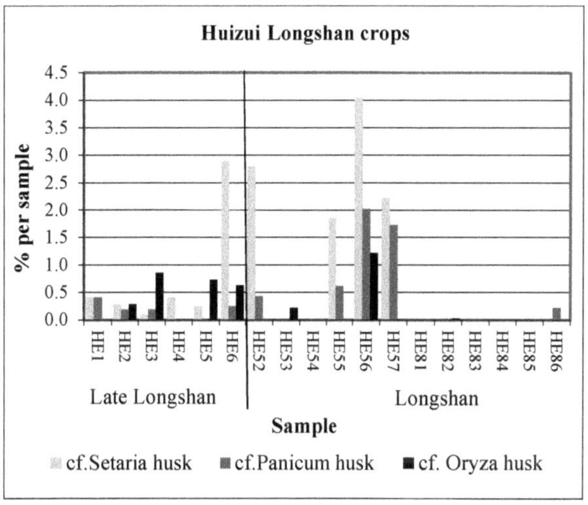

Figure 9.13 Huizui Longshan crop husks

Late Longshan (HE1-HE6) but are found independently in other Longshan samples (*Figure 9.14*). This might point to changes in when and where crop-processing stages are taking place. The presence of the leaves and husks suggests local cultivation is continuing, but the leaves make up lower proportions of the samples than the husk. This may be because the leaves have economic value as fodder or green manure so have been used elsewhere.

Unlike Baligang neither the *Oryza* husks nor leaves correlate with Cyperaceae husks suggesting a different cultivation or harvesting regime.

There are only five samples from Huizui for the Erlitou period; two of these contain husks from cereal crops. *Panicum* is predominant, followed by *Oryza* then *Setaria* (figure 9.15). *Oryza* leaves are present independently in two samples in higher proportions than the husk, suggesting *Oryza* was still being grown rather than traded in.

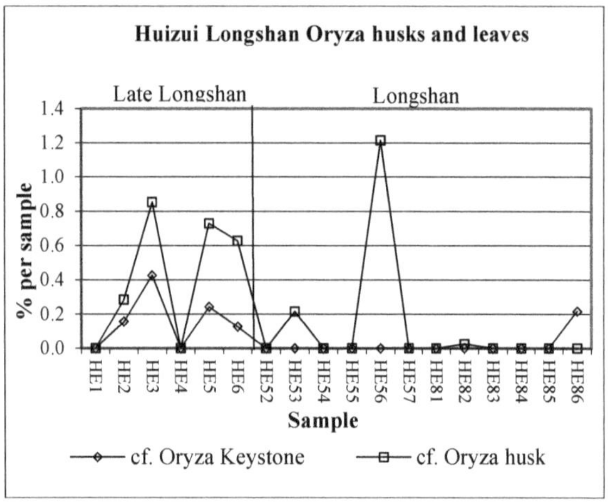

Figure 9.14 Huizui Longshan Oryza husks and leaves

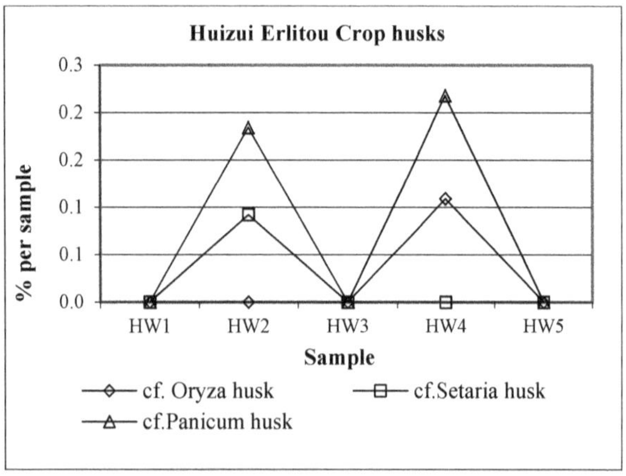

Figure 9.15 Huizui Erlitou Crop husks

SURVEY

The survey samples are from all three cultural periods, but only one is from the Yangshao and contains no cereal crop husks (Figure 9.16). One Longshan sample, Peichun B, contains relatively high levels of *Oryza* husk. There are no other rice morphotypes from leaves or stems in any of the other survey samples, which suggest the one sample containing *Oryza* husks may be anomalous and unreliable. Scooped bilobes are present in the samples as well as other phytoliths from plants that can be associated with rice farming, for example Cyperaceae, which raises the possibility of rice being farmed in the vicinity but not processed there. However, Cyperaceae, in particular, has other economic uses so could have been present in the samples for a number of other reasons. The levels of diatoms and sponge spicules are low which indicates that even if Oryza was being farmed it is unlikely to have been in paddy fields. The Longshan samples show *Setaria* as the predominant crop at these peripheral sites, with *Panicum* as a secondary crop (Figure 9.18).

Husk phytoliths from weedy millets were present in all the Longshan samples from the survey except one from NE Gaoya (S7).

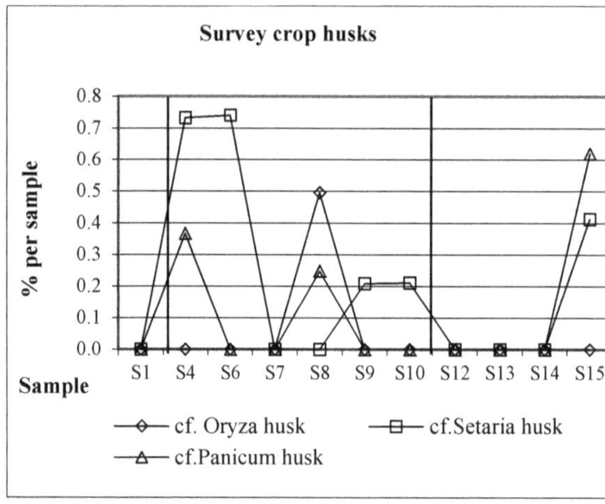

Figure 9.16 Survey crop husks

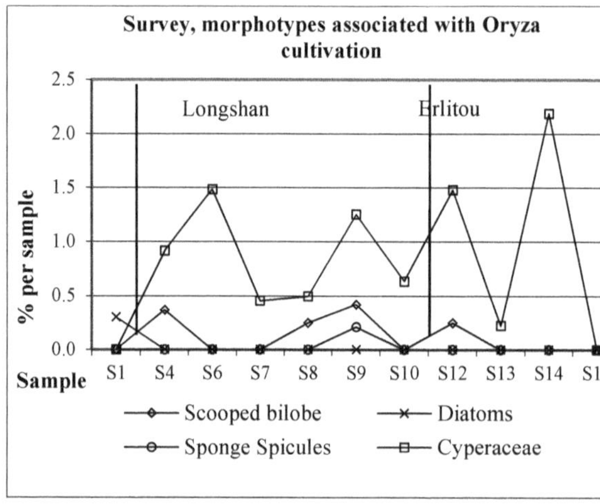

Figure 9.17 Survey, morphotypes associated with Oryza cultivation

Only one Erlitou sample, from South East Zhaiwan (S15), contains any crop husks, both *Setaria* and *Panicum*. Weedy millets were also found in both samples from SE Zhaiwan but were absent from the samples from the other Erlitou period survey site, Yuangou.

The survey samples follow a similar pattern to Huizui with generally higher proportions of crop husks during the Longshan and *Setaria* as the main crop. In the Erlitou crop husks are rare. Although the samples are mostly limited to one per site, the available evidence suggests a change to *Panicum* as the main crop, although as only one sample contained crop husks this is not a clear picture. The major contrast between the Huizui and the survey is the lack of *Oryza* in the smaller more peripheral survey sites. This might point to *Oryza* being an elite or special crop only available at larger sites where the greater labour organisation needed to farm *Oryza* is available.

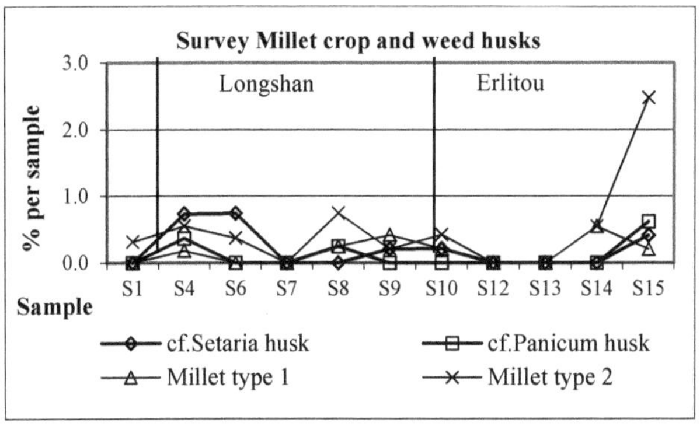

Figure 9.18 Survey Millet crop and weed husks

ERLITOU

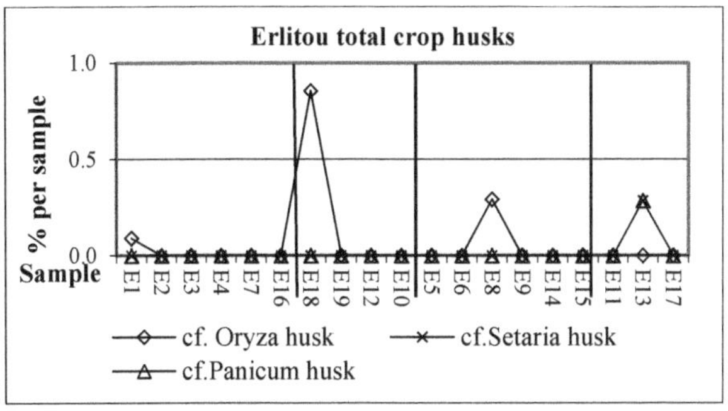

Figure 9.19 Erlitou site total % per sample crop husks

The divisions, from left to right, show the phases EII, EIII, EIV and the last is Erlitou but undated (Figure 9.19). As stated earlier the samples from Erlitou do not come from domestic contexts and are not necessarily representative of the site as a whole. There are few crop husks overall. The only crop husks from phase II are *Oryza*, there are none from phase III and again only *Oryza* in one sample from phase IV. One undated sample contains *Setaria* and *Panicum* husks.

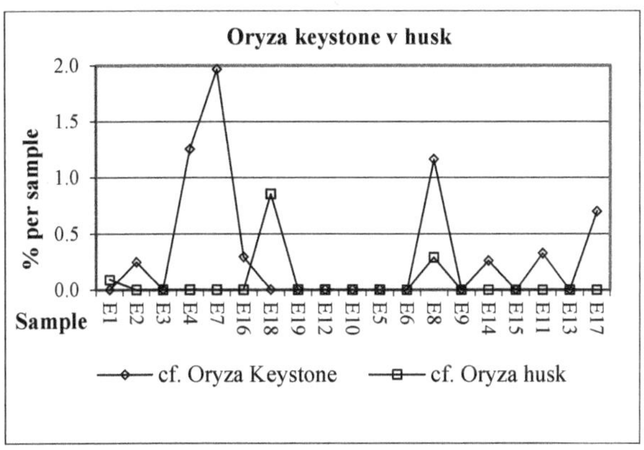

Figure 9.20 Erlitou site Oryza leaf v husk

Although there are few samples containing husk, there are more with *Oryza* leaf (Figure 9.20). *Oryza* keystones are usually in different samples from those containing husks.

9.3 CROP CHANGES, CULTURAL PERIODS

YANGSHAO

When looked at together the crop husks from the Yangshao present an interesting pattern. There is a clear difference in main crop choice between the sites. At Baligang rice is predominant. Apart from the one sample with *Panicum* husks, *Oryza* is the only crop. The single survey sample, from Gong Jia Yao, has no crop husk phytoliths. In contrast Xipo has husks from all three crops, principally *Panicum*, followed by *Oryza* and *Setaria*. Huizui is different again. Proportions are higher here. *Setaria* is clearly the most important crop but both *Panicum* and *Oryza* occur frequently and the *Oryza* husk in one sample, H08, has the highest percentage per sample of any crop in the Yangshao. In comparison with the Longshan (Figure 9.21) percentages are generally low, the majority fall below 0.5% per sample.

LONGSHAN

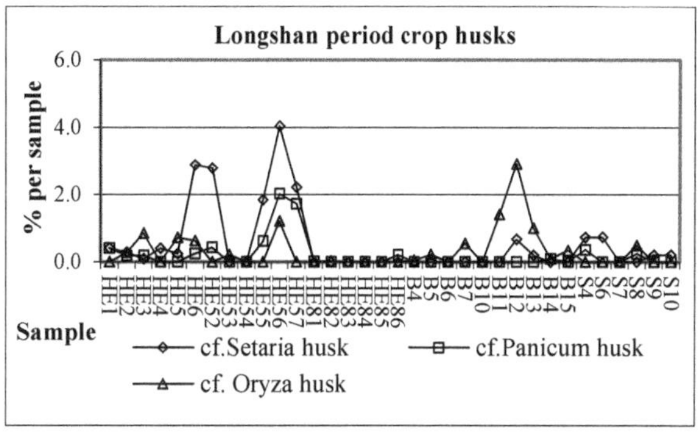

Figure 9.21 Longshan period crop husks

The Longshan sees a growing predominance of *Setaria* in the samples from the Yilou River Valley, Huizui and the survey samples. Baligang continues to favour *Oryza* although *Setaria* has been introduced.

ERLITOU

The relative frequencies of crop husk phytoliths decrease again in the Erlitou samples, falling to even lower levels than the Yangshao. The change from the Longshan to the Erlitou period is marked by a change from a reliance on *Setaria italica* as the main crop to a predominance of *Panicum* and to a lesser extent *Oryza*.

To summarise, crop selection seems more closely related to the position of the site than cultural period as can be seen by the contrast of predominantly rice at Baligang in the south and the millets, *Panicum* and *Setaria* at Xipo, Huizui and the survey sites to the north of the region.

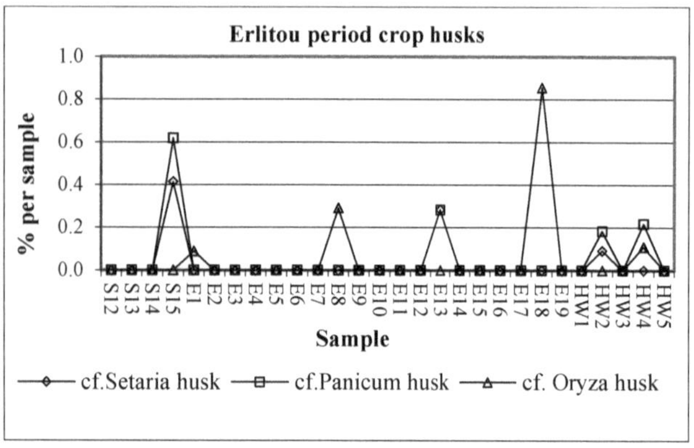

Figure 9.22 Erlitou period crop husks

However, although the main crop stays the same from the Yangshao to the Longshan at Huizui and Baligang there are changes to the secondary crops with *Oryza* gaining more importance at Huizui as well as *Panicum* suggesting an emphasis on mixed farming.

At Baligang the introduction of *Setaria* as a minor crop suggests a mixed crop repertoire, perhaps in response to demographic expansion or a changing environment creating the need to use previously uncultivated land.

The Erlitou period is more clouded. All crops are present at Huizui but *Panicum* seems to be the main crop, a situation reflected in the results from the survey sites, suggesting expansion into more marginal drier upland areas, a possible response to a cooler and drier local environment and needs of an increased population. At the Erlitou site there are few crop husks in any of the samples. *Oryza* occurs most frequently, perhaps highlighting the cultural importance of this particular crop as during this period the environment is far less suitable for rice agriculture than earlier.

9.4 CHANGING AGRICULTURAL STRATEGIES, THE INTRODUCTION AND EXPANSION OF RICE AGRICULTURE, IN PARTICULAR WET RICE FARMING

Increased proportions of distinctive phytoliths particular from *Oryza*, for example the husks (double peaked glume cells) and leaves (scoop edged *Oryza* keystones), might indicate the expansion of rice cultivation.

The average percentages per site show a marked increase in *Oryza* husks and leaves at Baligang between the Yangshao and the Longshan (Figure 9.24). At Huizui the levels of keystones from *Oryza* leaves drop in the Longshan and again in the Erlitou period, while the husks increase and then fall. At Xipo *Oryza* levels are low. The *Oryza* husk in the survey samples decreases from the Longshan to the Erlitou; there are no other *Oryza* phytoliths from these samples. The Erlitou site produced high proportions of *Oryza* keystone and some husk. These results suggest increasing *Oryza* agriculture at Baligang and Huizui from the Yangshao to Longshan.

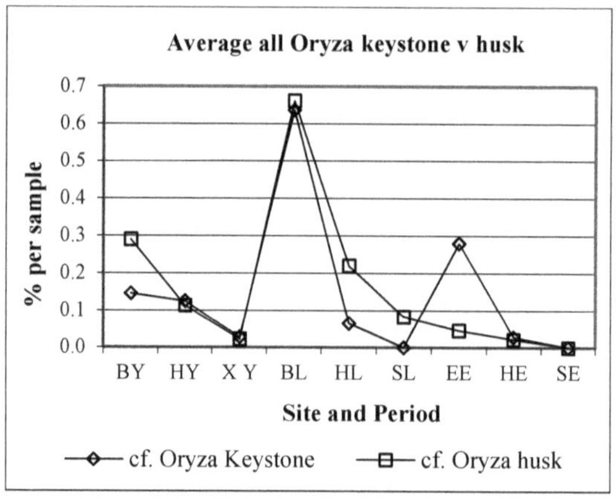

Key: B= Baligang, H= Huizui, X = Xipo, S = Survey, E = Erlitou, Y = Yangshao, L= Longshan, E = Erlitou

Figure 9.23 Average all Oryza keystone and husk

The scooped bilobes, Cyperaceae leaf and *Oryza* keystones generally correlate in the northern sites during the Yangshao and Longshan periods. Cyperaceae leaf and cones (from the husk) are high at Baligang during the Yangshao compared to the *Oryza* phytoliths.

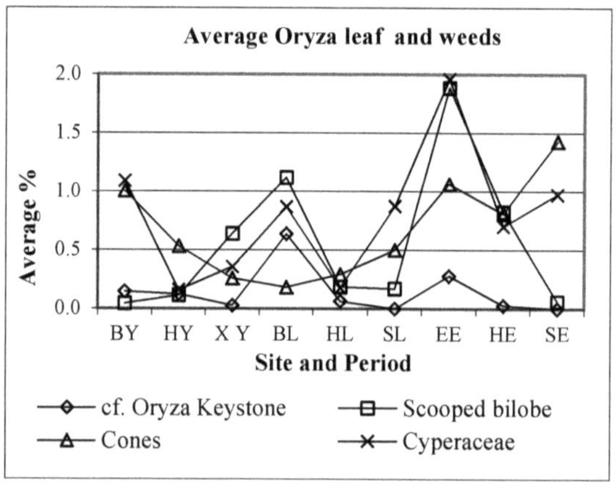

B=Baligang, H=Huizui, X=Xipo, S=Survey, E=Erlitou, Y=Yangshao, L=Longshan, E=Erlitou

Figure 9.24 Average Oryza leaf and weeds

The Erlitou period is difficult to interpret because of the contexts sampled at Erlitou but it would seem there is rice at the Erlitou site, less at Huizui and none at all at the survey sites.
The presence of different parts of rice crop weeds such as scooped bilobes from *Leersia* or *Zizania* or cones from Cyperaceae can also indicate cultivation (Figure 9.24).

The percentages of cones drop at Baligang during the Longshan but otherwise correspond. The Erlitou samples correlate although there is a drop in *Oryza* leaf at Huizui and high Cyperaceae to no *Oryza* in the survey samples.

BALIGANG

Baligang has well established rice agriculture by the Yangshao period. There are also *Oryza* keystones and husk multicells in the pre Yangshao sample (B16). Rice is the main crop throughout the Yangshao and Longshan. It would seem that *Oryza* was being treated differently in each period. During the Yangshao there are relatively high proportions of cones from Cyperaceae husks correlating with the *Oryza* husks. In contrast in the Longshan there are more morphotypes from weed leaves, suggesting a change in harvesting practices and maybe also a change in how the rice was being cultivated. Proportions of all *Oryza* morphotypes increase in the Longshan, as do the crop weeds seen by increasing scooped bilobes and Cyperaceae leaf (Figure 9.25). There is evidence of possible diversification with the introduction of *Setaria* italica (Figure 9.10).

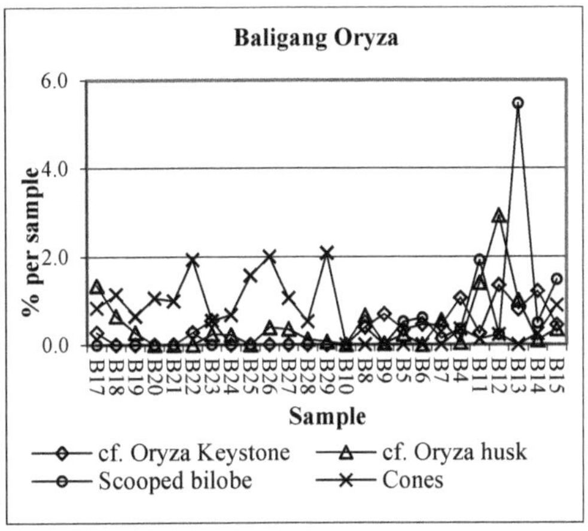

Figure 9.25 Baligang Oryza and weeds

THE YELLOW RIVER VALLEY SITES

XIPO

The phytolith evidence from Xipo points to a mixed agricultural economy (Figure 9.26) with *Panicum* as the predominant crop. There is evidence of rice agriculture. Like Baligang, there are comparatively high levels of Cyperaceae husk, but there are higher levels of *Oryza* keystones and scooped bilobes suggesting different harvesting practices. The overall level of *Oryza* phytoliths per sample is much lower than at Baligang. *Oryza* husks are not usually found in samples that produce *Panicum* and never with *Setaria* demonstrating it is being processed separately. There is little correlation between scooped bilobes from *Oryza* and *Oryza* weed leaves and Cyperaceae husk.

If the morphotypes from plants that need water, *Oryza* husk, Cyperaceae husk, and diatoms from Baligang and Xipo are compared, the difference is clear. While levels in all three are relatively high at Baligang there is a steep drop in the Xipo samples. This might suggest the rice cultivation regime at Xipo had not yet developed into paddy farming, which requires waterlogging. There is some Cyperaceae at Xipo but this genus can be found in dry and wet farming regimes and the scarcity of diatoms suggests the crops may have been dry farmed here.

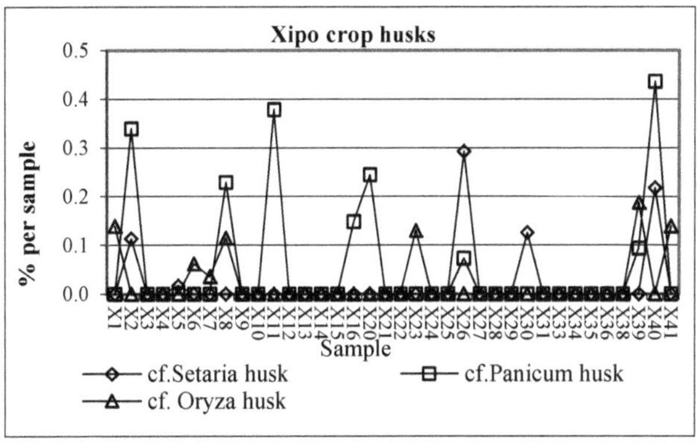

Figure 9.26 Xipo crop husks

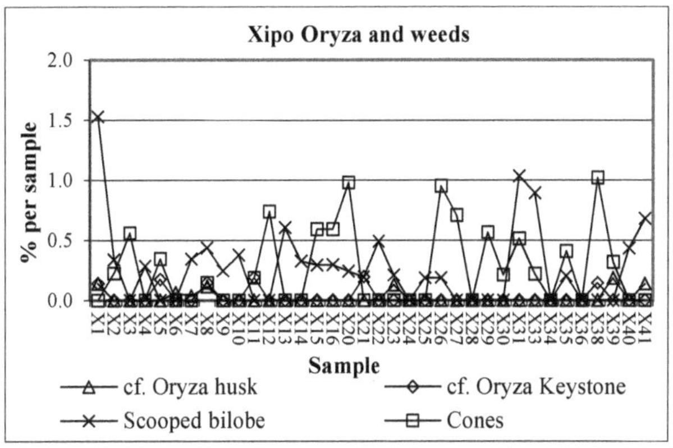

Figure 9.27 Xipo Oryza husk and leaf, Cyperaceae husk

HUIZUI

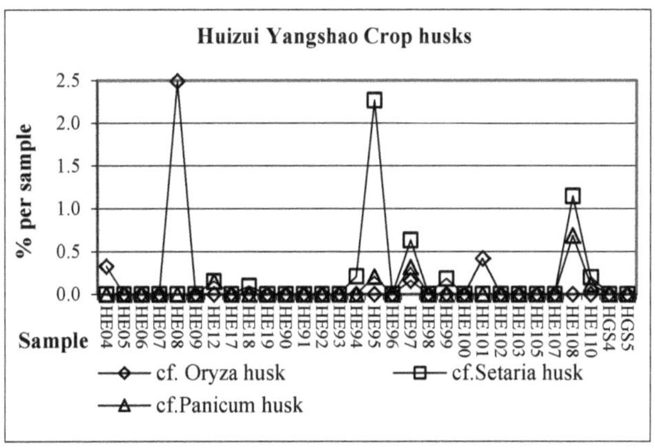

Figure 9.28 Huizui Yangshao crop husks

As at Xipo there is a mixed economy at Huizui during the Yangshao period, however *Setaria* is the predominant crop here highlighting Huizui's geographical position further east. Apart from one sample (HE97) rice is found separately from the millet crops.

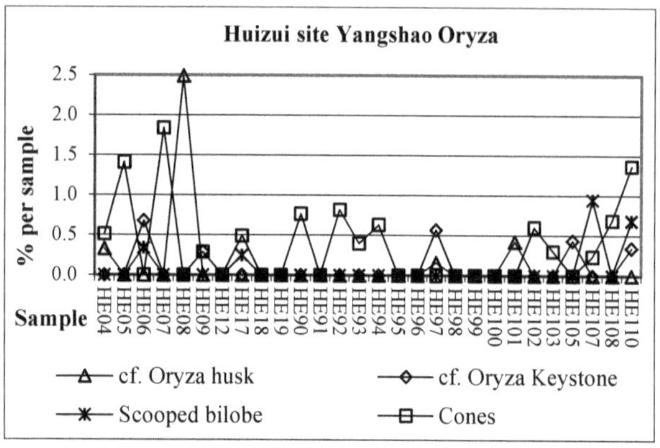

Figure 9.29 Huizui site only Yangshao Oryza and weeds

In contrast to the Yangshao results from Baligang, Cyperaceae husk and *Oryza* husk rarely show any correlation; neither does Cyperaceae husk correspond with *Oryza* keystones (Figure 9.29).

There seems to be little relationship between scooped bilobes and *Oryza* or Cyperaceae apart from in HE06 and HE17.

When the *Oryza* husks, Cyperaceae and diatoms at Baligang and Huizui are compared the contrast between the two sites is not as great as Baligang and Xipo but there is still a clear difference.

There is a significant increase in all crops, in particular *Setaria* in the Longshan period samples (Figure 9.30), indicating increasing diversification and possible expansion.

Oryza is not native to the Yellow River Valley and increasing rice agriculture suggests the adoption of new agricultural strategies.

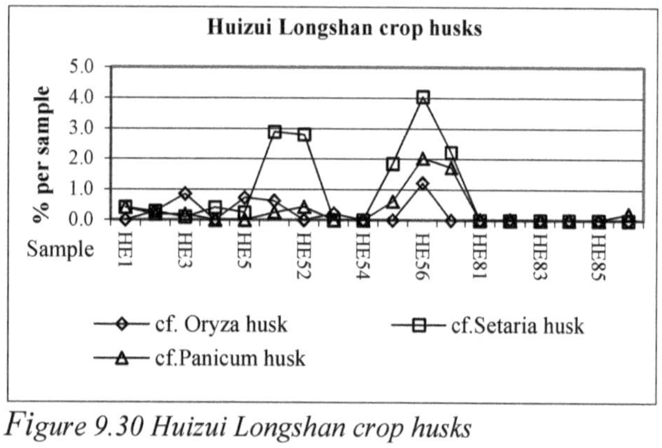

Figure 9.30 Huizui Longshan crop husks

The dividing lines describe separate features within the site, HE1-6 are from laminated layers, HE52-57 from another laminated pit fill, HE81-86 are from a section of floors and foundations. The *Oryza* husk levels increase in the Longshan (Figure 9.31). Cyperaceae husk decreases in samples from laminated pit fills but increases in samples HE82-HE86, this feature is a series of floors and foundations so the Cyperaceae found here is unlikely to be related to crop harvesting and processing, although

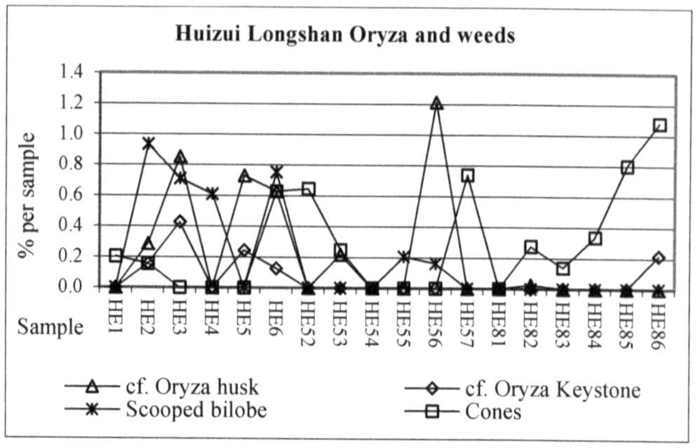

Figure 9.31 Huizui Longshan Oryza and weeds

there were *Oryza* keystones in HE86. It may possibly be packing material for the foundations. There are low levels of *Oryza* husk in HE82. The scooped bilobes do not appear to correspond with either *Oryza* husks or keystones.

The results from Erlitou period Huizui are limited to five samples; these demonstrate a change in main crop from *Setaria* to *Panicum* (Figure 9.31). *Oryza* husk and leaf are also present (Figure 9.32). *Oryza* weeds, represented by scooped bilobes from Oryzeae, and cones from Cyperaceae, are also found in these samples in higher proportions than the crops (Figure 9.33).

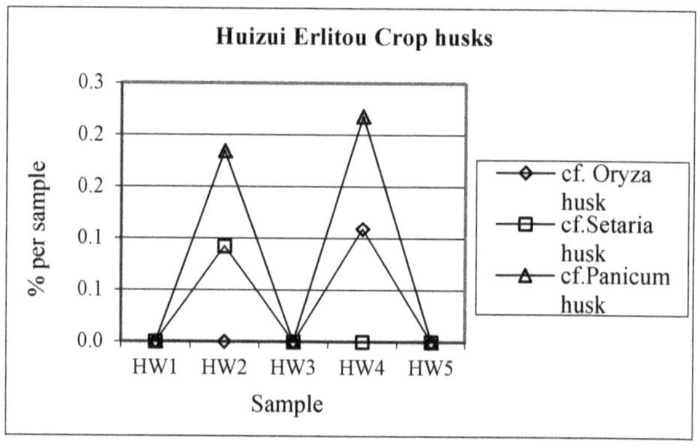

Figure 9.32 Huizui Erlitou crop husks

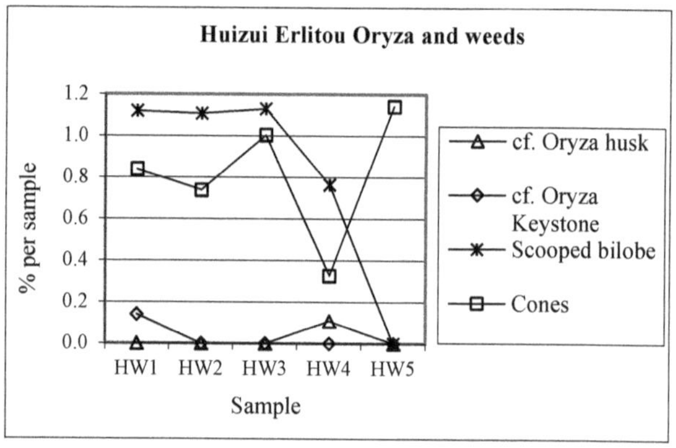

Figure 9.33 Huizui Oryza and weeds

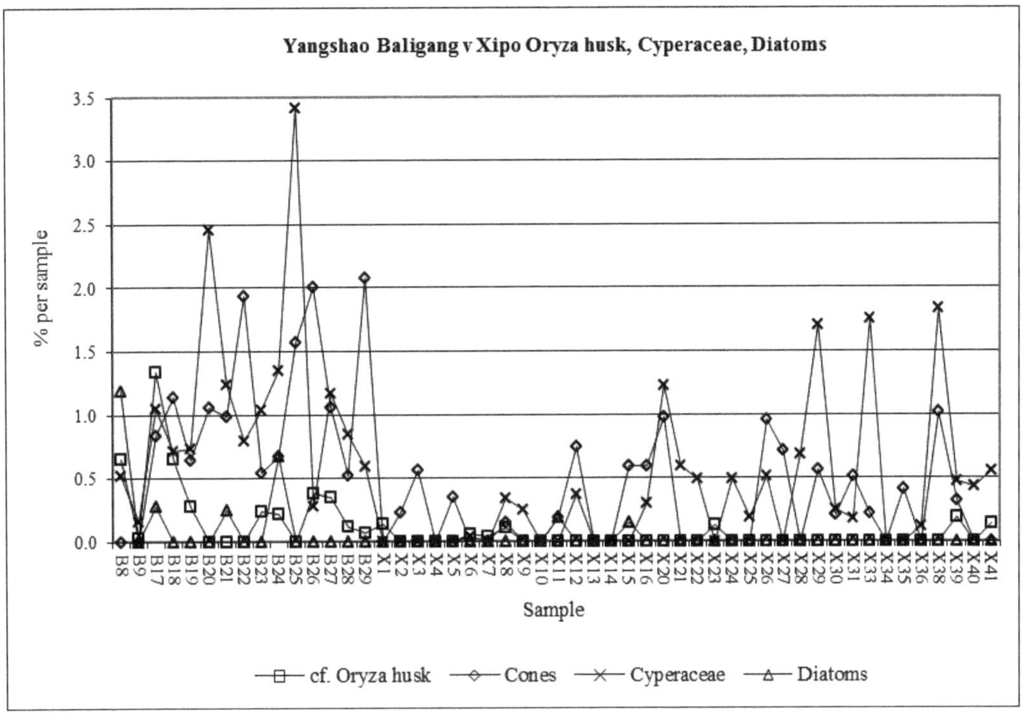

B=Baligang, X= Xipo

Figure 9.34 Yangshao Baligang and Xipo, Oryza husk, Cyperaceae husk and leaf, Diatoms

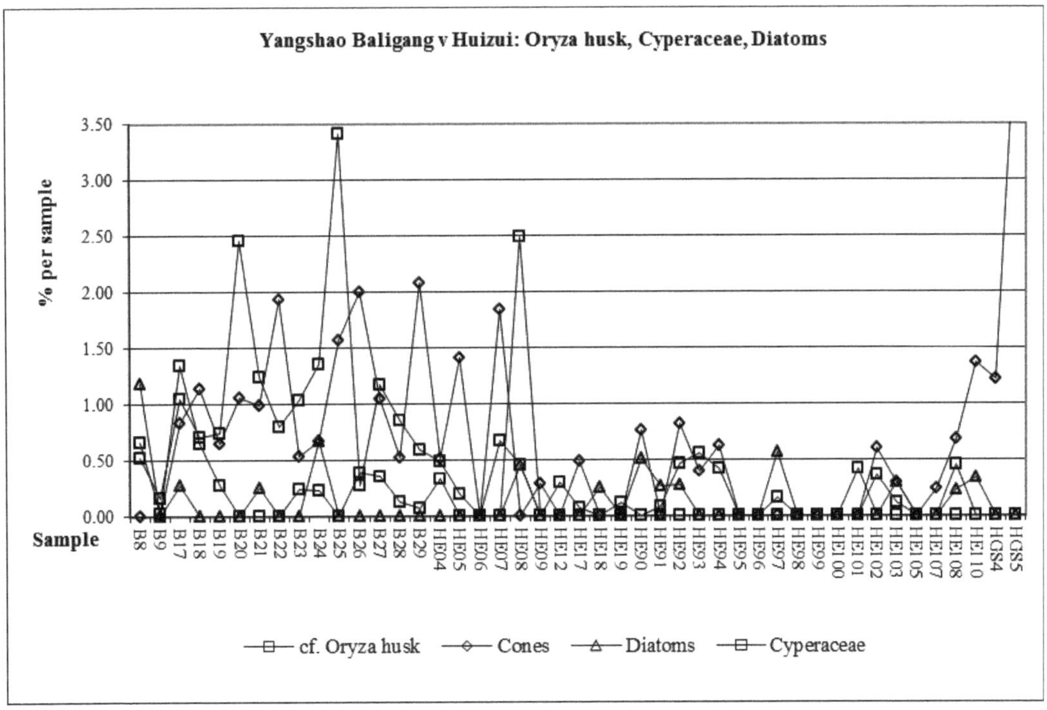

B = Baligang, HE = Huizui East, HGS = Huizui Geological Samples

Figure 9.35 Yangshao Baligang and Huizui: Oryza husk, Cyperaceae, Diatoms

SURVEY

The Yangshao sample has no crop phytoliths (Figure 9.36). The Longshan shows a similar mixed crop choice pattern to Huizui. The predominant crop is *Setaria. Panicum* is present in lesser proportions and one sample contains *Oryza* husk although it is the only *Oryza* of any kind in all the survey samples (Figure 9.37) so may not represent a locally grown crop. Again following a similar pattern to Huizui the predominant crop in the Erlitou period is *Panicum* with less *Setaria* and no *Oryza.* These crops are only found in one sample. However, it is worth including them in the analysis as they may add to the general picture of the period.

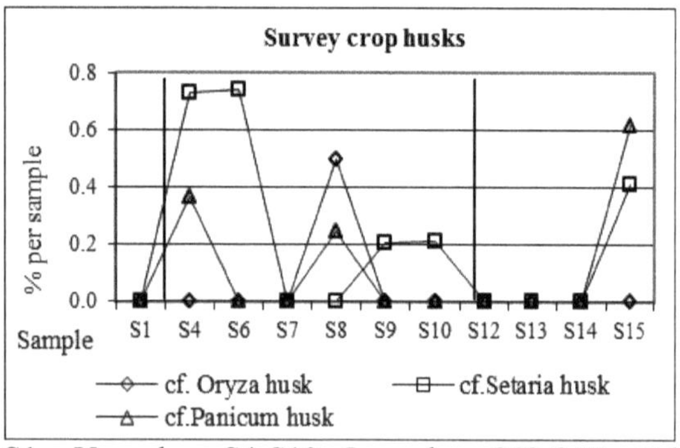

S1 = Yangshao, S4-S10= Longshan, S12-S15 = Erlitou

Figure 9.36 Survey crop husks

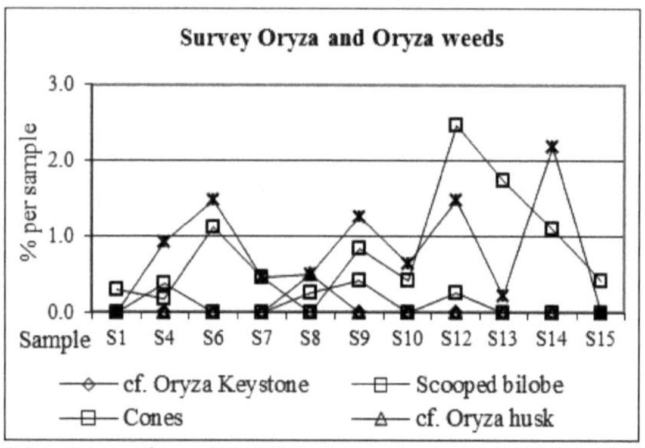

S1 = Yangshao, S4-S10= Longshan, S12-S15 = Erlitou

Figure 9.37 Survey Oryza and Oryza weeds

Erlitou

Unlike Huizui and the survey sites *Oryza* husks are the most common cereal crop of the few found in the Erlitou samples (Figure 9.38). One sample contains *Panicum* and *Setaria* in equal proportions. Erlitou follows the pattern of the other sites in having few crop remains and all at less than one percent per sample.

The proportions of scooped bilobe and Cyperaceae are relatively high in the Erlitou site samples, in particular during phase IV (Figure 9.39).

In conclusion, by the Yangshao period *Oryza* had been introduced to the Yellow River Valley sites, Xipo and Huizui, in northern Henan, although it was clearly a more important crop in the south at Baligang. While the evidence suggests rice cultivation close to the site at Xipo there is nothing to indicate the development of paddy farming. In contrast at Baligang the relatively high levels of Cyperaceae along with diatoms and sponge spicules in the samples containing rice points to at least wet farming. Huizui falls between the two, there is rice but no clear evidence of cultivation practices.

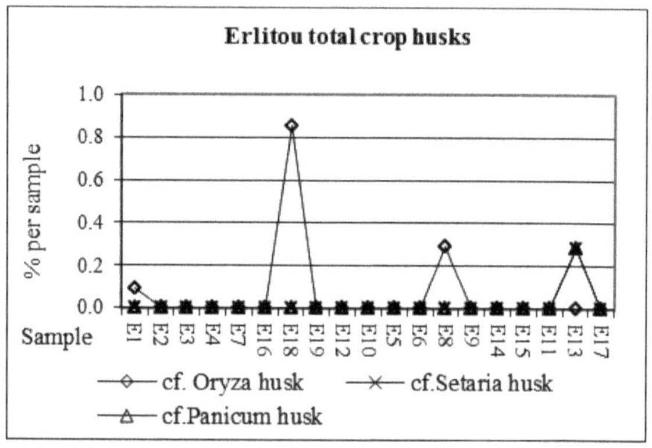

Figure 9.38 Erlitou crop husks

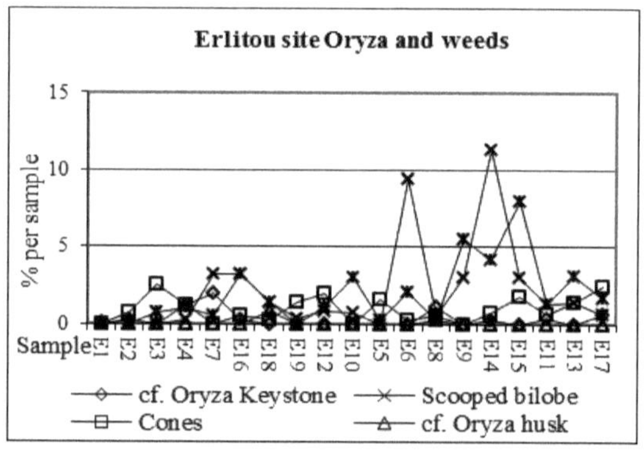

E= Erlitou, the numbers are the sample numbers
E1 – E18 = Phase II, E19-E10 = Phase III, E5 – E15 = Phase IV, E11-E17 = Erlitou general

Figure 9.39 Erlitou site Oryza and weeds

At Baligang in the late Yangshao and early Longshan Cyperaceae disappears from the samples, returning in low levels in the later Longshan samples. This could suggests a change in harvesting methods, cutting from the base of the plant would be more likely to include weeds, while selecting only the panicles would avoid collecting Cyperaceae along with the grain harvest. Alternatively, it could suggest more successful cultivation practices. Better paddy preparation and irrigation would mean the weeds would be waterlogged and if transplanting the rice seedlings would select against crop weeds too. This would be a clear indicator of intensification.

The increase in all crops in the Longshan points to a greater reliance on a suite of cultivated crops, including *Oryza, Panicum* and *Setaria.* The increased labour needed to construct and manage paddy farming and improved millet cultivation suggest demographic growth and the ability to organise labour for land clearance, harvesting and crop processing.

9.5 CROP PROCESSING

First, the model developed by Harvey and Fuller (2005) was tested on rice. Next, tri-plots were used to compare crop husk: weed husk: straw ratios, first for rice and then for millets. Finally the results from a grain storage pit at Huizui are described.

CROP PROCESSING MODEL

Harvey and Fuller (2005) propose a model for identifying crop-processing stages using phytolith data based on Hillman (1981,1984A, 1984B), Thompson (1992, 1996) and Reddy's, (1999) crop processing models for charred seed remains (see Chapter 7 for details).

This model has great potential for use in analysing labour and social organisation. However, Harvey and Fuller used a small dataset of 10 samples predominantly from ash middens (they do not state exactly how many or what other contexts) from the Neolithic site of Mahagara in central India. I initially selected a dataset of 60 samples from ash middens to test this model. Next, I added samples from 51 laminated pit fills because they are also likely to contain crop-processing waste. Relative frequencies were chosen rather than the absolute counts Harvey and Fuller used to enable comparison with correspondence analysis.

The Erlitou site samples were not included in the crop processing analysis because they were taken from palace wall contexts and despite some of the contexts being classified as ash middens it is not clear whether they represent the same types of feature as those from the other sites.

The Henan ash midden dataset was tested on *Oryza* in the same way as Harvey and Fuller's case study at Mahagara (2005:748-750). Millets were not tested at Mahagara due to the low numbers present and difficulties with identification (Harvey and Fuller, 2005:749).

USING THE MODEL TO TEST ORYZA

According to Harvey and Fuller, 2005 it should be possible to identify *Oryza* crop processing stages using phytolith data, (see Chapter 7). Fuller and Zhang, (2007: 955) suggest the ash middens are likely to contain residues from repetitive daily activities so can provide a picture of the routines of daily life. One problem with this model is the ash middens are composed of many deposits of secondary and tertiary waste disposed of over time. If the midden has received input from more than one stage of processing it will be difficult to unpick. However, if, as Fuller and Zhang suggest, the waste in the middens is from routine processing the stage that reaches them is likely to be the same. The laminated pit fills may prove clearer indicators, as they tend to be the result of single deposits. Following Harvey and Fuller 2005 (Figure 3) phytolith morphotypes used here also include all non-crop grasses.

HENAN CROP PROCESSING

Using Harvey and Fuller's (2005) case study at Mahagara as an example I examined the phytoliths produced by rice first, *Oryza* keystone, *Oryza* leaf/culm, *Oryza* husk and scooped bilobe. Scooped bilobes, distinctive short cells found in *Oryza* leaves, can also be found in other members of the Oryzeae family, for example *Leersia* and *Zizania*. Furthermore there is a strong likelihood the other plants that produce scooped bilobes in this region, are found growing with *Oryza sativa* so can be classed as *Oryza* crop weeds. As they are potential rice weeds and were part of the original model they were included in Figure 9.41.

Overall there are higher percentages of leaf to husk across all three periods, the levels of leaf increasing over time. The exception is Baligang during the Yangshao period.

The high percentages of *Oryza* husk to leaf at Baligang, particularly in the Yangshao samples, suggest the remains of late stage processing.

At Xipo there is far less rice but there are two samples with only husk, another two with only straw and one with equal proportions of both suggesting all crop processing stages may have been taking place on site.

At Huizui there are five samples that have only *Oryza* husk, seven with leaf only and two with husk and leaf, again suggesting both early and late processing stages taking place on site.

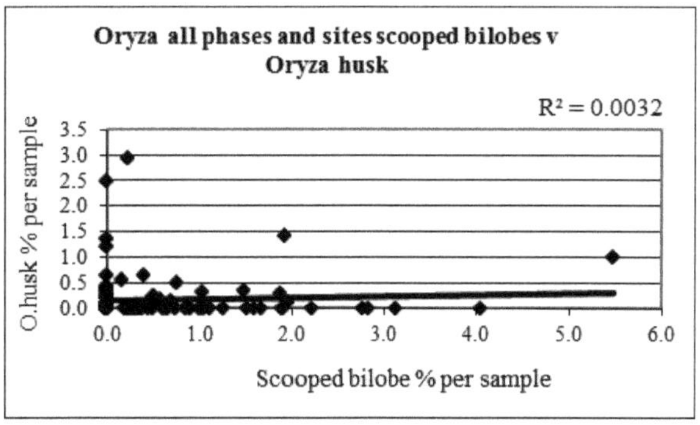

Figure 9.40 Scooped bilobes v Oryza husk

There are high percentages of scooped bilobes, with a peak at Baligang in the Longshan samples and high percentages throughout the Erlitou period samples; although during the Erlitou they are the only cf. *Oryza* morphotype in the majority of the samples. Unlike Mahagara (Harvey and Fuller 2005: Figure 8) the morphotypes in the dataset from Henan are rarely represented in equal quantities either as % per sample or number per gram. Neither do the scooped bilobes and *Oryza* husks correlate as a whole or when divided into

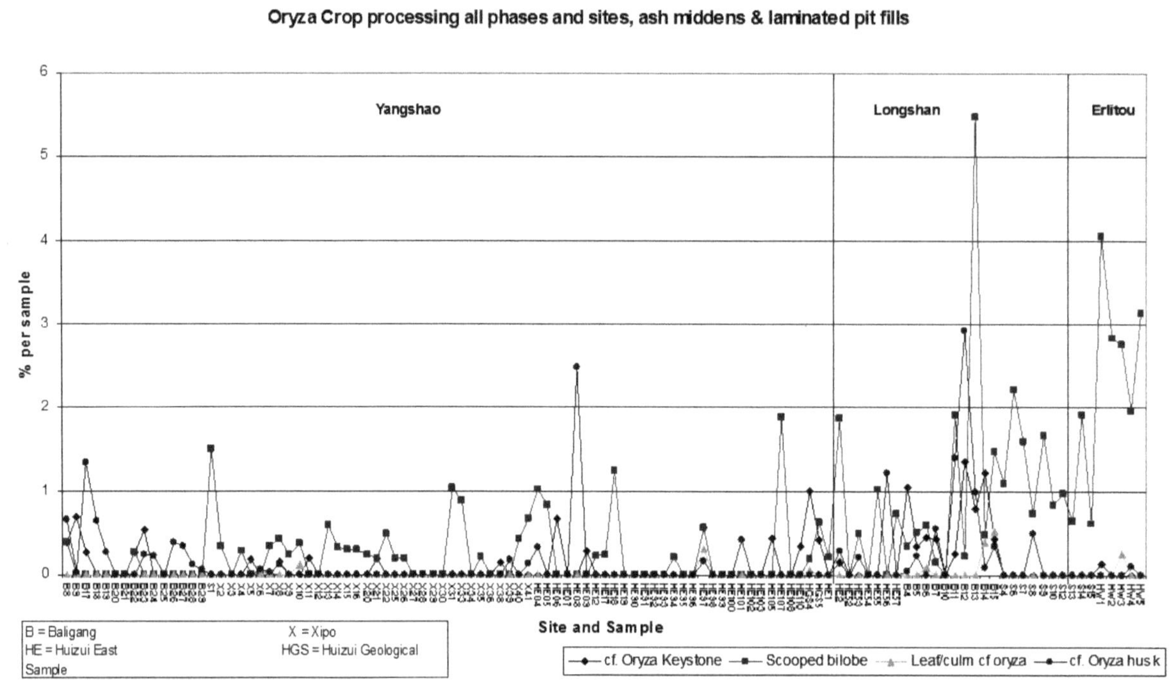

Figure 9.41 Oryza crop processing all phases and sites: Ash middens and laminated pit fills

phase or site (Figure 9.40). While the ratio of scooped bilobes to *Oryza* husk correlate in Harvey and Fuller's case study (2005:749 fig.9) the ratios for the Henan dataset as a whole do not (Figure 9.40). This may be due to the range of phases and sites but the ratios for individual sites and phases do not correlate either.

RESULTS CROP PROCESSING TRI-PLOTS

Possible crop processing residues were examined using the tri-plot template generated by Graham and Midgely (2000). These diagrams 'represent tri-variate data in which the three variables represent proportions of the whole'. In order to achieve this, the three data groups in each chart are represented as percentages of their total.

For the first plot all morphotypes both single and multicell from grass leaves, including both crops and wild plants, husks and crop husks were selected from a table of all ash pit and laminated pit fill samples. Samples from floor/foundation contexts were not included.

This plot shows the vast majority of phytolith remains are from grass leaves, in crop processing stage terms, from threshing residue. On a plant the volume of husk to leaf is greater so a larger proportion of leaf than husk phytoliths should be expected. However, this result does raise the question of whether leaves or husks produce more silica bodies, this is an area that should be explored in detail in the future.

For the second tri-plot morphotypes representing grass leaf, weed husk and crop husk were selected from the

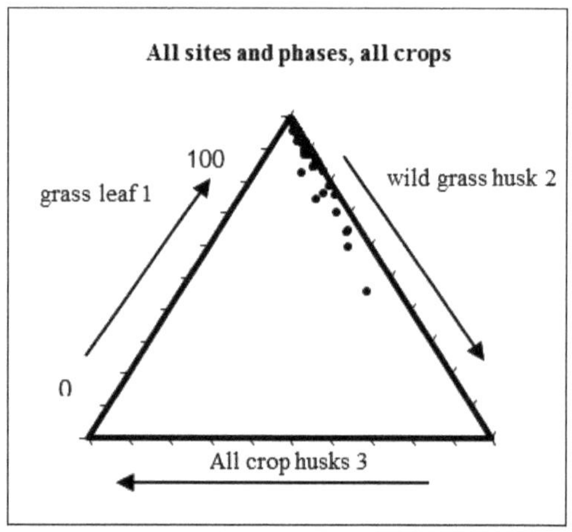

Figure 9.42 All sites and phases, all Crops

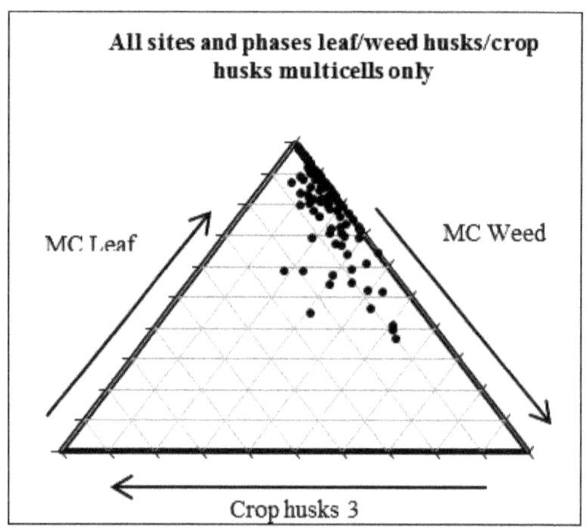

Figure 9.43 Multicells only

multicell forms only. Waste from the early stages of crop processing should be seen at the apex and towards the right of the tri-plot with high leaf values and low husk. Higher levels of wild grass husk, towards the lower right, suggest winnowing by products and crop husks towards the lower left should represent dehusking. In this plot all sites and phases are present and, like plot 1, the majority of the plant material found in the ash middens and laminated pit fills seems to represent threshing by-products.

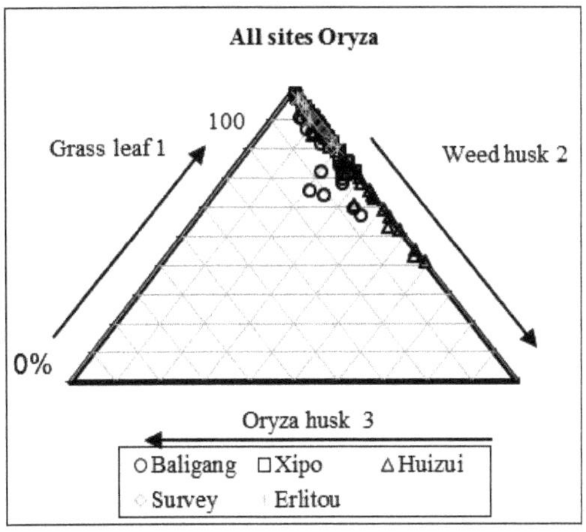

Figure 9.44 All sites Oryza

Figure 9.44 shows *Oryza* at all sites. There is a clear difference between Baligang and the Yellow River Valley sites. Xipo, Erlitou and the survey samples have low percentages of *Oryza* husk suggesting very little rice dehusking at these sites. For the survey samples and Xipo this probably results from low quantities of rice overall. The Erlitou samples are unclear and almost certainly not representative of the site as a whole, however they appear to contain mostly leaves and some weed husk with very little rice. Huizui has higher proportions of weed husks pointing to possible winnowing as well threshing input, although there seems little evidence of *Oryza* dehusking in these samples. The Baligang samples produce the highest levels of *Oryza* husk suggesting dehusking as well as threshing and winnowing waste.

BALIGANG

First the same dataset was used to look at the individual cultural phases. There does not seem much difference between the Yangshao and Longshan. Next, I refined the chart by using only morphotypes from *Oryza* and *Oryza* weeds. I included millet 1 and 2 husks as millets, such as *Paspalum scrobiculatum*, *Panicum cambogiense*, *Brachiaria reptans*, have been recorded as *Oryza* weeds in both deep water and upland rice. Cones are from Cyperaceae which have been recorded with all rice cultivation systems (Thompson, 1996: Table 32 Sources Kittipong, 1983, Vongsaroj, n.d).

***Oryza* and *Oryza* weed leaf**	***Oryza* weed husk**
Oryza keystone	Cones
Scooped bilobes	Millet 1
Leaf/culm Phragmites	Millet 2
Leaf culm reed	
Leaf/ culm square cell	
Cyperaceae	

Table 9.1 Morphotypes included as Oryza and Oryza weeds.

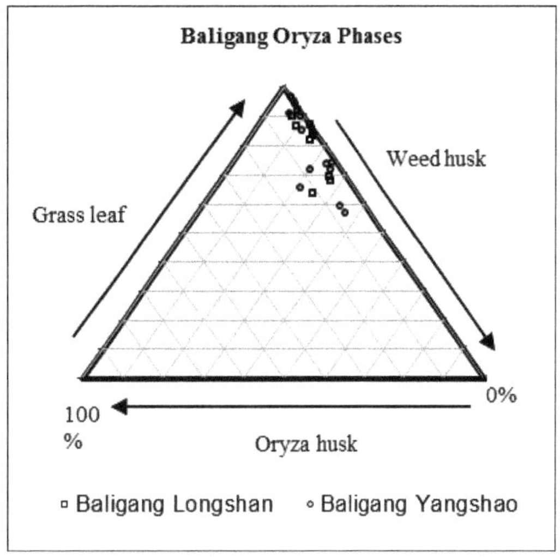

Figure 9.45 Baligang and Longshan Oryza only

The picture that emerges is quite different. A clear distinction can be made between the Yangshao and Longshan. The Longshan has generally higher proportions of *Oryza* and *Oryza* weed straw suggesting a possible change in harvesting practices in the Longshan. The higher proportions of weed and crop husk in the Yangshao could be related to harvest height, cutting higher on the plant to include some crop weed husks but fewer leaves, or it could mean there was more threshing waste generated in the Longshan.

Figures 9.46 and 9.47 show five of the 16 samples from the Yangshao contain *Oryza* leaf morphotypes while eleven have *Oryza* husk at mostly higher levels, in contrast to the Longshan where all of the nine samples contain leaf morphotypes and in all except one higher proportions than husk.

Figure 9.52 shows the contrast between Cyperaceae husk and Oryza leaf in the Yangshao and Longshan periods. A distinct change can be seen. The high proportions of Cyperaceae husk in the Yangshao fall dramatically in the Qujialing (Late Yangshao) samples (B10, B8, B9) and Longshan, while Oryza keystones increase.

If the same *Oryza* morphotypes only are applied to the three phases at Huizui, the Yangshao samples produced leaves, weed husks and one sample with comparatively high proportions of *Oryza* husk. The Longshan samples show another slightly different pattern, there are fewer leaves and more weed and Oryza husks than in the other samples. In contrast, the Erlitou period samples produce what appears to be mostly threshing waste. This could suggest different approaches to crop processing in each cultural period. Although different to Baligang the general trends are in the same direction.

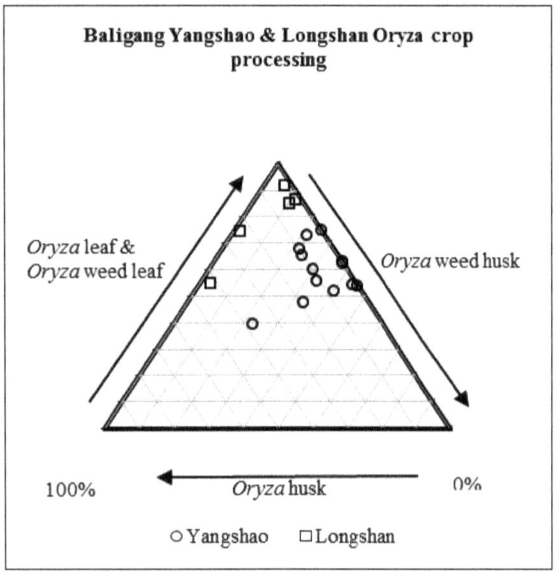

Figure 9.46 Baligang and Longshan Oryza only

XIPO

Despite there being limited rice remains at Xipo a distinction can be seen between samples with and without husk. The majority of samples have high leaf proportions, no crop husk and some weed husk suggesting input from threshing waste (Figure 9.44). As emphasised by the bar chart (Figure 9.51), there is some separation between samples containing *Oryza* leaf and husk. This could show different processing stages, threshing and dehusking, taking place on site but at different times or places. Three samples produce both husk and leaf, but as these samples are from ash middens this is likely to reflect the mixed nature of the deposits in this type of context.

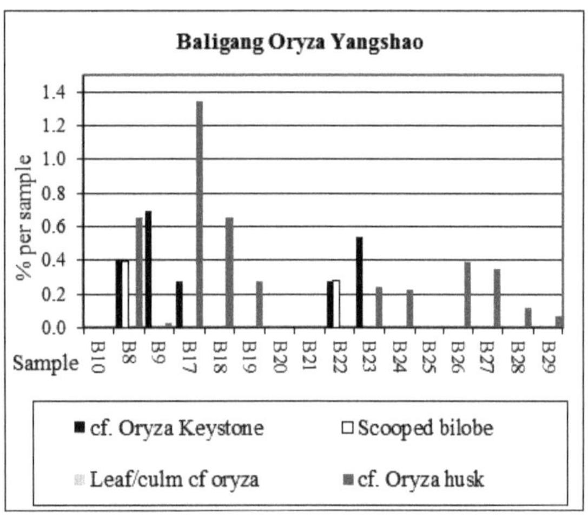

Figure 9.47 Baligang Oryza Yangshao

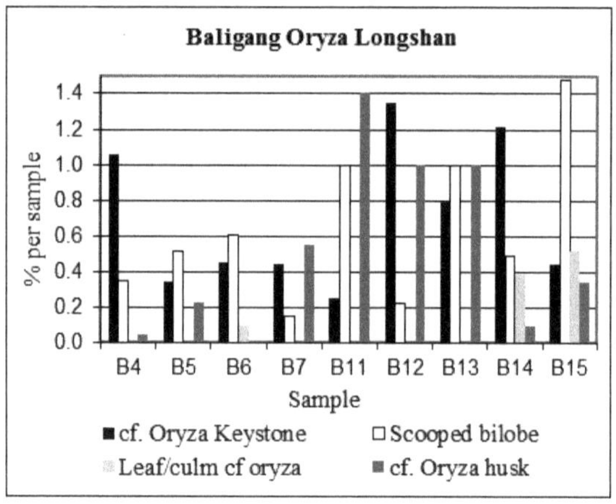

Figure 9.48 Baligang Oryza Longshan

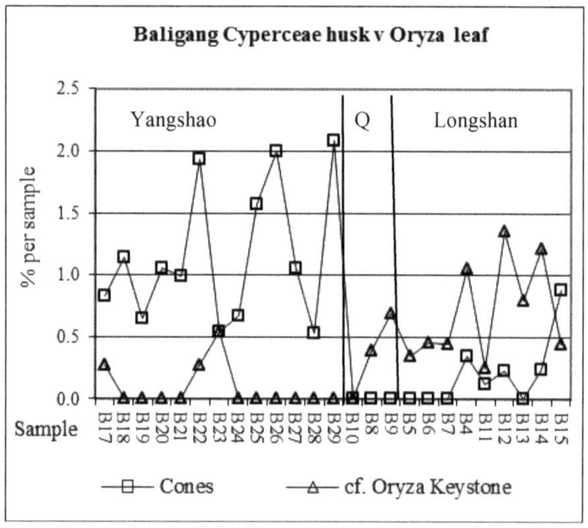

Yangshao and Longshan Q= Qijialing (Late Yangshao)

Figure 9.49 Baligang Cyperaceae husk (cones) v Oryza leaf (Oryza keystones)

HUIZUI

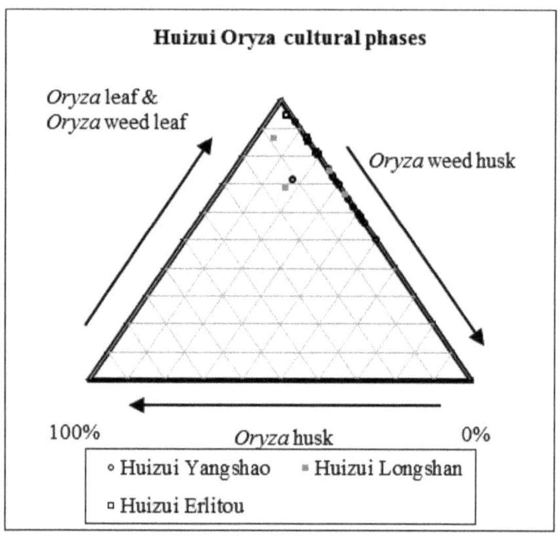

Figure 9.50 Huizui Oryza phases, Oryza only

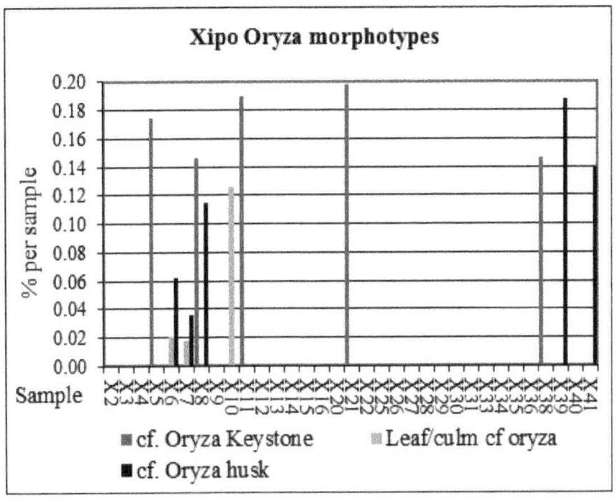

Figure 9.51 Xipo Oryza morphotypes

SURVEY

There are very few *Oryza* type morphotypes in any of the survey samples (Figure 9.47). There seems to be an increase in weed husk waste in the Erlitou period but the levels are still low in comparison to the leaf residues suggesting threshing waste. Scooped bilobes were the only morphotype that could have come from *Oryza* leaf, there were no *Oryza* keystones or leaf multicells suggesting rice was not being cultivated at these peripheral sites (Figure 9.52). The level of scooped bilobes drops in the Erlitou period. There is a relatively high level of *Oryza* husk in one sample (S8). This is the only *Oryza* in any of the survey samples.

When the results for all the cultural phases are looked at together there seems more activity during the mid and Late Neolithic, Yangshao and Longshan phases, than in the Early Bronze Age Erlitou. While the majority of samples from all sites and phases are composed of high proportions of grass leaves and stems suggesting threshing waste, there are higher levels of crop husk in the Yangshao samples from Baligang and higher percentages of weeds husk in both Yangshao and Longshan. This is in part a reflection of the southern geographic situation of Baligang, which seems to have produced higher rice husks than the other sites, but both Xipo and Huizui have Yangshao samples with relatively high levels of weed husks.

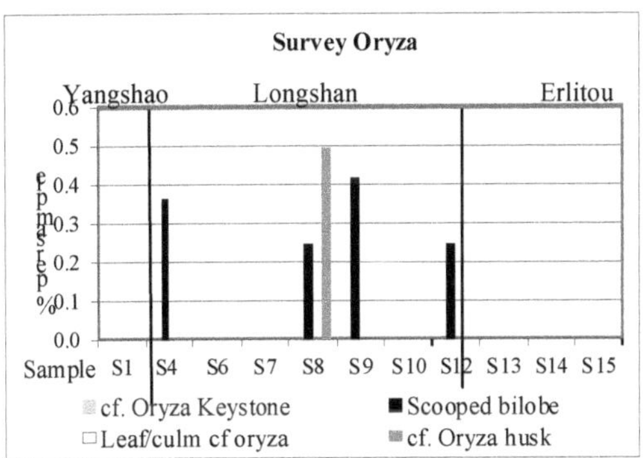

Figure 9.52 Survey Oryza, phases

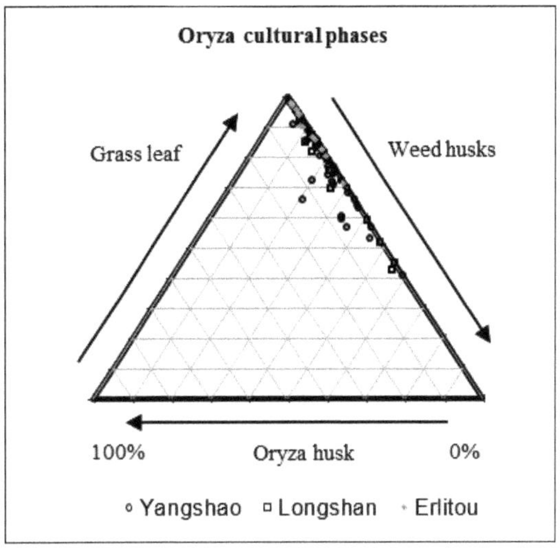

Figure 9.53 Oryza cultural phases

MILLETS

This chart (Figure 9.54) demonstrates the predominance of threshing waste at all sites. The Huizui samples stand out as they show higher percentages of weed and millet husk suggesting evidence for the later stages of processing taking place on site.

When looked at individually differences between sites become clearer (Figure 9.55). Baligang has very low proportions of millet crop husks compared to leaf and weed husks. This may be threshing waste or, more likely, represent the comparatively low levels of millet at this site.

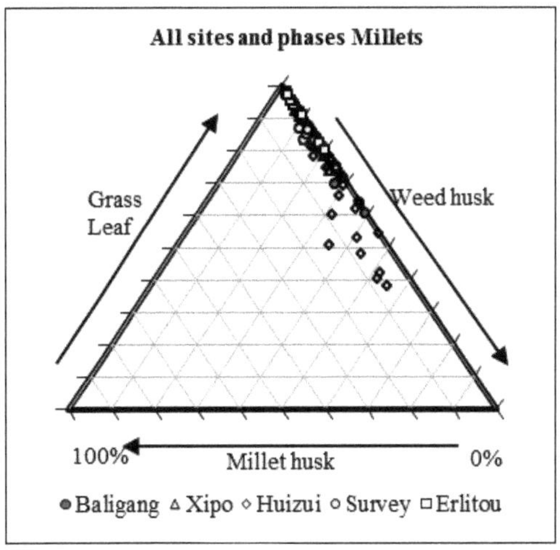

Figure 9.54 All sites and phases millet crops

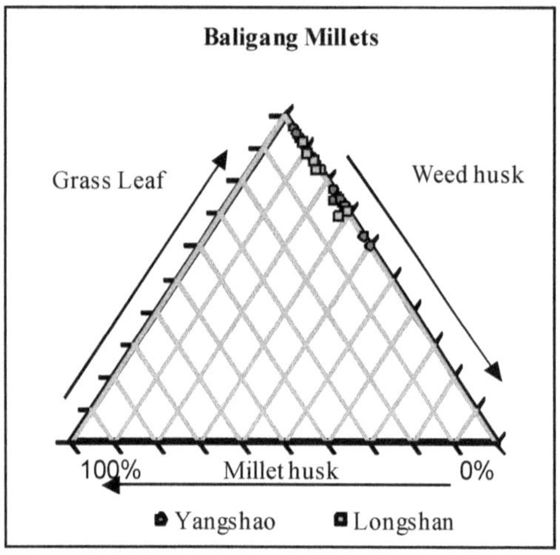

Figure 9.55 Baligang crop millets only

The crop processing by products at Baligang consist predominantly of threshing waste. Both Yangshao and Longshan deposits are mixed. *Panicum miliaceum* and *Setaria italica* levels are low at this site overall.

At Xipo (Figure 9.56) the results show high proportions of leaves pointing to threshing waste. There are slightly higher levels of millet husk here than at Baligang, but not enough to demonstrate separate dehusking.

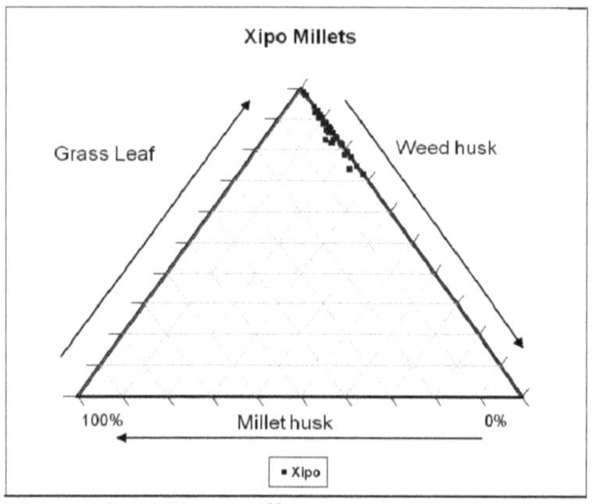

Figure 9.56 Xipo millets

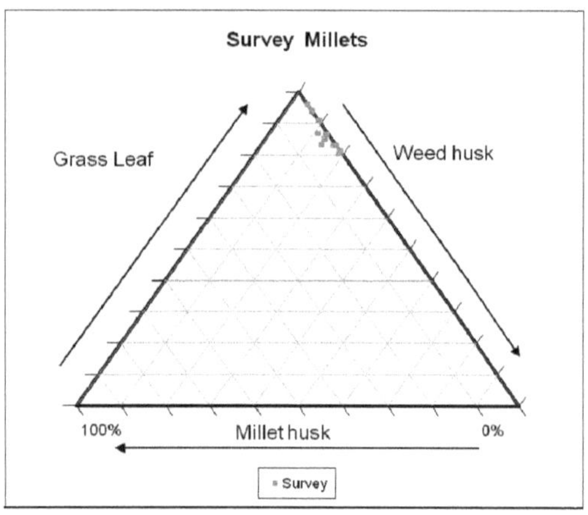

Figure 9.57 Survey Millets

The results suggest the survey samples contain mostly threshing by products, although there are some weed and crop husks (Figure 9.57). The samples are mixed throughout the phases.

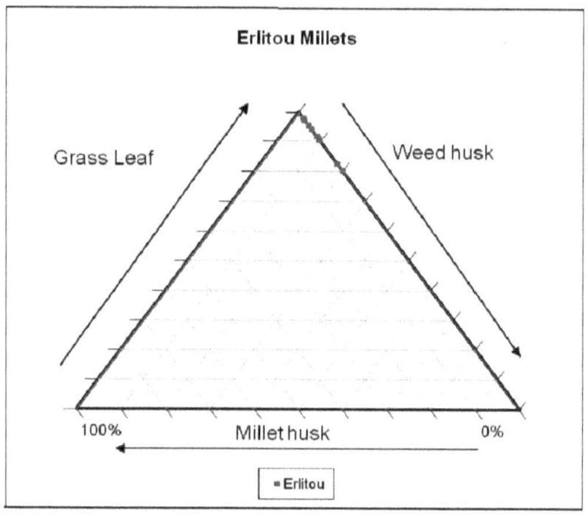

Figure 9.58 Erlitou Millets

The samples from the Erlitou (Figure 9.61) site are predominantly grass leaf with a few weed husks and no *Oryza* husks suggesting that if the samples contain crop-processing by-products they are from threshing.

At Huizui there is a greater range (Figure 9.59). While the highest levels are threshing, the weed husks suggest possible winnowing residue, and there are some relatively high proportions of crop husks suggesting dehusking. Threshing residues are predominant in all phases. The Yangshao and Longshan samples have higher levels of both weed and crop husk which points to higher levels of winnowing and dehusking input during these periods.

Although this is a very generalised picture some differences between phases can be noted. The majority of samples from all phases cluster at the right hand apex of the tri-plot suggesting predominantly threshing waste from leaves and stems with some weed seeds.

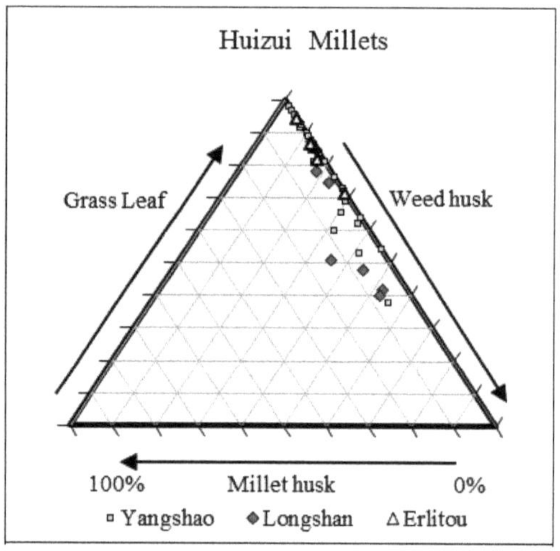

Figure 9.59 Huizui Millets

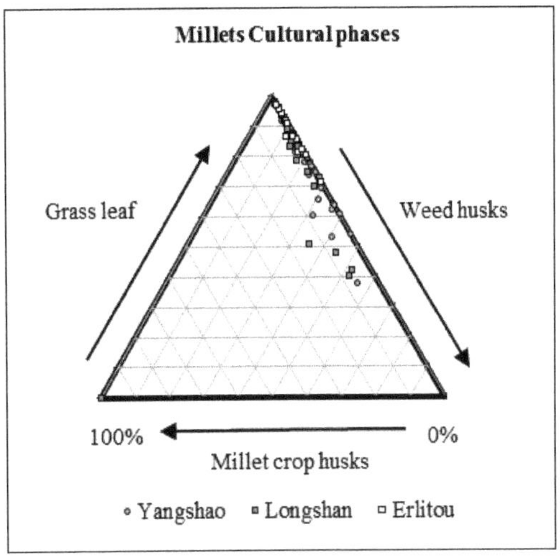

Figure 9.60 Millets, cultural phases

This may be partially influenced by some phytolith morphotypes, from leaves and stems, preserving in an identifiable state in greater numbers than those from husks, for example bulliforms and long smooth cells from leaves versus dendritic forms from husks.

Some samples from the Yangshao period, from Huizui, have relatively high proportions of weed husks and some crop husks. This increases at the same site in the Yangshao, before falling away in the Erlitou period.

Most Erlitou period samples are primarily composed of leaves and stems, with a few weed husks. There are very low percentages of millet crop husks from this period.

When both millet and rice charts for all cultural phases are looked at together a pattern begins to emerge.

9.6 HUIZUI GRAIN STORAGE (HE 90-96)

Seven samples were taken from the north wall and bottom of a large bell shaped pit dated to the Yangshao period. The walls were lined with burned earth containing impressions of what appeared to be grass or reed leaves lying at a diagonal to the top of the pit. In the corner was a dense layer of fine charcoal. At the bottom of the pit was a 5cm layer of mostly millet seeds (*Setaria italica*). These appear to be separate deposits.

Table 9.2 Huizui storage pit

HE90	White patch
HE91	Looser layer
HE92	White ashy
HE93	Fibres from imprints and charcoal
HE94	Yellow burned earth
HE95	Dense fine charcoal in corner
HE96	Charcoal from compacted grains at bottom

The charcoal layer (HE95) provided far higher densities of phytoliths than the other samples, total 7,315,339 per gram. The next highest was 849,442 per gram (HE92) and the lowest 11,952 per gram (HE96). The results are discussed using the relative frequency charts. Samples HE90-HE94 represent the pit lining, while HE95- HE96 are from the remains of the contents. There are no *Oryza* morphotypes in any of the samples from this pit suggesting rice and millets were stored separately.

The vast majority of phytoliths from the pit lining samples are from leaf/culms, followed by dicotyledons. There are few husks or inflorescences. The fibres from the lining of the pit (HE93) produced predominantly single cell (58.8% / 84079 per gm) phytoliths, unidentifiable to plant part. The burned earth sample (HE94) produced a relatively low density of phytoliths but has a wide variety of morphotypes.

The dense fine charcoal sample (HE95) contains abundant husks, in particular *Setaria* (2.3% /165915per gm). These are the highest *Setaria* crop densities of any sample collected. *Panicum* is present and crop weed husks, millets type 1 and 2 are also abundant. In contrast, (HE96) the 5cm layer of charred *Setaria italica* seeds covering the bottom of the pit produced no crop or weed husks. The most interesting here is the lack of identifiable phytoliths in the seed layer. The morphotypes with the highest proportions in this sample were indeterminate single cells 34% and hairs 28%. Starch rarely survives the method of processing used for these samples. This sample contains the highest percentage of starch of all samples analysed for the entire project (0.5%), which supports the presence of a high proportion of grain. The lack of husk phytoliths in this sample might suggest the whole grains in this pit were being stored dehusked.

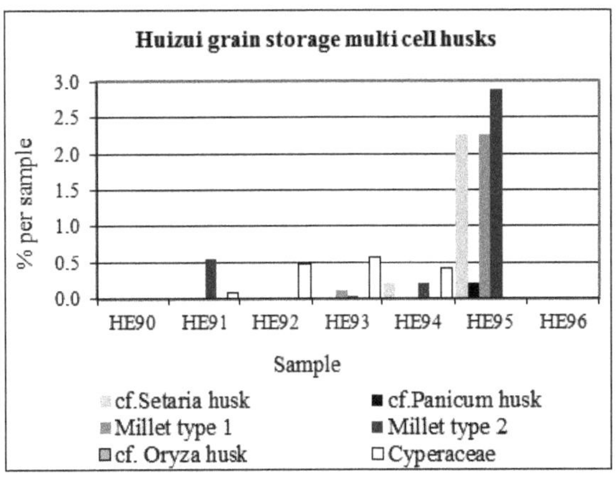

Figure 9.61 Huizui grain storage multicell crop and weed husks (not including indeterminate)

In conclusion, the results suggest it is possible to identify crop processing stages using phytolith data and the model developed by Harvey and Fuller. However, this model works far better with a crop such as rice where it is possible to identify leaf as well as husk rather than millets where this is not the case. Establishing which stages of processing are represented it is then possible to interpret differences in labour and social organisation over time, for example the changes at Baligang from the Yangshao to the Longshan suggest more intensified paddy farming and the need for organised labour to build and maintain paddy field systems.

9.7 RESULTS: NON-CROP MORPHOTYPES AND BROADER NON-AGRICULTURAL VEGETATION AROUND SITES.

Correspondence analysis of non-crop grasses and Cyperaceae provides a broad picture of the relationships between non-crop grasses present in the samples. Grasses were categorised according to Twiss (1992) Watson and Dallwitz (1992) and Metcalfe (1960). Using a dataset comprising of morphotypes from Panicoid, Chloridoid, Pooid, Bambusoid, Cyperaceae and water loving grasses, all samples were plotted and the three cultural periods highlighted.

NON-CROP GRASSES AND CYPERACEAE: CORRESPONDENCE ANALYSIS

The context type was tested but there was no differentiation visible. Rondels were included as part of the Pooid group, although occasionally rondels appear with bilobes so are not always from Pooids. The occurrence was judged sufficiently rare to place them with the Pooids.

There was little differentiation in the makeup of the samples over each cultural period (figure 9.65). Yangshao samples appear in every quadrant, although there are fewer in the top left and the Longshan samples follow a similar pattern. The Erlitou period samples also appear in three quadrants with none in the lower right.

When the samples are viewed by site rather than period some patterns begin to emerge (figure 9.63). Most Baligang samples fall to the right of axis one, contrasting with the survey samples to the left. Although Xipo samples appear in every quadrant they predominate in the lower part of the chart, while Erlitou site samples tend to fall into the upper two quadrants. The survey samples all fall towards the left of axis 1. Huizui samples are ubiquitous.

When the pie charts are examined, the major pull seems to be between panicoids to the left and Cyperaceae along with Pooid to the right and water loving towards the top of the chart (figure 9.64).

Next each cultural period was examined individually, beginning with Yangshao (Figure 9.65).

YANGSHAO

The samples from the Yangshao show a clear differentiation between sites according to the non-crop grasses (Figure 9.65). Baligang samples are to the right on axis 1 as are the majority of samples from Huizui.

Xipo samples are present on both sides but not in the lower right quadrant. The species plot shows panicoid, chloridoid and bambusoid all to the far left. Pooid and morphotypes common to water loving plants are in the centre of the plot while Cyperaceae cluster in the upper part of the chart. The pie charts for each sample demonstrate clearly that Cyperaceae is pulling samples from Baligang and Huizui to the upper right, while high proportions of panicoid grasses are what are separating many samples from Xipo (Figure 9.66).

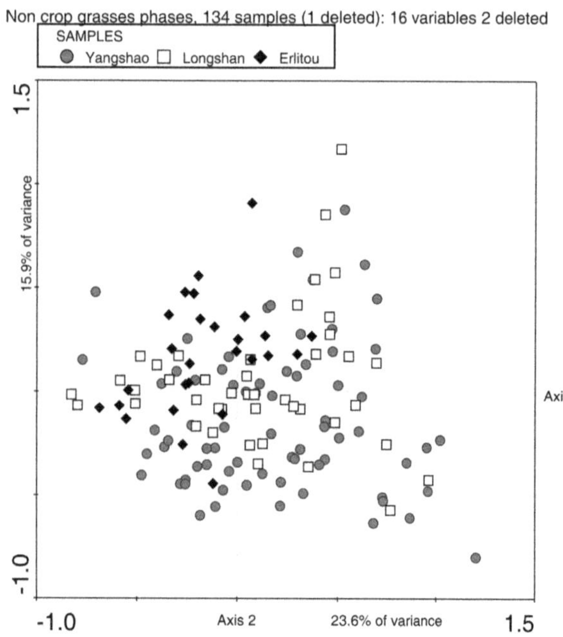

Figure 9.62 Non-crop grasses, phases (ubiquitous morphotypes removed Full dataset 144 samples, 17 variables)

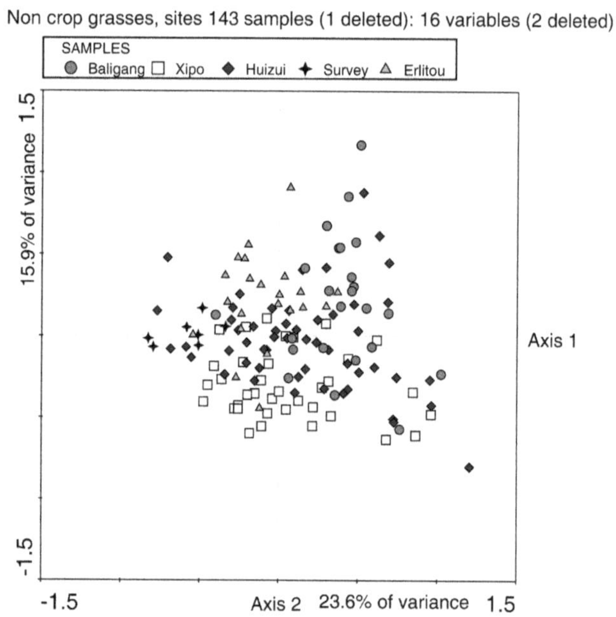

Figure 9.63 All non crop grasses and Cyperaceae, sites 143 samples (1 deleted): 16 variables (2 deleted)

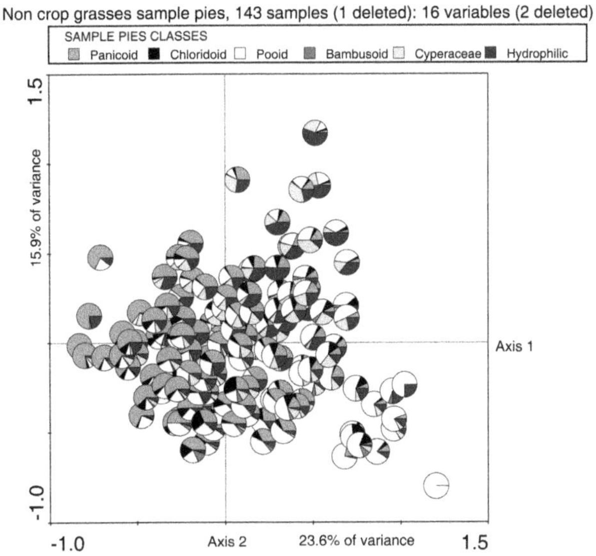

Figure 9.64 Non crop grasses and Cyperaceae, sample pie charts

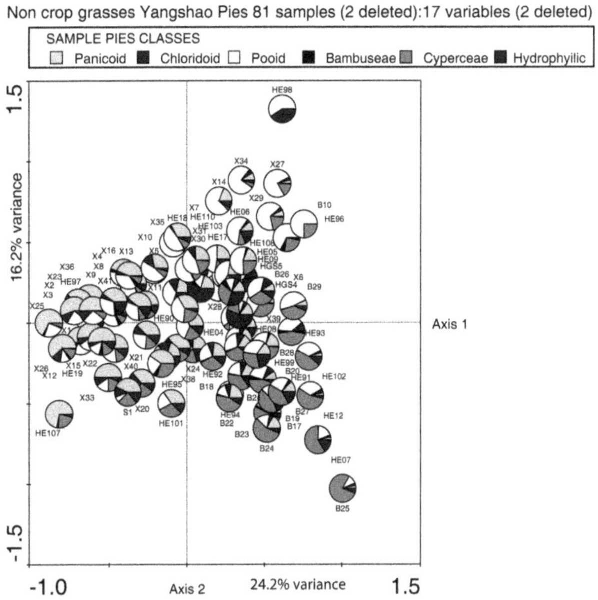

Figure 9.66 Yangshao non crop grasses and Cyperaceae, sample pie charts.

Differentiation between sites is even more defined in the samples from the Longshan period (Figure 9.67). All Baligang samples are confined to the right side of the chart so the proportions of their constituents are mostly different to the survey samples, which are all found in the bottom right quadrant. The Huizui samples cross the centre along axis 1 but the majority are to the left and upper part of the plot showing differentiation from Baligang along axis 1 and the survey along axis 2.

The pie charts show panicoids are predominant in the survey samples with some Cyperaceae but few Pooid grasses.

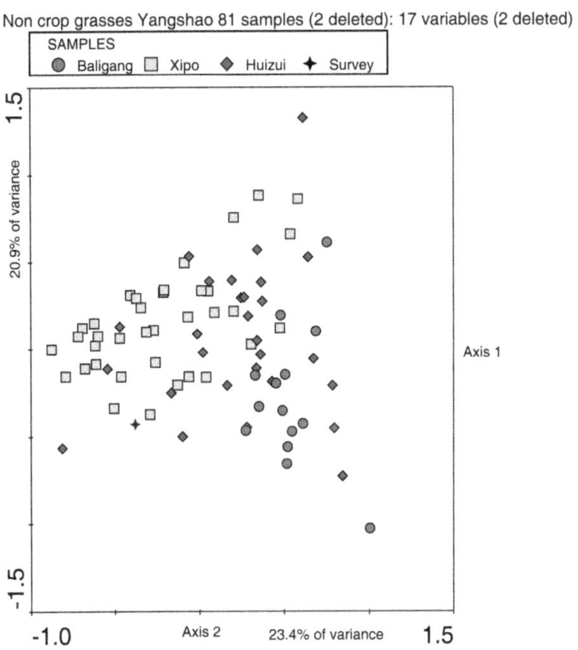

Figure 9.65 Yangshao, non crop grasses and Cyperaceae, sites (82 samples: 17 variables)

LONGSHAN

Baligang samples are the most mixed and have the highest proportions of Bambusoid of any site with high levels of Pooid alongside water loving morphotypes. Baligang also has the highest proportions of Chloridoids. The Huizui samples are mixed with predominant Pooid, Cyperaceae and panicoid grasses. They contain far higher levels of Pooid than the survey samples but look closer to these than Baligang. between floors at Huizui may be the result of floor covering.

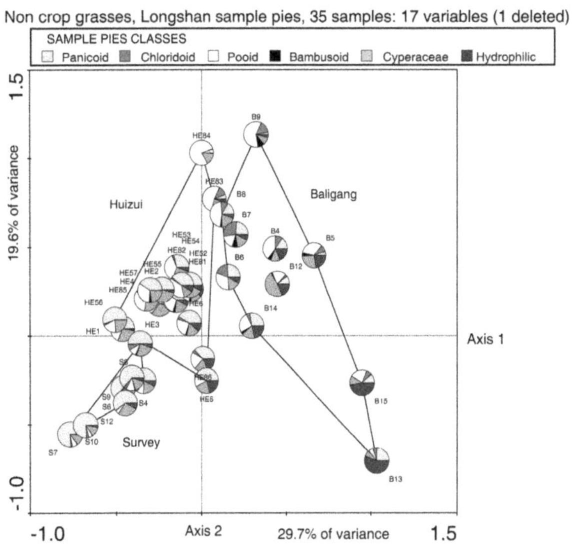

Figure 9.67 Longshan non crop grasses and Cyperaceae, sites and pie charts

The context types may provide insight here as *Phragmites*, could have been used as either a floor covering or packing material. McPhail (2007: 107) suggests the lack occupation soils.

When the context types were examined, perhaps unsurprisingly, there was a contrast in results between ash middens and floors and foundations on axis two highlighting the different components found in each (Figure 9.68). The laminated pit fill samples were predominantly in the upper part of axis 2. The greatest difference is between the Survey and Baligang ash middens suggesting the main cause of variability is the location of the site rather than the context.

ERLITOU

The Erlitou period sites are more mixed than the Longshan (Figure 9.69). This may in part be due to the lack of samples from Baligang in the south of Henan. Erlitou site samples are found in every quadrant mixed with the survey samples. Huizui is only in the upper part of the plot.

C3 V C4 PHOTOSYNTHETIC PATHWAYS

Phytolith analysis can be used to suggest patterns in grass subfamilies that often relate to whether species with a C3 or C4 photosynthetic pathway dominate grasslands.

AVERAGES C3 V C4 GRASSES

Using Twiss' 1992 classification of phytolith short cells from Graminae subfamilies short cells were selected to represent grasses with C3 and C4 photosynthetic pathways. First, I looked at rondel versus bilobe to represent Pooid versus Panicoid However, some panicoids; represented by bilobes can be either C3 or C4

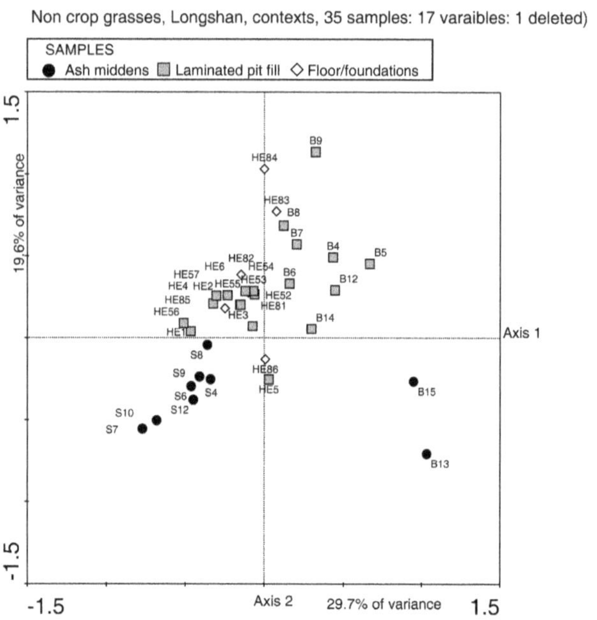

Figure 9.68 Longshan, non crop grasses and Cyperaceae, contexts

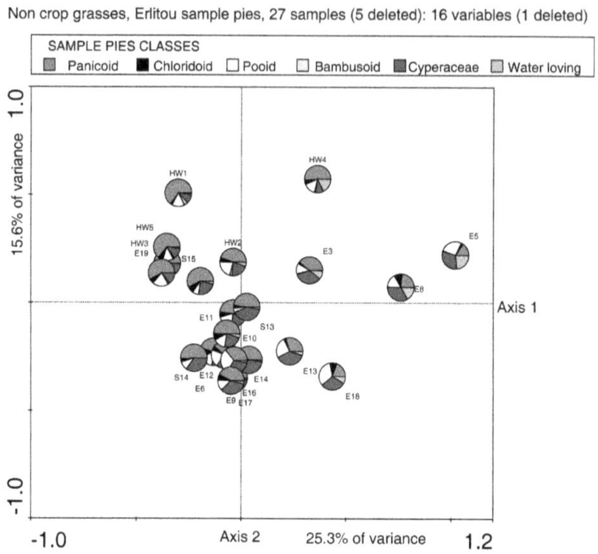

Figure 9.69 Erlitou non crop grasses and Cyperaceae, pie charts

,for example *Panicum*. In some of the archaeological samples examined for this project rondels occasionally appeared in the same multicells as bilobes, although this was very rare.

Figure 9.70 shows the samples from the Yangshao are dominated by C3 Pooid grasses at all sites. At Baligang this continued into the Longshan, at northern sites there was a shift towards C4 (a higher percentage of panicoids). The high proportions of panicoid continue into the Erlitou period. Pooids also increase at the Erlitou site and Huizui but not in the survey samples.

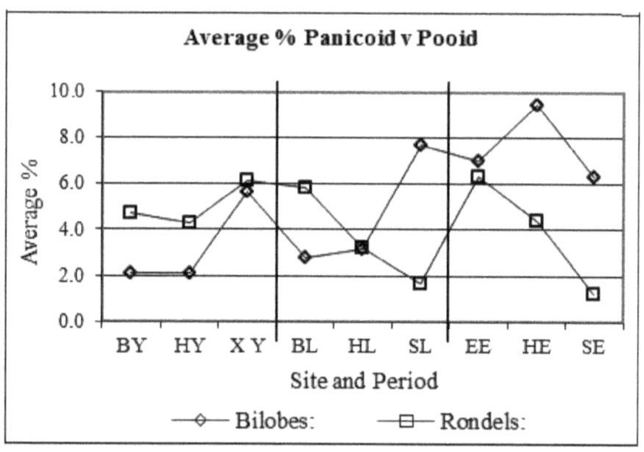

B= Baligang, H= Huizui, X=Xipo, S=survey, E= Erlitou
Y= Yangshao, L= Longshan, E= Erlitou site

Figure 9.70 Average% Panicoid v Pooid

This suggests the local environment around Baligang, in the southern part of the province, may have remained similar throughout the Yangshao and Longshan and indicates wetter grassland. In contrast in the Yellow River valley there is a change from higher proportions of pooid in the Yangshao to increasing levels of panicoid in the Longshan, especially in the survey samples, suggesting an environmental change either human induced or natural. This may reflect the third millennium drying, which was felt more strongly in the north than south (Xiao *et al*, 2004:1675, Li *et al*, 2004: 198-200 Figure 3.2). This trend continues into the Erlitou period with panicoid in consistently higher proportions than Pooid, although they are close at the Erlitou site.

However, to test this hypothesis fully well dated palaeosols from around the sites need to be sampled and unfortunately these were unavailable.

Next, *Stipa* rondel and scooped bilobe were plotted for C3 and cross and saddle for C4 (Figure 9.71). Using these morphotypes the chloridoids (saddles) follow a similar pattern to the Pooids in Figure 9.70. During the Yangshao the morphotypes from C4 grasses, saddles and crosses, are higher at Baligang but drop at Huizui, while those from grasses with a C3 pathway, scooped bilobes and *Stipa* rondels increase slightly, however saddles still dominate. At Xipo there are higher levels of all with cross falling to below the C3 short cells.

This fits with a cooler drier climate in the north west of Henan at that time, as chloridoid grasses prefer a dry upland environment (Iriate and Paz, 2009:118).

In the Longshan at Baligang the scooped bilobes increase but there is still a majority of saddles, which does not correspond with a humid local environment. This increase in scooped bilobes is likely to be related to rice cultivation. At Huizui there is an increase in panicoids, which is mirrored and amplified in the survey samples.

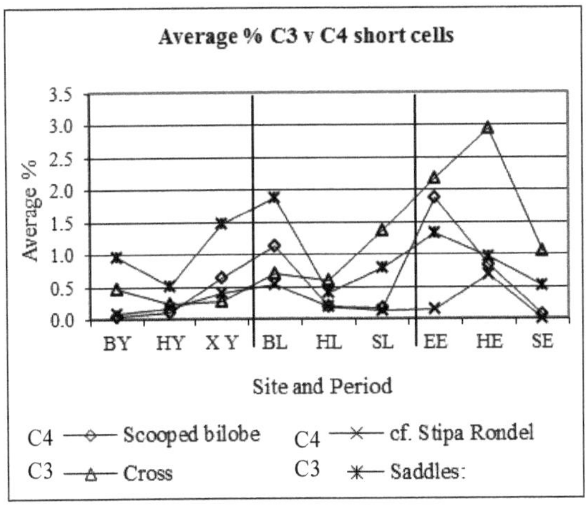

B= Baligang, H= Huizui, X=Xipo, S=survey, E= Erlitou Y= Yangshao, L= Longshan, E= Erlitou site

Figure 9.71 Average% C3 v C4 short cells, scooped bilobe, Stipa rondel, cross, saddles

The Erlitou period sees increased levels of cross at the Erlitou site and Huizui and although there is a fall in the survey samples this is the dominant morphotype. Scooped bilobes are high at Erlitou and increased at Huizui although both morphotypes representing C3 grasses are scant in the survey samples.

All diagnostic short cells were classified as C3, (scooped bilobe, rondel, *Stipa* rondel) or C4, (Bilobe, *Setaria* bilobe, cross, polylobe, saddle) (Figure 9.75). As in the earlier charts (Figures 9.70 and 9.71)

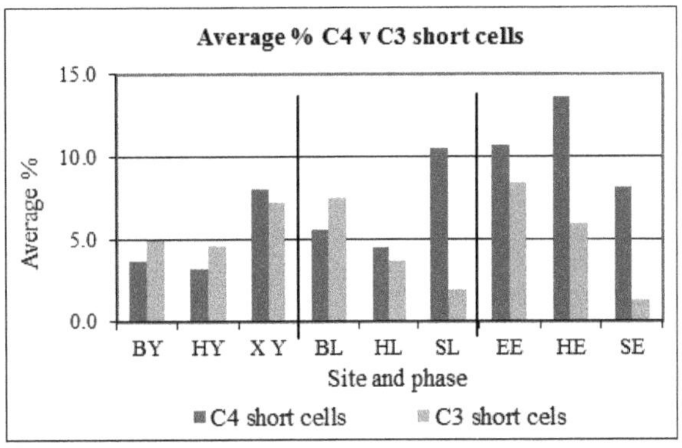

Figure 9.72 Average % C4 v C3 short cells

C3 levels are higher than C4 at Baligang in both Yangshao and Longshan. This seems likely to be a reflection of the rice crop as much as local environment particularly given Baligang's southern position in relation to the other sites. Huizui has a similar result to Baligang in the Yangshao but this may be less related to rice agriculture and more to local environment. Rice is the second crop after *Setaria italica* in a mixed agricultural economy at Huizui. At Xipo there are higher proportions of C3 grasses possibly due to its position in the more arid west of the region. In the Longshan Baligang has a higher average percentage of both C3 and C4 grass short cells but in similar relative proportions to the Yangshao. In contrast the levels of short cells from C4 grasses have increased at Huizui. The mixed agricultural economy continues into the Longshan but *Panicum* as the second crop has overtaken rice. The neighbouring survey samples have far higher proportions of C4 in this period too. These levels of C4 may in part be due to the higher levels of millet crops than earlier but could also reflect a generally drier climate.

During the Erlitou period C4 short cells dominate the assemblages from all three sites, suggesting a predominantly drier environment than in the Yangshao. This is in part borne out by the crop choices; both Huizui and the survey area show increased *Panicum*. However, *Oryza* is still being grown which points to social rather than environmental motivation.

BULLIFORMS

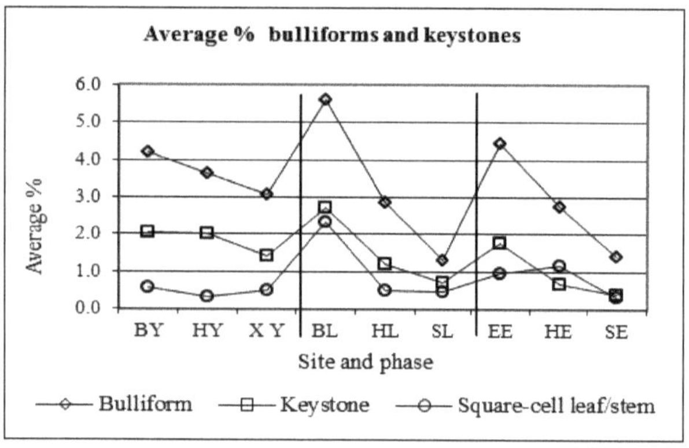

B= Baligang, H= Huizui, X=Xipo, S=survey, E= Erlitou
Y= Yangshao, L= Longshan, E= Erlitou site

Figure 9.73 Average % bulliforms

How far do the results relate to the crops or environmental conditions according to proxy data?

Phytoliths from wetland plants such as Cyperaceae (Ollendorf 1992) and *Phragmites*, a ubiquitous reedy grass that grows in marshy areas, can be used to point to a damp habitat (Figure 9.74). The average percentage of total bulliforms generally correlates with the *Phragmites* and Cyperaceae, although the reed multicells provide a slightly different pattern, possibly due to the uncertain identification. This category consists of morphotypes that look like *Phragmites* but do not possess all the identification criteria to be classified as such. During the Yangshao and Longshan bulliforms and Cyperaceae are highest at Baligang. In the Yangshao *Phragmites* levels at Baligang are relatively low, both Xipo and Huizui have higher proportions. Reed type morphotypes are higher than *Phragmites* at all Yangshao sites.

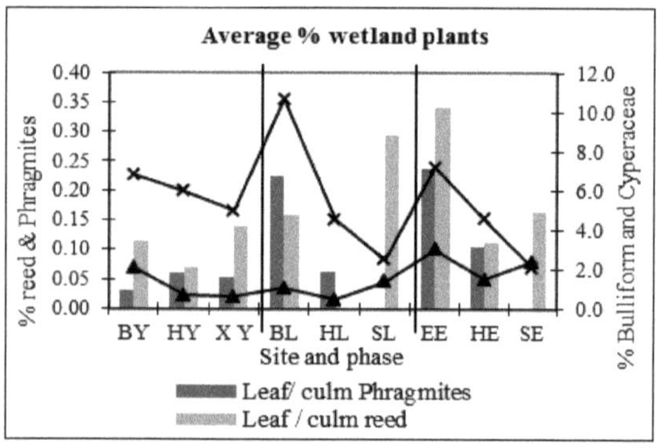

B= Baligang, H= Huizui, X=Xipo, S=survey, E= Erlitou
Y= Yangshao, L= Longshan, E= Erlitou site
Total Bulliform = Bulliform, Keystone and Leaf/culm square multicells
Total Cyperaceae = Cones and Cyperaceae multicells

Figure 9.74 Average % wetland plants

The Longshan appears less consistent than the Yangshao. While Baligang has the highest average percentage of all morphotypes except reed, Huizui sees a drop in *Phragmites* and Cyperaceae and has no reed morphotypes. The survey sites have a high percentage of reed forms and the highest average for Cyperaceae but no *Phragmites* and lower total bulliforms. In the Erlitou period the Erlitou site has high average percentages for all the morphotypes. These fall sharply at Huizui although they are still greater than in the Longshan and reeds are present. The survey samples still have no *Phragmites* and low bulliforms but Cyperaceae has increased.

From these results it would seem the environment around Yangshao sites was drier the further north and west they were positioned.

These results are reinforced by the diatoms and sponge spicules (Figure 9.75). The presence of diatoms and sponge spicules can provide indications of the local environment (Imosieke *et al*, 1995, Gist Gee, 1931). Both would be expected in water so are likely to be found in rice paddies, although some diatoms and sponge spicules are found in soils (Wilding and Drees, 1968, Pearsall and Piperno, 1993:97).

The total bulliform (excluding *Oryza* keystones) result parallels that of *Oryza*, especially *Oryza* keystones, suggesting the crops heavily influence the samples and that keystones come from grasses that enjoy a similar environment (Figure 9.76). The total wetland related morphotypes show a more humid environment at Baligang in the Yangshao and Longshan (Figure 9.77). All grasses increase at Baligang in the Longshan. C3 predominate in both periods. This is likely to be connected at least in part to rice agriculture. However, while C3 grasses and wetland plants increase during the Longshan at Baligang, levels fall at Huizui which also cultivates rice. C3 grasses increase at Huizui, while wetland plants fall. The survey samples produce high percentages of C3 grasses and even lower levels of wetland plants. These results may reflect the increasingly arid environment at this time, especially to the north of the region in the Yilou River Valley.

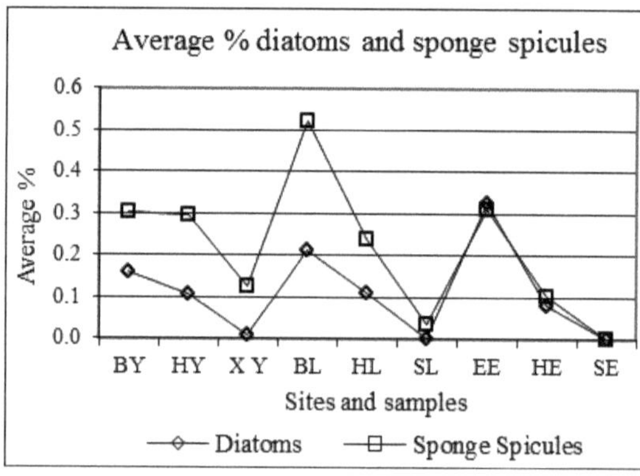

Figure 9.75 Average % diatoms and sponge spicules

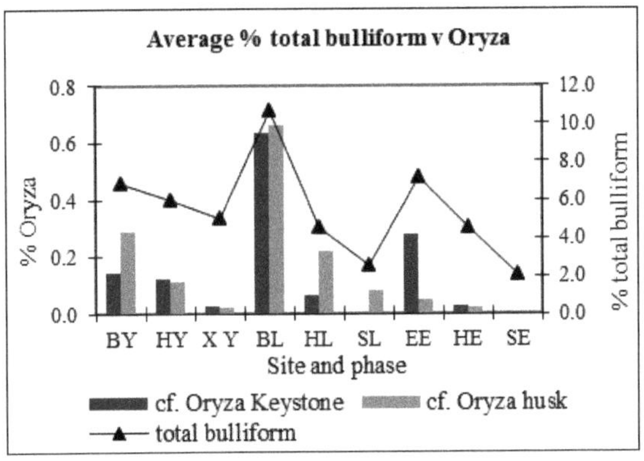

B= Baligang, H= Huizui, X=Xipo, S=survey, E= Erlitou
Y= Yangshao, L= Longshan, E= Erlitou site

Figure 9.76 Average % total bulliform v Oryza

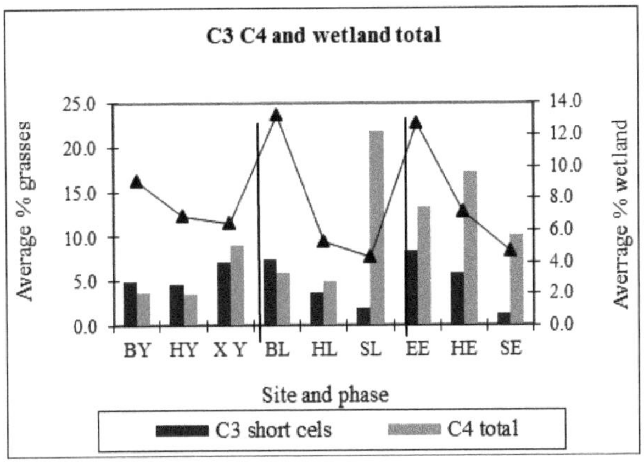

Figure 9.77 Average % C3, C4 and wetland totals

During the Erlitou C3 grasses increase dramatically at Huizui while wetland plants increase to the Yangshao levels. The survey samples see a fall in both C3 and C4 grasses and a slight increase in wetland plants. The Erlitou site produced high levels of wetland plants and C3 and C4 grasses.

9.8 COMPARATIVE DATASET

In Figure 9.81A the Chinese sites are represented on the right, India on the left. The separation between the two is clear. When first tested the main cause of the initial discrepancy seemed to be the presence of dicotyledons in many of the Chinese samples and very few in the Indian. Because of possible differences in the way the samples were counted the dicotyledons were removed from both datasets. The Indian and Chinese samples are still separate although the majority of samples have high proportions of grass leaf (Figure 9.78B). The Chinese samples are on the right of the chart. The major differences are seen in the four quadrants. The millet crops can be seen in the Chinese samples but are absent from the Indian. Chinese samples contain rice with water loving species, whereas the Indian samples in the top left quadrant have high levels of rice but mostly accompanied by grass husk. This might suggest dry versus wet farming and is something that could be explore further in the future. The lower left quadrant shows the Indian samples made up of wild grass and palms showing a very different environment from that of Neolithic Henan.

The comparative datasets show a clear division between Neolithic China and India. The Chinese samples are circled. Within this division it is possible to see a possible difference between millet and wet rice farming in China and dry rice and no farming in India.

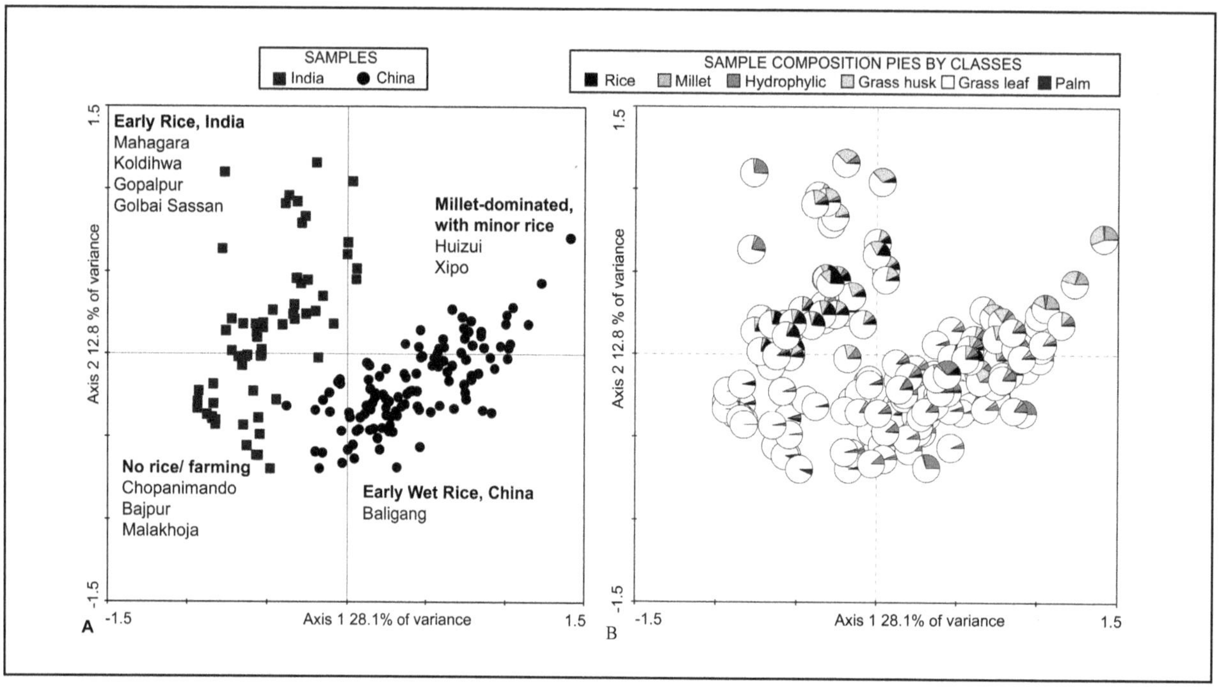

Figure 9.78A China and India Neolithic sites and samples, Figure 9.78B Chinese and Indian Neolithic sites without dicotyledons

CHAPTER 10

INTERPRETATIONS AND FINAL THOUGHTS

10 .1 INTRODUCTION

This chapter will bring together issues raised in the introduction with interpretation of the phytolith data. Variation in crop choices, both temporally and spatially, will be discussed first, followed by possible evidence for agricultural intensification, extensification or expansion and how this may relate to social change. Whether phytolith data can be used to distinguish crop processing stages and what this can reveal about changes in labour scales and social organisation will be considered. Next, phytoliths from non-crop plants and whether it is possible to glean evidence for environmental change from human influenced archaeological contexts will be discussed and finally what the comparative dataset can add to the greater picture.

10. 2 CROP CHOICES

Crop choices made by the people living in north central China during the Yangshao, Longshan and Erlitou periods were based on a number of factors and there is evidence for change throughout the three cultural periods. Correspondence analysis of the samples from all sites and periods does not immediately show a clear distinction between cultural periods in relation to phytoliths from crop plants, suggesting the same suite of crops, *Panicum*, *Setaria* and *Oryza,* were used throughout Henan in all three periods, but in different proportions at different times and places. This may be related to elements of local environment and social organisation. However, when the cultural periods are examined individually variation in crop choices between the Yangshao and Longshan periods is clear. The Erlitou period is more complicated to unpick, in part because the samples from the Erlitou site are not from clearly domestic features and in part because there are fewer samples and unlike samples from the Yangshao and Longshan none from the southern part of the region (Baligang).

The results of this study show there were both temporal and spatial variations in crop choices in Henan during the Neolithic and Early Bronze Age. While during the Yangshao densities are generally low, the levels of the three main crops, *Setaria italica*, *Panicum miliaceum* and *Oryza,* are fairly close. *Setaria* is higher in the relative percentages, while *Oryza* predominates in the absolute densities. This difference is related to site and is discussed below. The proportions of all crops are much higher in the Longshan, suggesting higher input into the sites, pointing to increased agricultural production and possible changes in how crop processing was organised. Overall, *Setaria* is the dominant crop during this period. The Erlitou sees a fall in crop remains recovered. *Setaria* still dominates in the relative percentages per sample but in the absolute densities *Panicum* is the principal crop. *Oryza* is present at the larger sites, Erlitou and Huizui, but not in any of the small village sites sampled in the survey.

YANGSHAO CROP CHOICES

The lower densities and proportions of all cereal crop husks in the Yangshao samples may in part be a greater use of gathered wild foods than in later periods, as at Banpo (Chang, 1986:112), and also demonstrated in the Upper Ying Valley where wild fruit and soybeans (*Glycine*) were common during the Yangshao but absent from the early Longshan samples (Fuller and Zhang, 2007: figure3).

Three large sites with Yangshao contexts were sampled; Xipo, in the drier cooler north west of the province, Huizui, in the Yilou river valley in the north and centre, and Baligang in the warmer more humid south. One small village from the survey, Gong Jia Yao, was also sampled but produced no cereal crop remains. The results suggest the main staple crop choices were related to the sites' geographical positions. The three large sites with samples from the Yangshao period all produce different results. Xipo and Huizui have mixed economies but at Xipo the emphasis leans towards *Panicum*, while Huizui there is more *Setaria* and rice. Baligang seems completely dependent on rice agriculture at this time.

XIPO YANGSHAO CROP CHOICES

According to the phytolith data *Panicum miliaceum* is the most frequent crop at Xipo and usually in the highest proportions. As *Panicum* is a crop that prefers dry conditions this fits with Xipo's position in the cooler more arid northwest corner of Henan. *Setaria italica* and *Oryza* are both present suggesting a mixed economy, which may be a result of the slightly more marginal environment of Xipo compared to the other sites. C-4 millets (*Setaria* and *Panicum*) were certainly important at Xipo as can be seen in the isotopic evidence from the faunal remains from pig and dog; demonstrating that grain (or grain products, e.g. from human faeces) was a major component of animal fodder (Pechenkina *et al*, 2005: 1185). Xipo is a middle Neolithic large village ca 40ha, the second largest in the Zhudingyuan area of western Henan, the largest being Beiyangping ca 85ha. Although there is no evidence for an elite at this time, Xipo's size and the presence of large elaborate buildings segregated from ordinary houses suggest Xipo may be a central settlement with public buildings (Ma, 2005: 15, 46). There is evidence for feasting at Xipo (Ma, 2005), an important means of enhancing and cementing social relations. The isotope evidence has demonstrated the importance of millets as food for pigs, an important element of feasting (Pechenkina *et al*, 2005:1185,). *Oryza* has been used for fermented drinks for funerals and feasts during the Yangshao (McGovern *et al* 2004:249). This may be one of the reasons why *Oryza* may have had a special status. It is found at Xipo despite the local environment being better suited to dry farmed crops. The presence of *Oryza* so far from its natural habitat implies that there were strong cultural motivations for the adoption of this crop, but it is also the case that environmental conditions in the alluvial valleys in this period made irrigated rice cultivation feasible (Rosen 2008).

HUIZUI YANGSHAO CROP CHOICES

Huizui, (Liu *et al* 2007:93), further east along the Yellow River Valley, also has a mixed economy in the Yangshao, however *Setaria italica* predominates here, followed by relatively high levels of *Oryza*. *Panicum miliaceum* is fairly frequent but in low proportions. This reflects the site's easterly position in Henan, which would have been expected to be more humid than Xipo so also more suitable for farming *Setaria italica*. The predominance of *Setaria* during the Yangshao at Huizui when compared to Xipo suggests a local environment well suited to *Setaria* cultivation.

BALIGANG YANGSHAO CROP CHOICES

The southernmost site, Baligang, also follows what might be expected from its geographical position in southern Henan within the Yangtze catchment. *Oryza* is by far the most common cereal crop, there is only one sample containing *Panicum*, which has been interpreted at a possible *Panicoid* crop weed, (see results) and no *Setaria*.

The Yangshao samples tend to demonstrate choice of main crop according to what is suitable for the local environment and growing conditions.

LONGSHAN CROP CHOICES

The crop choices in the Longshan seem to primarily reflect the local environment, for example rice at Baligang, but the agricultural economies at both Huizui and Baligang appear more mixed than in the Yangshao period. The uptake of an extra crop or increasing quantities of a secondary crop, such as *Panicum*, may be a response to escalating populations requiring more food and insurance or buffering against crop failure or shortages. Geoarchaeological evidence points to local environmental degradation due in part to climate change at Huizui during the Late Yangshao and early Longshan (Rosen, 2008:305), which could have provided incentive for further developing a mixed economy. Another possibility is over exploitation of the land and natural surrounding environment around the sites as suggested by Ma (2005: 26). There is a drop in remains from wild foods exploited in the Ying Valley from the Yangshao to Longshan (Fuller and Zhang, 2007: fig 3). A fall in available wild foods would also provide motivation for cultivating a broader range of crops and extending into more marginal areas.

HUIZUI LONGSHAN CROP CHOICES

Huizui is a regional centre during the Longshan period (Lee *et al*, 2007:93). At Huizui, the crop pattern is similar to the Yangshao but there is an increase in proportions of all phytoliths from crop plants. The emphasis in the Yellow River Valley is on millets, in particular *Setaria* although at Huizui there is a mixed agricultural economy and rice is also cultivated. The levels of millet crop husk increase. *Setaria* is clearly the most important crop; *Oryza* and *Panicum* are also common. The millet weeds increase in proportion to the millet crops. *Oryza* husk occurs more frequently, in contrast to the *Oryza* leaves, which appear in fewer samples. None match the level of the densest sample from the Yangshao. This does not correspond with the expected drop in *Oryza* due to the reduction of available land for rice cultivation. Having developed paddy technology and given the labour organisation and energy invested to construct and maintain paddy fields, the people at Huizui may have been reluctant to give up rice farming and continued cultivating *Oryza* despite difficulties created by a changing local environment.

BALIGANG LONGSHAN CROP CHOICES

The results from the Longshan show a similar general increase in proportions of crop husks can be seen at Baligang in all the samples. There is a corresponding increase in *Oryza* leaves pointing to continued local cultivation. There is a fall in some of the crop weeds, namely Cyperaceae husk, which may be connected to the agricultural practices discussed below. *Oryza* remains the staple crop but low levels of *Setaria* and *Panicum* are now both present, suggesting a more mixed economy than previously. There is also a suggestion of expansion into more marginal areas evidenced by the presence of *Panicum*. The uptake of *Setaria* could be due to an increasingly cooler and drier environment, making it a more suitable crop than previously. It could also be due to an increased population and heavier demands on food production. Increased inputs suggest more focussed farming practices and the introduction of diversity of crop choices points to risk insurance.

SURVEY LONGSHAN CROP CHOICES

The survey samples were gathered from small sites peripheral to Erlitou in the same region as Huizui, the Yilou River Valley. All are from the Late Longshan period. Apart from one, possibly anomalous, sample containing *Oryza* husk, the only crop husks in the survey samples are millets, predominantly *Setaria,* all in proportions of less than one percent. The only site to produce *Oryza* is Peichun B, which does not have any *Setaria* but is one of only two Longshan period sites that yield *Panicum*. The picture the results present is not unlike Huizui but at lower levels. These peripheral sites may not have had the technology or available labour to produce *Oryza*, suggesting *Oryza* may have been considered a special or elite crop. Another possibility suggested by the presence of phytoliths from *Oryza* crop weeds, but very scarce evidence of *Oryza* itself, is that rice was being grown around some of these sites for consumption elsewhere. Peichun B and NE Gaoya are below 150m but the other sites, Mazhai and NE Zhaiwan are above 200m so probably too marginal for rice. However, there is little evidence to suggest this. The villages are small so there may not have been a sufficiently large labour force for rice farming. If the same change in hydrology and alluviation, making less land accessible for rice agriculture from the Late Yangshao/Early Longshan, applies at these sites as in the Liujian valley at Huizui, rice cultivation may not have been worth the investment of time and labour. These sites could represent the development of some smaller specialised sites in more marginal areas during this and subsequent periods (Liu, 2004:245).

ERLITOU CROP CHOICES

The Erlitou period sees a further change in emphasis on particular crops in the Yellow River Valley. There is a sharp fall in phytoliths from all crops, particularly *Setaria* and *Oryza*, while *Panicum* increases at the survey sites. All crop husk proportions from this period are low, less than one percent per sample, the majority less than half a percent pointing to lower agricultural input. The most striking change is the general fall in *Setaria* levels at Huizui and in the survey samples. The increase in proportions of *Panicum* may suggest expansion of farming into higher, more marginal land. It is possible there was less arable land available to support a burgeoning population and increased reliance on *Panicum* could, in part, be a response to the continuing cooling and drying of the climate.

HUIZUI ERLITOU CROP CHOICES

By this time Huizui had become a secondary centre to Erlitou and specialised in stone tool production. The dataset from Erlitou period Huizui is small so may not give full picture. There is an overall fall in the levels of all crop husks, which may be a result of a change of focus from subsistence farming to manufacture for trade. Grain could have been traded in from outside as well as produced locally. However, agricultural activity continued at Huizui as evidenced by the phytoliths, which are predominantly *Panicum*, with some *Setaria* and *Oryza* husk and leaves. The charred seed remains show higher levels of *Setaria italica* than *Panicum miliaceum* as well as *Oryza* and also the introduction of *Triticum* (wheat) (Lee *et al*, 2009: table 1, Lee and Bestel, 2007: table 2).

ERLITOU SITE CROP CHOICES

The Erlitou site samples contain *Oryza* husk and one sample with *Panicum* but no *Setaria*. These few samples from close to the palace walls cannot be representative of a city of up to 100,000 people.

SURVEY ERLITOU CROP CHOICES

The samples from the villages peripheral to Erlitou, suggest a change from *Setaria* as a main crop to *Panicum*. There is no *Oryza* in the Erlitou period samples from these sites. As in the Longshan there is a possibility the survey villages could be growing *Setaria* and or *Oryza* but exporting it while keeping *Panicum* for local consumption. Cyperaceae phytoliths were found in high levels and scooped bilobes were also present, both indicators of rice crop weeds.

10.3 TESTING EXISTING HYPOTHESIS

While the phytolith data in this project does not show changes in settlement patterns, it can show changes in crop choices at different sites and cultural periods that may be related to changes the local environment and social organisation. Ma's proposition is that changes in settlement patterns during the Yangshao are linked to changes in natural resources, a changed social landscape of larger more integrated settlements developed in response to burgeoning populations, and the effect of increasing land clearance on a local environment. Such shifts may be reflected by the phytolith data not only for changes in the Yangshao but also for corresponding developments between the Yangshao and Longshan and Longshan and Erlitou.

Overall, higher proportions and densities of phytoliths from *Oryza* leaves and culms can be seen in Longshan samples compared to those from the Yangshao period. At Baligang *Oryza* was already the main crop and at Huizui, although *Setaria* is the main crop, there are increased levels of *Oryza*. There is a decline in phytoliths from *Setaria* and *Oryza* crop husks from the Longshan to the Erlitou period. The fall in *Oryza* may be due in part to the lack of samples from the south of the region from this period, but the changing climate and local environment also has an effect. Although *Panicum* husk decreases slightly in percentage per sample, in comparison to *Setaria* and *Oryza* there is an increase. There is a clear increase in *Panicum* husk density during the Erlitou period. This could suggest expansion into more marginal land either because of increased demand due to demographic expansion or a change in local environment. Both seem possible. The climate at this time had cooled and dried making *Setaria* a less suitable staple. The land previously used for *Oryza* at Huizui was no longer available due to the down cutting of the river. The arable land available for *Setaria* may have been degraded or the demand for *Oryza* as a special or high status crop may have meant some of the land previously used for *Setaria* was being used to cultivate rice.

10.4 EXPANSION, INTENSIFICATION, EXTENSIFICATION

RICE

The remains of *Oryza*, in particular, may provide insight into agricultural expansion, intensification and extensification. *Oryza* is not indigenous to the Yellow River valley and requires very different cultivation methods from the established millet crops in this region. The development and improvement of the technology needed to grow wet rice needs social organisation and investment of organised labour for the initial construction and maintenance of rice paddies, paddies, terraces, dams and irrigation systems (Allen and Ballard 2001:158, Brookfield, 2001:184, Leach, 1999:315)

Oryza was present at all the Yangshao sites investigated, except Gong Jia Yao. While it was the main crop at Baligang, both Xipo and Huizui had mixed economies and millets were more common than rice. Neither of these sites is within the plausible natural range for wild rice, even considering the warm humid environment during the Mid Holocene Climatic Optimum, so what could be the motivation for introducing a new crop that needs a constant water supply and can be labour intensive? The introduction and expansion of a new crop, such as *Oryza*, may provide insight into social responses to changing environmental conditions. Rosen, (2008:305) suggests rice production may have been made easier by the expansion of alluvial floodplains in the Yilou valley during the middle Holocene (from the Early Yangshao) as a result of the changing climate. This could certainly have been a stimulus to begin rice farming. But, environmental conditions had been stable and suitable for paddy farming for a long time before *Oryza* was adopted (Rosen 2007, 45) so the introduction of *Oryza* as a crop at this time could point to social motivation, possibly demographic pressure, or developing social integration (Rosen 2007, 46), or alternatively that *Oryza* was only a relatively new domesticate, with recent arguments from the Lower Yangtze placing full morphological domestication after 4500 BC (Fuller *et al.* 2009), with intensification and expansion around 4000 BC (Fuller and Qin, 2007), i.e. during the latter part of the Early Yangshao period. It is not clear whether paddy farming was necessarily the first, nor the only, method used for rice farming in Henan. Both dry and wet farming are a possibility. Evidence from the Yangtze at this time points to primitive paddy fields (Fuller and Qin, 2009). However, later the site of Liangchengzhen in Shandong (from 2500 BC) may have had dry rice cultivation (cf. Crawford *et al.* 2005). Also, some scholars argue that there was limited dry rice cultivation in Late Jomon Japan, introduced from China around 2000 BC (e.g. D'Andrea 1999; Toyama 2002: Table 1; Fujio 2004). Using phytolith data for identification of different farming regimes is area for future research. Paddy farming is labour intensive and requires the social organisation and ability to mobilize a labour force to build and maintain. There would also be differences in the length of cropping season, requiring agricultural labour for longer. Chinese historical sources suggest that early ripening (short growth season rice) was largely introduced to China in the past 1000 years (Ho 1956). Most rice varieties and those of prehistory are expected to have had much longer growing seasons (~ 6 months) compared to 3 or 4 months for the millets. All of these would have set seed after summer (e.g. September/ October).

Oryza was being cultivated in the Yellow River Valley by the mid Yangshao Period, and may have included dry-cropped varieties. This can be seen by the results from Xipo (Middle Yangshao), which show both husks and leaves as well as phytoliths from crop weeds in the form of scooped bilobes. However, there is no evidence of paddy farming from Xipo. The crop weeds follow a different pattern from that at Yangshao Baligang. There are no diatoms in the samples containing *Oryza* at Xipo and few Cyperaceae suggesting the *Oryza* may have been dry farmed at this time, or fields were dry for a significant portion of their life cycle. It is doubtful that rainfall in itself would have been sufficient, but rice might have been sown on the alluvial plains after receding floods. One possibility is wet farming in the flood waters of the local rivers, similar to décrue farming in West Africa as described by Harlan and Pasquereau (1969). The seeds are planted before the water rises and submerges the fields and harvested as the waters recede. Another is dry farming, there are few sponge spicules and fewer diatoms in the Xipo samples especially compared to Baligang and Huizui. Near Huizui the palaeosol of a paddy field has been identified based on sedimentary characters and the presence of bulliforms (A. Rosen, pers. comm.).

At Huizui, further east, the picture is different. Here, diatoms, Cyperaceae and phytoliths from other rice crop weeds, for example scooped bilobes, appear frequently with both *Oryza* leaf and husk suggesting wet or paddy farming, although the samples dated to the Late Yangshao contain no *Oryza* husks and only one produced *Oryza* leaf. During the Longshan *Oryza* ubiquity increases in the Huizui samples suggesting it has gained in importance as a crop. The climate was becoming cooler and more arid so less suitable for *Oryza* cultivation so new and improved farming methods may have been developed (in order to protect the investment made in the Yangshao), such as improved water control created by the construction of bunds, and dams.

At the northern sites it would seem that rice is a supplementary crop during the Yangshao period, while at Baligang it is the main crop. However, the high levels of cone shaped phytoliths from Cyperaceae husks suggest the rice at Baligang is heavily weed infested so the rice fields do not seem particularly well managed at this time. Macro-remains evidence, at least from Longshan samples, include abundant nutlets of *Fimbristylis miliacea* (D. Q. Fuller, pers. comm.), which suggests low standing water levels early in the growth cycle, since better managed (slightly deeper) water should prevent establishment of this weed (Reissig *et al* 1986: 292).

The survey has one Late Longshan site, Peichun B, that produced a few *Oryza* phytoliths, but there is no evidence for any rice at all in the rest of the survey samples from any period.

The sharp increase in *Oryza* husks at both Huizui and Baligang in the Longshan highlights the growing importance of this crop. This increase in *Oryza* might be related to the demands of an increasing population and developing social competition. The advent of chiefdom level polities and with the ability to mobilise labour for large construction projects such as building rammed earth town walls suggests the organisation required to build and maintain paddy fields was in place by the Longshan. The climate fluctuated then dried and cooled throughout the Longshan period. Along with an increasing population, this may have provided motivation to improve paddy technology, as evidenced at Baligang by the dramatically falling levels of Cyperaceae husk. Improved water management would have drowned rice weeds, such as many Cyperaceae. The introduction of the sowing method of transplanting rice seedlings into paddies would also lead to fewer crop weeds, although there is no evidence this had occurred in Neolithic or Bronze Age times. There is a corresponding increase in phytoliths from rice leaves in the Longshan, suggesting a change in harvesting practices from cutting the top only to harvesting lower down the plant to include the leaves. Crop leaves can be a valuable processing by-product as fodder; although it is not clear how early in the Longshan period sheep and cattle were adopted in China (Flad *et al.* 2007, Yuan and Blench 2008). Further dating evidence is needed to assess whether the increase in rice correlates with the advent of domesticate ruminants in what had been only pig-keeping.

The expansion and improvement of paddy farming suggests both intensification and extensification were taking place at the end of the Yangshao and during the Longshan period at Baligang. In addition to changes in the proportions of crop weed husks this notion is supported by the increased input of distinctive phytoliths from *Oryza*.

At Huizui the change is not as emphatic but the pattern is similar. The introduction of *Oryza*, which needs a consistent water supply, into the Yellow River Valley region may have been facilitated by the creation of steadily alluviating floodplains during the early Neolithic such as those Rosen (2008:305) describes at Huizui. At around 2400-2000 cal. BC, during the Longshan period, climatic conditions became markedly cooler and drier. At Huizui increased stream flow at this time cut through the sediments deposited during the Yangshao and early Longshan periods drastically reducing the land available for irrigation and paddy farming (Feng *et al*, 2004; Rosen 2007, 46, Shi *et al.* 1993).

There are discrepancies in the phytolith remains related to rice cultivation from Baligang and Huizui in the Yangshao and Longshan suggesting divergences in the agricultural practices at each site. This could be a reflection of the relative importance of rice as the main crop at Baligang and a secondary crop at Huizui. The difference in the relative proportions of Cyperaceae in samples containing *Oryza* husk at the two sites over the Yangshao and Longshan could also be related to the cooler drier local environment at Huizui. There were far fewer diatoms and sponge spicules in the Huizui samples than those from Baligang.

There is a fall in *Oryza* input at Huizui in the Erlitou period samples. At the Erlitou site *Oryza* seems comparatively high; particularly *Oryza* keystones from leaves and levels of phytoliths from *Oryza* weeds are high at Erlitou, increasing from phases II to IV. Rice weeds are also high in the survey samples. One possibility is that rice was being imported into Erlitou in sheaves and processed there, or was being specially grown in quantity near this site. Evidence of macro-remains also includes rice grains from sites in the Yilou Valley (Lee, *et al*, 2007: table 1, Chen *et al*, 2003).

MILLETS

The evidence of millets should be considered in relation to the expansion of agriculture, in particular for *Panicum* agriculture. Millet crops were negligible at Baligang during the Late Yangshao. *Panicum* was the predominant crop at Xipo with *Setaria italica* and *Oryza* as secondary crops. At Huizui the economy was also mixed but *Setaria* was the main crop, fitting with Huizui's more north-easterly position. During the Longshan increased *Setaria* was also found at Baligang. The relationship of *Setaria* to rice at Baligang may be similar to the relationship of *Panicum* to *Setaria* at Huizui. The secondary crop played a role in buffering risk and expanding agriculture onto soils that were less conducive to the primary crop. In northern Henan, increases in rice and *Panicum* in the Longshan period can be suggested to represent a combination of extensification, into more marginal land with *Panicum,* with expansion and intensification through rice in selected lands. This is seen in the Ying Valley evidence from macro-remains (Fuller and Zhang 2007), and again at Huizui, where Longshan levels of *Panicum* and *Oryza* increase. The survey samples produced evidence for increasing *Oryza* agriculture but *Panicum* is still secondary to *Setaria* at all sites until the Erlitou period when it comes to the fore at Huizui and the survey sites.

This is suggested to be a response to demographic pressure. It may be possible that demand for surplus was strong enough for *Panicum* expansion into marginal areas and incentive for improvements to *Oryza* agricultural technology. The possibility that a similar trend at Baligang with the intensification of rice but the expansion of *Setaria* took place for similar regions requires further confirmation.

The Erlitou period sees a sharp decrease in all crop husks from all sites. At Huizui there are few crop husk phytoliths, while the majority are *Panicum* there is little phytolith evidence of *Setaria* or *Oryza* cultivation. This may be a result of the decrease in arable land for rice agriculture and the cooler, drier environment that set in from ca. 2200-2000 BC. Huizui was a specialist stone tool production site and there is a possibility that the people were importing dehusked grain from outside in exchange for stone tools. However, the macro results from an earlier study at Huizui (Lee and Bestel, 2007:52) present a different picture to this project's phytolith evidence. The macros produce far higher levels of *Setaria italica* (511 *Setaria* grains to 70 *Panicum* and 11 *Oryza*). The evidence for a shift towards *Panicum* during the Erlitou period at Huizui where *Setaria* seems to all but disappear in the phytolith record, contrasts with the macro-remains (Lee and Bestel, 2007) which may suggest that different processing practices were applied to these two millets. One possible reason for the low levels of *Setaria* husk is that they may have been dehusked elsewhere. As communities became larger, and agricultural land increased, including expansion to marginal lands for millet, it may be farmers themselves became more specialized and those focusing on the different millets developed differing practices. In addition, as the earlier paddy fields had been left high and dry by the Liujian River cutting down into the floodplain (Rosen, 2007:354), some of the land previously used for *Setaria italica* may have been transferred to use for rice agriculture although exactly where this might have happened is unclear at the moment. Samples with charred grains processed for phytoliths from the Longshan storage pit at Huizui produced few husk phytoliths. The macros also show free-threshing *Triticum* (wheat), which is not evident in any of the phytolith samples, although is present in very small quantities. In addition, as this wheat is free-threshing chaff would be disposed of with straw after initial threshing, which is likely to take place off site. Apart from SE Zhaiwan, the survey sites yield no crop husks at all from the Erlitou period. *Panicum* and *Setaria* are both present at Zhaiwan with *Panicum* in higher proportions.

EXPANSION, INTENSIFICATION, EXTENSIFICATION

Although the underlying causes of intensification are under debate. Boserup's (1965) model for agricultural intensification as a unilineal path from swiddening to multicropping seems too simplistic for what was happening in north central China at this time. Firstly, there is no direct evidence of swiddening and while demographic expansion has been established in the region through regional survey of settlement patterns (Liu *et al*, 2004), the cereal crops discussed were all being cultivated at the sites in northern Henan by the mid Neolithic and at Baligang by the Late Neolithic. There may have been increasing demographic pressure on limited resources leading to innovations, such as the development of paddy farming, the adoption of new crops, for instance millets at Baligang, technological advances seen by improved water management and a higher labour input to construct and maintain new field systems and paddies and then harvest, process and distribute increased crops. All these are criteria for intensification described by Allen, (2001:204), Brookfield and Hart, (1977:332) and Netting, (1974:36).

The multiple pathways suggested by Kirch (1994 in Leach, 1999) seem more likely. The components of Kirsch's diverse productive strategies: intensification, specialisation, diversification and expansion at certain social and spatial scales can all be found at Baligang in the Longshan. Improvement of agricultural techniques, to increase rice production, demonstrates intensification. The phytolith evidence of changes in Cyperaceae husk (levels) suggests developments in the way *Oryza* was being cultivated at Baligang pointing to improved paddy technology where weeds were being drowned out by a more consistent water control and waterlogging. Improved water management suggests higher labour input. Specialisation in paddy farming is shown by increased yields. Nevertheless, a diversified and risk-buffering strategy was maintained through growing *Setaria* and *Panicum*, probably on more marginal land. Improved rice agriculture at Baligang in the Longshan also probably involves the expansion of land used for cultivation. Landesque in the investment of labour in physical evidence of intensification in long term improvements to the land, such as paddies fields and dams and irrigation systems can also be perceived at both Baligang and to a lesser extent at Huizui (Allen and Ballard 2001:158, Brookfield, 2001:184, Leach, 1999:315), although preserved paddy fields have not been found. Agricultural intensification has been related to socio-political development (Leach, 2003:31, Brookfield, 1972:38). Socially generated demands for produce (Bender, 1985, Leach 1999, Gilman, 1981, Sand 1999, Brookfield, 1972:38) might be seen in the continuation of rice farming at Huizui despite the local landscape becoming less suitable, as well as continuity of rice production into the Erlitou period, with rice grains from Erlitou period site contexts (Lee *et al*, 2007: Table 1). Evidence that *Oryza* was used in fermented beverages from the Yangshao and into the Longshan, (McGovern *et al*. 2004:) and the importance of ritual drinking vessels (Underhill:2004, Barnes, 1993) suggest that at the very least rice had an important role in feasting and mortuary ritual and could be regarded as a 'special' crop. The escalating difficulty in growing rice at Huizui as the arable land suitable for rice agriculture decreased is likely to have put pressure on market forces, encouraging intensification of rice production in other areas where it remained possible (Netting, 1993).

The social ability to mobilise labour suggests central political and social organisation, highlighting the question of whether social coercion or population pressure are the driving forces behind agricultural intensification or whether it is a result of the production of goods for social purposes, for example 'special' or elite crops such as rice. In the case of north central China the main motivation would seem to be increased demand from the pressure of an expanding population, in face of a drying climate, along with a social response to a cooling and drying local environment, although an element of hankering for the socially desirable is also likely to play a part.

10.5 NON-CROP PHYTOLITHS: INPUTS FROM CROP-PROCESSING AND SOCIAL ORGANIZATION

The results suggest that the bulk of the non-crop phytoliths in the ash middens and a large proportion of the phytoliths from the laminated pit fills are from crop processing residues. The majority of non diagnostic phytoliths from monocotyledons and those from the non crop plants represent either weed seeds gathered with the harvest or the leaves and stems, the straw, from the crops and weeds that have been harvested with the cereal crop and removed during crop processing as suggested by M. Jones (1985) G. Jones (1987) and Fuller and Stevens (2009) for macro remains.

If, as Fuller and Zhang (2007) suggest, the ash middens contain the residues of routine daily activities, spatial and temporal differences can be identified.

The dividing line between bulk and daily processing may suggest varying levels of organisation and labour mobilisation during harvest period when there is likely to be high demand for labour (Stevens, 2003: 72). Understanding whether such organisation is small scale, kinship based family units, or large scale even centrally organised can provide insight into changes in social organisation. Evidence of varying scales of crop processing products and residues can suggest how bulk and daily processing were organised. Overall comparison of leaves/stems, weed husks and crop husks from all sites and phases demonstrates higher levels of the early stages of processing, threshing waste, than any other stage.

CROP AND NON-CROP PHYTOLITHS: BALIGANG

Baligang has higher rice levels than the sites in the north and differentiation between the Yangshao and Longshan samples is clear. During the Yangshao the majority of samples contain *Oryza* and weed husks but no *Oryza* leaf, suggesting dehusking was a routine activity, taking place in a domestic context possibly as the rice was used on a daily basis. This points to small-scale family level organisation. The crops threshed and winnowed away from the domestic context, but stored in the husk and processed as needed. In contrast the Longshan samples are predominantly mixed, leaf and husk. The levels of weed straw are correspondingly high, indicating a change in crop processing practices. The Longshan samples suggest bulk-processing waste from all stages discarded in the same place. This could imply all processing stages were taking place at the same time pointing to centralised rather than household organisation. More centralised labour organisation during the Longshan corresponds with intensification of rice agriculture. Organised labour would be needed for construction and maintenance of paddies. Greater rice production could require large groups of organised labour for the harvest of increased yields. The results from Baligang suggest a change from household to centralised organisation from the Yangshao to the Longshan. These results are reinforced by the remains of Late Yangshao longhouses at Baligang, which suggest corporate organisation during the Yangshao (Underhill, 2002:144). The changes in how crop processing was organised from the Yangshao to the Longshan point to changes in social organisation.

CROP AND NON-CROP PHYTOLITHS: HUIZUI

At Huizui, all periods have high levels of threshing waste from the millet crops. *Oryza* husk is found alongside the millet crops more frequently during the Longshan at Huizui, suggesting waste from dehusking is discarded in the same place whatever the crop. This may be a change in how processing and storage are organised. A difference can be seen in the Late Longshan where *Panicum* levels fall and *Oryza* husk and leaf/culm appear in the same samples suggesting a change in where and when *Oryza* is being processed. There is also an increase in weed and *Oryza* husks, possible winnowing by products. Waste from threshing and dehusking in the same samples could point to full processing at the same time. If there is less rice being grown or if it is being imported into the site in sheaves it may well have become more valuable and only processed as it was used.

Many samples contain only millet crop husks suggesting *Panicum* and *Setaria* were being processed either together and/or at the same time and their crop processing residues disposed of in the same place. There were slightly higher levels of millet weed husk in the Longshan than the Yangshao, possibly pointing to winnowing waste but it could also be a reflection of the greater inputs of millet crops overall during the Longshan period.

STORAGE

Storage is in evidence at Huizui. The remains of a large bell shaped storage, pit dated to the Longshan, produced a 4cm layer of whole *Setaria* grains. However, the sample from this level yielded scant husk phytoliths suggesting the grain was stored dehusked. This pit also contained a charcoal level rich in remains from *Panicum* and *Oryza* as well as *Setaria* husk. This demonstrated not all *Setaria* was stored dehusked, and suggests that bulk dehusking may have sometimes been used as fuel and thus found mixed with wood charcoal.

Fully processed grain is very difficult to see with phytoliths as they are produced in the husk not the caryopsis. This lack of visibility may account for the discrepancy between the phytoliths and the charred seed evidence for *Setaria* at Huizui during the Erlitou period. While *Setaria* has become rare in the phytolith data, the macro-botanical evidence suggests it continues as the main staple at Huizui during the Erlitou period. Dehusking before storage points to mobilisation of large organised groups for processing. Harvest can be a time of labour bottlenecks, fully processed grain in a large storage pit suggests large organised groups mobilised to process and store the bulk of the crop. Smaller groups, such as households, have to deal with seasonal demands by partly processing before storage and completing the task on a day-to-day basis as needed (Fuller and Stevens, 2009).

The crop-to-crop weed ratios of the macro results for the Yangshao in the Ying Valley suggest a consistent input of waste from primary winnowing pointing to uniform organisation of crop processing (Fuller & Zhang, 2007). The phytolith results of millet processing from the Yellow River Valley for the same period suggest consistent waste from threshing, an earlier stage also pointing to uniform organisation.

While this is true of the Yangshao settlements, Xipo and Huizui, the Longshan settlements show greater diversity between settlements. The villages from survey have a different pattern from Huizui. The survey samples produce mostly threshing waste in a similar manner to the Yangshao samples. Huizui has higher levels of crop and weed husks suggesting later stages of processing, winnowing, storage and dehusking. This may demonstrate a continuation of uniformity, demonstrating a large-scale communal processing pattern at the small survey settlements. In contrast at Huizui the different stages of processing point to some small household scales of processing for daily use, which Fuller and Stevens (2009) term a "focused" scale of labour mobilization. The large storage pit containing fully processed *Setaria* at Huizui also suggests diversity of crop processing within the site. This might point to changes in scales of social organisation from household or kinship groups to more centralised systems and also a contrast in differentiation between settlements to within settlements. Diversification between settlements can be interpreted as part of the increase in social complexity in Central China at this time.

10.6 PRODUCTION AND CONSUMPTION

YANGSHAO PRODUCTION AND CONSUMPTION

The phytolith evidence from three sites with Yangshao contexts, Xipo, Baligang and Huizui, suggests that during this period production and consumption were local. Both leaves and husks from *Oryza* are present at the three sites and husks from crop weeds also point to crop processing taking place on or near site. Geoarchaeological and phytolith evidence point to *Oryza* cultivation around Huizui during the Yangshao period (Rosen *et al*, in press). The emphasis on *Oryza* at Baligang also suggests local cultivation and processing. From the Xipo samples I studied the presence of keystone phytoliths from *Oryza* leaves, as well as crop weeds such as Cyperaceae suggest input from early rice crop-processing waste, pointing to local cultivation in wetter fields. As suggested above this might have been a décrue system, involving sowing into receding river floodwaters. The evidence for paddy-field cultivation of Huizui could represent either a later development of more intensive rice cultivation or a regional variation.

The mid Yangshao saw the beginning of tiered settlements and variation in settlement size. By the mid Yangshao small settlements were being developed on higher and more remote ground, although most sites were still clustering on arable land close to rivers. In this landscape context small-scale décrue rice cultivation would have been easy. Some of these small settlements are thought to be specialist sites with agriculture taking place around the larger villages on more fertile ground. The lone Yangshao sample from the survey at Gong Jia Yao may be one of these specialist settlements. It produced no crop remains and is sited between 150-200m above sea level.

Expansion of rice cultivation outside its natural range during the Yangshao period indicates the role of interregional cultural contact. The earliest hard evidence for rice in the Yangshao region comes from Early Yangshao grains from Nanjiaokou in northwest Henan, with a direct AMS date of 3900-3800 BC (Ling Qin, personal communication), while by Middle Yangshao times (after 3800 BC) rice finds are more frequent, for example at Xipo. An increased dependence on domestic animals, such as the pig in the Yangshao period (see, e.g. Yuan and Flad 1992; Barton *et al.* 2009), may also provide an additional use for grain surplus in pig feed. Pigs had an important role in feasting and burials so ownership may have afforded status. *Setaria* may have been more valuable on the hoof than as grain. A need for surplus may have motivated land clearance and an insurance crop such as *Panicum*, especially if related to expansion to more marginal drier land. Evidence of non-agricultural economic activities at this time, such as kilns and two styles of pottery, suggest the beginnings of specialisation.

LONGSHAN PRODUCTION AND CONSUMPTION

The Longshan period provides evidence of increases in population density and settlement size. Population growth meant aggregation to larger settlements and this was often accompanied by warfare (Barnes, 1999, Liu 2004) and the construction of large town walls, demonstrating the ability to mobilise large labour forces for communal projects (Liu 2004). Growing demand for food may be understood in the archaeobotanical record by increased destinies of crops and changes in how crop related activities are organised, as well as expansion and intensification of existing agricultural economies.

The introduction of sheep and cattle shows a heavier dependence on domestic animals, which in turn might make demands on the agricultural economy for fodder and also expansion for grazing. The value of crop processing residues, such as leaves and stems would be augmented by such demands. The shift towards more focused, small-scale processing and the more routine evidence for threshing residues on sites could be seen as contributing to the foddering of cattle and sheep.

Elite and non-elite houses within settlements and social grouping within cemeteries provide evidence of more complex social stratification. Residues left by fermented beverages (McGovern *et al* 2004) demonstrate they were made from rice rather than millets as in the Yangshao, providing further evidence of rice as a special or elite crop and demonstrating motivation for continued rice cultivation. Part time ceramic specialisation (as described by Underhill, 2004, Liu, 2004) may have meant available labour during harvest time. Evidence of Qujialing ceramics, for example, in western Henan during the Yangshao points to could also suggest regional interaction. There is a suggestion of large-scale immigration into Henan from southeast regions at this time (Li *et al*, 2005:85:89).

ERLITOU PRODUCTION AND CONSUMPTION

The Erlitou period saw rapid population growth in the Yellow River Valley, accompanied by increased specialisation and the monopoly of bronze prestige items by a ruling class (Liu and Chen 2003). A dense distribution of sites with a four-tier settlement hierarchy developed with palatial structures at Erlitou. Increased stratification saw social polarisation and political centralisation. Key central settlements were sited on fertile land with easy access to water, for example Erlitou on the Yilou river floodplain. The phytolith evidence, although extremely limited, offers a suggestion of possible production and consumption sites. Huizui is known to have been a site specialising in stone tool manufacture (Ford, 2001, 2004, Webb *et al*, 2007). However, as a large secondary centre in a fertile valley it is likely Huizui was producing at least a portion of its own food. The evidence for millets and *Oryza* suggests the bulk of crop remains are threshing waste, which suggest local produce. Only one survey settlement produced any crop phytoliths; however other sites do have *Oryza* and millet crop weeds so there is a possibility they were producing grain for the larger sites although again more evidence is needed.

Isotopic evidence demonstrates the diet consisted of mostly C4 plants (millets) but some traces C3 (rice/ other) were found at Erlitou (Zhang *et al*, 2007, Wu *et al* 2007, in Jing and Campbell, 2009:101). The very restricted phytolith evidence shows rice and millets were present and suggest some rice processing was taking place. With so few samples from such a limited area it is not possible to say much about what was happening at the Erlitou site regarding production or consumption.

10.7 NON-CROP PHYTOLITHS: EVIDENCE FOR ENVIRONMENTAL CHANGE?

Some morphotypes do not come from crop plants. Although most palaeoecological studies are made with off- site samples, for example (Osterreith *et al*, 2009, Lu *et al*, 2006, Lu *et al*, 2007, Alexandre *et al*, 1997), the non-crop phytoliths from the archaeological samples collected for this project can provide insight into the environment around the sites sampled. The issue here is to unpick the environmental from cultural influences. Using non-crop phytoliths removes the crops but inevitably the associated weed flora remains. Another area of cultural influence is the context the phytoliths were retrieved from. To gain a broad representation correspondence analysis was used first, followed by site and phase average percentages. The non-crop phytoliths from grass short cells were selected and categorised according to grass subfamily, Pooid, Panicoid, Chloridoid and Bambusoid, and then by photosynthetic pathway to reflect patterns of vegetation change. The initial correspondence analysis shows the grass subfamilies cluster in samples according to site rather than period, supporting the notion previously suggested that, as the cultural results demonstrate, the geographical position of individual sites has primary influence over the vegetation contained in the samples. This is not surprising; even if phytoliths from the crop weeds are influencing the content of the samples the crops need to be able to grow in the local environment, although this still could be related to the anthropic action.

When the cultural periods are examined individually clear distinctions can also be made between them, both cultural and environmental. The Longshan sites show greater diversity than the Yangshao. This may be a cultural influence. The agricultural economies were more mixed at all sites during Longshan so a corresponding increasing diversity in crop weeds should be expected. Differences according to context type can also be noted, in particular between ash middens and floor foundations. This is also culture-related variation.

The average Panicoid versus Pooid for each site in each cultural period suggests changes in the local environment that broadly parallel the environmental and climatic changes described in Chapter 3, as do the results from examining the patterns produced by species with C3 or C4 photosynthetic pathways. The climate gradually warmed and ameliorated to a peak during the Yangshao period at the Mid-Holocene Climatic Optimum, followed by a more arid period that cooled during the Longshan and subsequent climatic fluctuations in the Erlitou period. C3 grasses dominated during the Yangshao; however the high levels of C4 grasses at Xipo reflect its position in the driest region during the Climatic Optimum. A change to predominantly Panicoid C4 grasses at the northern sites during the Longshan could be connected to an increase in density of millet farming but can also indicate a change to more arid conditions. This continues and increases into the Erlitou period, reinforcing the evidence for drier environmental conditions.

Phytoliths from wetland plants can provide an indication of nearby damp habitats, but as rice was cultivated around most of these sites there will inevitably be input from water loving plants. Morphotypes from wetland grasses tend to follow the same pattern as the C3 grasses, but peak at Baligang in both Yangshao and Longshan periods. This seems connected to rice agriculture. If the bulk of the grasses are from crop-processing waste and rice is the main crop the majority of grasses in the samples are likely to be water-loving species. There is another peak at Erlitou which also corresponds with high rice levels so again is very probably cultural rather than environmental. The Yangshao and Longshan are easier to read than the Erlitou period. There are fewer samples from the Erlitou and it may be that by this time cultural and contextual influences are too strong to be able to unpick the environmental evidence.

The results can be related to proxy environmental data from other studies (see Chapter 3) to give picture of vegetation variation over time and in different parts of the region. Whether this is a genuine reflection of climate driven changes or of relative inputs of crop plants and their associated weed floras is not clear. This study demonstrates that some non-crop phytoliths from archaeological contexts can be used successfully to provide insight into the local environment and broad climate- driven changes. But, this information is a rough guide and is only useful in conjunction with firmer proxy data to support rather than propose. A palaeoecological study would be better served using off-site samples. Changes in inputs of non-crop phytoliths from archaeological samples are inevitably culturally influenced by changes in crops, weed flora, crop processing, as well as contextual influences, for example, how the plant found its way into the archaeological record.

10.8 COMPARATIVE DATASET

Comparison with the phytolith dataset from Neolithic India (of Harvey 2006) provided a broader context in which to situate the results of the Henan study. The results provided clear distinctions between the two datasets. It would seem from the correspondence analysis that the bulk of the Indian samples were being grown as a dry crop, in sharp contrast to the Chinese samples. Climate and water availability obviously play a part in the choice of farming methods. As a whole the Indian samples have higher proportions of *Oryza* husk per sample. This may be because the Chinese were also growing millets. *Oryza* appears to be the only crop in the Indian dataset. The evidence for wet farming and secondary crops from China might also be related to increased social complexity. In China there is evidence of risk management, buffer crops, mixed agricultural economies, and paddy farming suggesting extensification and intensification, from the Late Yangshao through the Erlitou periods (3000-1600 BC). The available evidence from India, mainly from Late Neolithic and Koldihwa (1800-1500 BC) and Chalcolithic (1500-1000 BC), lacks evidence for these, but suggests low-intensity dry cropping and routine crop processing of rice. With time, and more sites, these contrasts could be investigated further.

10.9 CONCLUSIONS AND FINAL REMARKS

The major aims of this project were to determine the levels to which phytolith data collected from varying archaeological contexts could be used to inform questions on agricultural economic diversity, past environmental change and processes that underlie the rise of social complexity in Late Neolithic China. There were a number of methodological questions, however, that needed to be addressed.

The first methodological issue was identification. Work has been done on the identification and classification of phytoliths produced in rice and millet husks (Pearsall, 1995, Zhao *et al*, 1998, Harvey, 2006, Lu, *et al*, 2009), but more was, and is still, needed. The production of a comprehensive reference collection and the importance of isolating identification criteria are issues that should be considered in any work involving phytoliths. References of phytoliths in leaves and culms and more importantly of associated crop weed husks are vital, especially the weedy grasses. This is an area that once further developed could provide real insight into the development of changing agricultural systems and from there a clearer picture of the relationship between developing agricultural economies and the social organisation required to produce staple and prestige food crops.

The next issue was how to deal with the quantities of data produced by such a large, rich dataset as the one generated from the samples collected for this project. Correspondence analysis proved a valuable analytical tool and a useful starting point, providing broad patterns in the data, which could then be refined to answer more precise questions. This method was particularly useful for determining differences in vegetation input into varying context types; thereby helping establish which contexts might provide remains appropriate for tackling specific questions.

As with most other forms of archaeological data, phytoliths work better when used with other types of proxy data, such as charred seeds. This is illustrated here by the results from the grain found in the storage pit at Huizui, which provided scarcely any husk phytoliths despite being taken from a thick layer of *Setaria italica* seeds. The phytolith results alone could have been misleading, just as the seeds alone do not provide all the answers. A separate layer from the same pit contained no visible seed remains but vast quantities of phytoliths from *Setaria* husks.

The results of this study demonstrate how phytolith data can be used to interpret archaeobotanical questions. Changes in crop choices between sites and throughout the cultural period were identifiable and provided information on agricultural economic diversity. Analysis of the data provided insight to levels of expansion, extensification and intensification of rice farming although this could be extended using more samples from each site and period. The millets are more complicated to interpret. The leaves are difficult to identify and there were not enough indefinable criteria to establish crop from weed leaves with *Panicum* and *Setaria*. This may also be an area where future work could provide better understanding.

Using the macro archaeobotanical crop processing model on this phytolith dataset, following Harvey and Fuller

(2005), suggested changes in how processing was organised over time at the different sites, from extended family/ semi-communal scale organization in the Yangshao to mostly more, focused, small-scale family organization in the more divided hierarchical society of the Longshan. This suggests that archaeobotanical evidence can provide insight into developing social complexity, as compared to other cultural traits, such as the building of large rammed earth city walls, or craft specialisation. Such specialization and public projects inevitably make demands on the labour of some or most households and will impact the scheduling of household agricultural activities, as will expansion of household production for surpluses aimed to support specialists and elites.

Another key aim of this study was to investigate whether it is possible to unpick the environmental from cultural signatures in a phytolith dataset collected from archaeological contexts. As with other questions phytolith data is better used alongside other proxy data, but nevertheless key environmental changes are apparent in the patterns generated by the non- crop phytoliths, but these patterns should only be used as a starting point.

Appendix 1 Holocene Sequence table

Date BP CAL	Climate	Environment	Cultural activities	Proxy evidence	Source
22 -18,000 (un cal) LGM	dry	Extremely variable		Isotopic	GRIP, Barton *et al* 2005
14-12 BP		Daihai Lake Sparse wood/grassland		Pollen	Li et al 2004
12900- 11600 (cal) Younger Dryas		Disappearance rice phytoliths			Li et al 2004, Lu *et al* 2002
12955-8855 Holocene	Climatic amelioration	Qinghai Lake, Tibetan plateau first maximum level		Pollen, stable isotopes, Palaeosols	Lister et al, 1991, An *et al* 2000
12955-11485 – 8855-7825	Warming and dry early phase	Xingjiang region		Geochemical indices, pollen, magnetic susceptibility	Feng *et al*, 2006
12750 –11485	Abrupt transition from dry to humid			Magnetic susceptibility of loess palaeosol sequence	An *et al* 2000,
12750-11485- 5775-4455					
11200-10800 BP	Warm and wet early phase	Tibetan Plateau		Geochemical indices, pollen, magnetic susceptibility	Feng *et al*, 2006
11-10,000 Holocene		Dahai Lake steppe		Pollen	Li et al 2004
10500-10 BP	Peak in precipitation	Seasonal contrast amplified northern frontal zone retreats		Insolation, solar radiation maximum	An *et al*, 2000
10180		Dahai Lake Sparse wood/grassland			Li et al 2004
		First tree pollen peak, Picea & Abies Rice phytoliths Yangzte		Pollen	Feng *et al* 2006, Lu *et al* 2002
10180-8855 Mid Holocene climatic optimum, Increased solar radiation	Warm, and wet in many areas			Insolation peak	Kutzbach & Gallimore, 1988, An *et al* 2000, Feng *et al* 2006
10250-7900	Dry and mild	Dahai Lake Herbs and shrubs mixed conifer/broad leaf		Pollen	Xiao et al 2004

Date BP CAL	Climate	Environment	Cultural activities	Proxy evidence	Source
9985		Dahai Lake Betula, Pinus Quercus, mixed conifer/ broadleaf forest		Pollen	Li et al 2004
9800-7800	Wet	End of Loess deposition, Well vegetated Dinxi Western Loess Plateau		Sratigraphic contrasts, wetland swamp/loess, Pollen, molluscs, organic matter, carbonate concentration	Feng *et al* 2006
9500-8000	Summer monsoon rainfall maximum			Biogenic silica	Schettler *et al*, 2006
9475 –3175 Holocene Megathermal	Summer monsoon rainfall decreases				Schettler *et al,* 2006
9-7000					Huang, *et* al, 2000
8230-7825			Emergence of early Neolithic cultures Yellow River Valley	Farming villages, domestic animals, graves, millet & rice	Lui, 2004
8855	Weakening of Asian monsoon				
8855-7825–5125-3175			*Setaria* and *Panicum* domesticated		An, 1989, Zhou, 1985
8855-5775 Holocene Optimum	Warm and wet	Temperate forest		Geochemical indices, pollen, magnetic susceptibility	Feng *et al*, 2006
79-7250 cal				Geochemical indices, pollen magnetic susceptibility,	Feng et al, 2006
7900-4450cal	Cool and humid			Pollen	Xiao *et al*, 2004
7825	Warm and humid with fluctuations warmer	Large scale mixed coniferous & broadleaved forest		Pollen	Xiao et al 2004
7800-4000 cal.	High precipitation, warm conditions	2nd tree pollen peak, temperate forest	Crop cultivation		Huang *et al* 2000, Feng *et al* 2006
7-4800 BP	wet	Stable	Rain fed agricultural cultures	Palaeosols, pollen	An *et al*, 2005
6850-5775		Daihai Lake Mixed conifer/broad leaf forest		Pollen	Li et al 2004
65-5100 cal Holocene Optimum Maximum	Arid phase		Decline in Neolithic culture	Pollen	Huang *et al* 2000

Date BP CAL	Climate	Environment	Cultural activities	Proxy evidence	Source
6400 varve yr BP	Warm and humid			Pollen	Xiao *et al*, 2004
5775-4455 - 3175	Summer monsoon minima				Schettler *et al*, 2006
5125 Attenuation of East Asian monsoon	Variable drying and probably warm stage	Tibetan plateau		Geochemical indices, pollen, magnetic susceptibility	Feng *et* al, 2006
5125- 4895 End of Holocene optimum	End of wetter climate than today			Pollen, diatom	Tarasov *et al*, 2000
5100-4800	Coldest period during Holocene	Drop in lake levels on the Tibetan plateau		Pollen, decrease in Magnetic susceptibility, particle size analysis, chemical composition	Jian *et al* 2000 Chen *et al*, 1999, Zhu & Chen, 1994, Lister *et al* 1991 Yan *et al* 1999
4900 varve yr BP	Mild and slightly humid			Pollen	Xiao *et al*, 2004
48-4450	Summer monsoon minima				Schettler *et al*, 2006
48-3400	Mild and humid			Pollen	Xiao *et al*, 2004
4670 -3915		Daihai Lake conifer Pine forest			Li et al 2004
4450-3950	Dry interval			Aeolian siliciclastic influx	Schettler *et al*, 2006
4450-2900	Cold dry episode			Pollen	Xiao *et al*, 2004
4385-3770	Cooler and drier	Woody plants declined		Pollen	Xiao *et al*, 2004
4,000 –3,600	Warm humid interval			Pollen	Xiao *et al*, 2004
3700 varve yr BP		Devastated ecologically	Longshan decline Appearance of Erlitou	Palaeosols, pollen,	An *et al*, 2005, Lui,2004
3500-2900	Summer optimum minima				Schettler *et al*, 2006
34-1950BP	Mild, slightly dry			Pollen	Xiao *et al*, 2004
3175 - Katathermal	Dry	Dahai Lake conifer, broad leaf – more open		Pollen	Xiao et al 2004
3175-	Gradual climatic decline	Regional precipitation peak retreats to southern China			Huang *et al*, 2000, An *et al*, 2000

Date BP CAL	Climate	Environment	Cultural activities	Proxy evidence	Source
2900 -	Fluctuating cool and dry late stage,	Xinjiang region		Geochemical indices, pollen magnetic susceptibility indices, pollen	Feng *et al* 2006
2200 varve yr BP	Cool dry	Forests disappear, vegetation density decreases		Pollen	Xiao *et al*, 2004
	Summer monsoon minima				Schettler *et al,* 2006

Appendix 2 North China crops and economic plants

From: Fuller, Qin, Harvey, 2007, Lee *et al*, 2007, Crawford *et al*, 2005, DF unpublished data (Ying), Banpo Museum, 1992, Underhill, 1997, Liu *et al* 2005 Fuller unpub. Data

CULTIGEN	Common name	Chinese name	Region of Origin	Archaeobotanical evidence	Economic use	Produces Phytoliths	In reference collection
WEEDS							
Amaranthus sp	Amaranth			Longshan, Lianchengzhen, Shandong, Ying Valley Longshan & Erlitou			
Asteraceae	Aster			Longshan, Lianchengzhen, Shandong, Yangshao Longshan Ying Valley		X	
Boragainaceae	Borage			Late Yangshao, Early Longshan Yellow River Basin (Yilou)		X	X
Brassicaceae	Wild mustard			Erligang Yellow River Basin (Yilou) Ying Valley Late Longshan & Shang		X	not identifiable
Chenopoduim sp.	Chenopod, lambsquarter			Late Longshan, Erlitou, Erligang sites Yellow River Basin (Yilou), Ying Valley, Yangshao, Longshan, Erlitou	dryland crop weed, gathered food,	X	
Cyperaceae,	Sedge			Erlitou & Erligang Yellow River Basin (Yilou), Yiáng Valley, Yangsho, Longshan, Erlitou	basketry, food, wet rice weed	X	
Cyperaceae, Cyperus	Sedge			Ying Valley, L Yangshao, E Longshan,	basketry, food, wet rice weed	X	
Cyperaceae, Scirpus	Sedge			Ying Valley,Yangshao, E Longshan,	basketry, food, wet rice weed	X	X
Euphorbarceae	Spurge			Late Yangshao, Early Longshan Yellow River Basin (Yilou) Yangshao Ying Valley	dryland crop weed,	X	
Glycera sp.	Panic mannagrass			Neolithic sites,Northern China and Yellow River Basin (6500-5000BC) (Yilou)		?	
Malva	Mallow			Ying Valley,			
Oxalis				Ying Valley, Longshan			

CULTIGEN	Common name	Chinese name	Region of Origin	Archaeobotanical evidence	Economic use	Produces Phytoliths	In reference collection
WEEDS							
Phragmites	Reed			Yangshao, Longshan, Erlitou, Yellow River Basin, Lianchengzhen, Shandong	thatching, basketry, food	X	X
Pologonaceae, *Polygonum sensu lato*				Ying Valley,			
Pologonaceae, *Rumex*				Ying Valley			
Polygonum sp.	Knotweed			Late Longshan, Erlitou sites Yellow River Basin (Yilou)			
Portulaca sp	Purslane			Longshan, Lianchengzhen, Shandong			
Ranunculus	Buttercup			Ying Valley, Yangshao			
Silene spp. Caryophyllaceae				Ying Valley, Late Longshan			
Solanaceae	Nightshade			Early Longshan Yellow River Basin (Yilou), Shandong, Ying valley			
Typha sp	Cattail/ Bullrush			Longshan, Lianchengzhen, Shandong	thatching, basketry, food		
	Minature beefsteak plant			Early Longshan,Erlitou Yellow River Basin (Yilou)			
FRUIT							
	Bramble			Erligang, Yellow River Basin (Yilou)	food		
Prunus sp	Plum			Erligang, Yellow River Basin (Yilou)	food		
Celtis bungeana	Hackberry			Longshan, Lianchengzhen, Shandong	food		
Prunus humillis	Cherry			Longshan, Lianchengzhen, Shandong	food		
Ziziphus jujube	Jubjub			Upper Ying Valley	food		
GOURDS AND MELONS							
cf Cucumis melo	melon			Upper Ying Valley	food		

Appendix 3 Samples
Baligang Samples

Site	Lab no	Sample	Period	Homogenous (ash layer/midden)	Laminate pit fill
Baligang	B1	CT1307 3:S1	Eastern Zhou		CT1307 3:S1
Baligang	B10	CT1307 12:S1	Yangshao (Early Qujialing)		CT1307 12:S1
Baligang	B11	H1634 3:S1	Longshan	H1634 3:S1	
Baligang	B12	H1632 3: S3	Longshan		H1632 3: S3
Baligang	B13	H1656 2: S3	Longshan	H1656 2: S3	
Baligang	B14	H1608 2: S2	Longshan		H1608 2: S2
Baligang	B15	H1646 1: S2	Longshan	H1646 1: S2	
Baligang	B2	CT1307 4:S1	Eastern Zhou		CT1307 4:S1
Baligang	B3	CT1307 5:S1	Eastern Zhou		CT1307 5:S1
Baligang	B4	CT1307 6:S1	Longshan		CT1307 6:S1
Baligang	B5	CT1307 7:S1	Longshan (early or just before) Shijiahe period		CT1307 7:S1
Baligang	B6	CT1307 8:S1	Longshan (early or just before) Shijiahe period		CT1307 8:S1
Baligang	B7	CT1307 9:S1	Longshan (early or just before) Shijiahe period		CT1307 9:S1
Baligang	B8	CT1307 10 (b) S2	Yangshao (Qujialing)		CT1307 10 (b) S2
Baligang	B9	CT1307 11 :S1	Yangshao (Qujialing)		CT1307 11 :S1
Baligang	B16	H1985	pre Yangshao	H1985	
Baligang	B17	H1977	Early Yangshao	H1977	
Baligang	B18	H1959	Mid Yangshao (earlier)	H1959	
Baligang	B19	H1959 -2	Mid Yangshao (earlier)	H1959 -2	
Baligang	B20	CT601 ZS:1	Mid Yangshao (later)		CT601 ZS:1
Baligang	B21	CT601 ZS:2	Mid Yangshao (later)		CT601 ZS:2
Baligang	B22	CT601 ZS:3	Mid Yangshao (later)		CT601 ZS:3
Baligang	B23	CT701 ZS:1	Mid Yangshao (later)		CT701 ZS:1
Baligang	B24	H1906 -4- A	Mid Yangshao (later)	H1906 -4- A	
Baligang	B25	H1906 -4- B	Mid Yangshao (later)	H1906 -4- B	
Baligang	B26	DT 506 -5	Mid Yangshao (later)		DT 506 -5
Baligang	B27	DT 506 -3	Late Yangshao		DT 506 -3
Baligang	B28	DT 506 -4	Late Yangshao		DT 506 -4
Baligang	B29	DT 506 -4- C	Late Yangshao		DT 506 -4- C

Erlitou samples

Site	Lab no	Sample	Period	Homogenous ash layer/ midden	Laminate pit fill	Floor /foundation
Erlitou	E01	T107:1	Erlitou II	T107:1		
Erlitou	E02	T107:2 L5B	Erlitou II	T107:2 L5B		
Erlitou	E03	T104:3L3	Erlitou II	T104:3L3		
Erlitou	E04	T104:3L2	Erlitou II	T104:3L2		
Erlitou	E05	T104:3L1	Erlitou IV	T104:3L1		
Erlitou	E06	T102:4AR	Erlitou IV			T102:4AR
Erlitou	E07	T102:4W	Erlitou II	T102:4W		

Erlitou	E08	T102:417	Erlitou IV	T102:417		

Site	Lab no	Sample	Period	Homogenous ash Layer/ midden	Laminate pit fill	Floor/ foundation
Erlitou	E09	T115:5L2	Erlitou IV	T115:5L2		
Erlitou	E10	T115:5L1	Erlitou earlier than IV	T115:5L1		
Erlitou	E11	T116:6L3	Erlitou			T116:6L3
Erlitou	E12	T117:7ALF	Erlitou III>IV			T117:7ALF
Erlitou	E13	T117:7AL2	Erlitou			T117:7AL2
Erlitou	E14	T114:8L3	Erlitou IV		T114:8L3	
Erlitou	E15	T114:8L2	Erlitou IV		T114:8L2	
Erlitou	E16	T114:8L1	Erlitou II		T114:8L1	
Erlitou	E17	T113:9AL	Erlitou	T113:9AL		
Erlitou	E18	T117:10AL2	Erlitou II			T117:10AL2
Erlitou	E19	T110:11L1	Erlitou III?			T110:11L1

Huizui East samples

Site	Lab no	Sample	Period	homogenous (ash layer/midden)	laminate pit fill
Huizui East	HE1	T2:H8L1	Longshan (Late)		T2:H8L1
Huizui East	HE2	T2:H8L5	Longshan (Late)		T2:H8L5
Huizui East	HE3	T2:H8L6	Longshan (Late)		T2:H8L6
Huizui East	HE4	T2:H8L7	Longshan (Late)		T2:H8L7
Huizui East	HE6	T2:F5L2	Longshan (Late)		T2:F5L1
Huizui East	HE52	T3 H4 NS 1	Longshan		T3 H4 NS 1
Huizui East	HE53	T3 H4 NS 2	Longshan		T3 H4 NS 2
Huizui East	HE54	T3 H4 NS 3	Longshan		T3 H4 NS 3
Huizui East	HE55	T3 H4 NS 6	Longshan		T3 H4 NS 6
Huizui East	HE56	T3 H4 NS 8	Longshan		T3 H4 NS 8
Huizui East	HE57	T3 H4 NS 9	Longshan		T3 H4 NS 9
Huizui East	HE5	T2:F5L1	Longshan (Late)		T2:F5L2
Huizui East	HE7	T5:RL2	Erlitou	T5:RL2	
Huizui East	HE90	T7 H9 W2	Yangshao	T7 H9 W2	
Huizui East	HE91	T7 H9 L	Yangshao	T7 H9 L	
Huizui East	HE92	T7 H9 P	Yangshao	T7 H9 P	
Huizui East	HE93	T7 H9 W	Yangshao	T7 H9 W	
Huizui East	HE94	T7 H9 Y	Yangshao	T7 H9 Y	
Huizui East	HE95	T7 H9 C	Yangshao	T7 H9 C	
Huizui East	HE96	T7 H9 CS	Yangshao	T7 H9 CS	
Huizui East	HE97	T5 H1 2	Yangshao	06HYHE T5 H1 2	
Huizui East	HE98	T5 H1 3a	Yangshao	06HYHE T5 H1 3a	
Huizui East	HE99	T5 H1 3b	Yangshao	06HYHE T5 H1 3b	
Huizui East	HE100	T5 H1 3ci	Yangshao	06HYHE T5 H1 3ci	
Huizui East	HE101	T5 H1 3cii	Yangshao	06HYHE T5 H1 3cii	
Huizui East	HE102	T5 H1 3ciii	Yangshao	06HYHE T5 H1 3ciii	
Huizui East	HE103	T5 H1 3di	Yangshao	06HYHE T5 H1 3di	
Huizui East	HE105	T5 H1 3e	Yangshao	06HYHE T5 H1 3e	
Huizui East	HE107	T5 H1 3jii	Yangshao	06HYHE T5 H1 3ji	
Huizui East	HE108	T5 H1 4a	Yangshao	06HYHE T5 H1 4a	
Huizui East	HE110	T5 H1 5	Yangshao	06HYHE T5 H1 5	

Huizui East, Geological and Huizui West samples

Site	Lab no	Sample	Period	Homogenous ash layer/ midden	laminate pit fill	floor /foundation
Huizui East	HE09	T4 H12 FA	Yangshao (Late)			06HYHE T4 H12 FA
Huizui East	HE12	T4 H12 FB	Yangshao (Late)			06HYHE T4 H12 FB
Huizui East	HE17	T4 H12 FC /A	Yangshao (Late)			06HYHE T4 H12 FC /A
Huizui East	HE18	T4 H12 FC /B W	Yangshao (Late)			06HYHE T4 H12 FC /B W
Huizui East	HE19	T4 H12 FC/ C W	Yangshao (Late)			06HYHE T4 H12 FC/ C W
Huizui East	HE08	T1 H17 LD	Yangshao			06HYHE T1 H17 BF4U
Huizui East	HE04	T1 H17 BF4U	Yangshao			06HYHE T1 H17 BF4L
Huizui East	HE05	T1 H17 BF4L	Yangshao			06HYHE T1 H17
Huizui East	HE06	T1 H17 FA	Yangshao			06HYHE T1 H17 FA
Huizui East	HE07	T1 H17 FB	Yangshao			06HYHE T1 H17 FB
Huizui East	HE81	Terrace F1 1	Longshan			06HYHE Terrace F1 1
Huizui East	HE82	Terrace F1 2	Longshan			06HYHE Terrace F1 2
Huizui East	HE83	Terrace F1 3	Longshan			06HYHE Terrace F1 3
Huizui East	HE84	Terrace F1 4	Longshan			06HYHE Terrace F1 4
Huizui East	HE85	Terrace F1 5	Longshan			06HYHE Terrace F1 5
Huizui East	HE86	Terrace F1 6	Longshan			06HYHE Terrace F1 6
Huizui GS	HGS4	T3 N1 L6	Yangshao	T3 N1 L6		
Huizui GS	HGS5	N1 L8	Yangshao	N1 L8		
Huizui West	HW1	T3H2:L3	Erlitou	T3H2:L3		
Huizui West	HW2	T3H2:L2	Erlitou	T3H2:L2		
Huizui West	HW3	T3H4:L1	Erlitou	T3H4:L1		
Huizui West	HW4	T3H4:L4	Erlitou	T3H4:L4		
Huizui West	HW5	T3H4:L3	Erlitou	T3H4:L3		

Xipo Samples

Site	Lab no	Sample	Period	homogenous (ash layer/midden)	laminate pit fill	floor /foundation
Xipo	X01	HLX06 X005	Yangshao (Miaodigoou)			X005
Xipo	X02	HLX06 X006	Yangshao (Miaodigoou)	X006		
Xipo	X03	HLX06 X007	Yangshao (Miaodigoou)	X007		
Xipo	X04	HLX06 X011 (5)	Yangshao (Miaodigoou)	X011 (5)		
Xipo	X05	HLX06 X060 (54)	Yangshao (Miaodigoou)	X060 (54)		
Xipo	X06	HLX06 X064A (60)	Yangshao (Miaodigoou)	X064A (60)		
Xipo	X07	HLX06 X079 (100)	Yangshao (Miaodigoou)	X079 (100)		
Xipo	X08	HLX06 X103 (90)	Yangshao (Miaodigoou)	X103 (90)		
Xipo	X09	HLX06 X105 (92)	Yangshao (Miaodigoou)	X105 (93)		
Xipo	X10	HLX06 X107 (94)	Yangshao (Miaodigoou)	X107 (95)		

Site	Lab no	Sample	Period	Homogenous ash layer /midden	Laminate pit fill	Floor/ foundation
Xipo	X11	HLX06 X074A (70)	Yangshao (Miaodigoou)		HLX06 X074A (70)	
Xipo	X12	HLX06 X052 (45)	Yangshao (Miaodigoou)	HLX06 X052 (45)		
Xipo	X14	HLX06 X053 (46)	Yangshao (Miaodigoou)	X053 (46)		
Xipo	X15	HLX06 X015 (4)	Yangshao (Miaodigoou)	X015 (4)		
Xipo	X16	HLX06 X041 (34)	Yangshao (Miaodigoou)	X041 (34)		
Xipo	X20	HLX06 X020 (10)	Yangshao (Miaodigoou)	X020 (10)		
Xipo	X21	HLX06 X099 (86)	Yangshao (Miaodigoou)	X099 (86)		
Xipo	X22	HLX06 X104 (91)	Yangshao (Miaodigoou)	X104 (91)		
Xipo	X25	HLX06 X096 (83)	Yangshao (Miaodigoou)	HLX06 X096 (83)		
Xipo	X26	HLX06 X024 (14)	Yangshao (Miaodigoou)	HLX06 X024 (14)		
Xipo	X27	HLX06 X008	Yangshao (Miaodigoou)		HLX06 X008	
Xipo	X28	HLX06 X004 F4	Yangshao (Miaodigoou)		HLX06 X004	
Xipo	X29	HLX06 X002 F3	Yangshao (Miaodigoou)		HLX06 X002	
Xipo	X30	HLX06 X049 (43)	Yangshao (Miaodigoou)		HLX06 X049 (43)	
Xipo	X31	HLX06 X037 (30)	Yangshao (Miaodigoou)	HLX06 X037 (30)		
Xipo	X32	HLX06 X109 96	Yangshao	HLX06 X109 96		
Xipo	X33	HLX06 X059 (53)	Yangshao (Miaodigoou)	HLX06 X059 (53)		
Xipo	X34	HLX06 X051 (44)	Yangshao (Miaodigoou)		HLX06 X051 (44)	
Xipo	X35	HLX06 X009	Yangshao (Miaodigoou)	HLX06 X009		
Xipo	X36	HLX06 X0100 87	Yangshao	HLX06 XO100 87		
Xipo	X37	HLX06 X019 (9)	Yangshao (Miaodigoou)	HLX06 X019 (9)		
Xipo	X38	HLX06 X065A (61)	Yangshao (Miaodigoou)	HLX06 X065A (61)		
Xipo	X39	HLX06 X035 (28)	Yangshao (Miaodigoou)	HLX06 X035 (28)		
Xipo	X40	HLX06 X096 (83)	Yangshao (Miaodigoou)	X096 (83)		
Xipo	X41	HLX06 X109 (96)	Yangshao (Miaodigoou)	HLX06 X109 (97)		
Xipo	X23	HLX06 X081 F4	Yangshao (Miaodigoou)		Hearth	X081 F4
Xipo	X24	HLX06 X088 (75)	Yangshao (Miaodigoou)		Hearth	X088 (75)

Survey samples

Site	Lab no	Sample	Period	Homogenous (ash layer/midden)
Jing-Yanggong	S11	03-041	Erlitou	03-041
Yuangou A Erlitou	S12	03-165	Erlitou	03-165

Site	Lab no	Sample	Period	Homogenous (ash layer/midden)
Yuangou A	S13	03-165	Erlitou	03-165
SE Zhaiwan upper	S14	03-216	Erlitou	03-216
SE Zhaiwan lower	S15	03-218	Erlitou	03-218
Mazhai site	S4	03-213	Late Longshan	03-213
Jueshan	S5	03-147	Late Longshan	03-147
NE Gaoya lower	S6	03-132	Late Longshan	03-132
NE Gaoya upper	S7	03-132	Late Longshan	03-132
Peichun B	S8	03-119	Late Longshan	03-119
NE Zhaiwan upper	S9	03-217	Late Longshan	03-217
NE Zhaiwan	S10	03-217	Late Longshan	03-217
Jiuligou upper	S2	03-159	Late Yangshao	03-159
Jiuliugou W	S3	03-159	Late Yangshao	03-159
Gong Jia yao	S1	03-183	Yangshao	03-183

Appendix 4 Description of terms used for phytoliths in study

	ICPN nomenclature	*Notes*
Monocot single cells		
Long smooth	Elongate psilate /– epidermal long cell	Leaf/culm grasses
Long sinuate	Elongate sinuate – epidermal long cell	Leaf/culm grasses
Long rods	Elongate psilate/ Cylindric smooth Epidermal long cell	Long and narrow sometimes cylindrical– Cyperaceae
Long dendritic	Elongate dendritic– epidermal long cell	Dendroform Gramminae inflorescence*
Stomata		Anatomical term
Papillae	Papillae cell Hair cell	Anatomical term – Hair on glume
Hair	Hair cell	Anatomical term *
Bulliform	Parallepipedal bulliform – (see Metcalfe)	Anatomical term Grass leaf
Keystone	Cuneiform bulliform –	Sometimes called fan shaped bulliform Grass leaf
Oryza type keystone	Cuneiform bulliform	Fan shaped keystone with distinctive scalloping around fan *Oryza* leaf
Crenate	Epidermal short cell	-
Bilobe	Bilobate – epidermal short cell	Leaf/culm many Panicoid grasses
Setaria type bilobe		Distinctive short bilobe found in *Setaria* leaf/culm and glume*
Scooped bilobe	Bilobate - concave apexes - epidermal short cell	Bilobe - arranged horizontally often found in Oryzeae, including *Oryza Sativa* and *Zizania*
Cross	Quadra-lobate – epidermal short cell	Found in some Panicoid grass leaf/culm (ex: *Coix lachryma jobi)*
Rondels	Rondel - epidermal short cell	-
Stipa type rondel		
Saddles	Saddle - epidermal short cell	Found in Chloridoid grass leaf/culm
Collapsed saddle	epidermal short cell	Bambusoideae
Cones	Conical	Hairs. Found on Cyperaceae inflorescence

Dicot single cells	*ICPN nomenclature*	*Notes*
Double-peaked glume cell	Bi-echinate - epidermal cell	Specific to rice husks
Rugulose spheroid	Globular echinate	From the Arecaceae family
Smooth Spheroid	Globular psilate	
Verrucate		
Crescent		*Equisetum*
Dicot Elongate	Elongate laterally irregular	-
Tracheid	Tracheid	Anatomical term from the vascular parts of plants
Two-tiered		
Blocks	Square regular psilate	Regular flat block shapes from
Platey		
Sheet	Irregular shape psilate	Irregular shapes of flat silica panel
Single Polyhedron	Polyhedron shape psilate	From dicot leaves
Scalloped	Irregular shape	Sometimes from Cucurbit rind
Single Jigsaw	Multi-lobate irregular pattern	- From dicot leaves

	ICPN nomenclature	*Notes*
Multicells - Leaves		
Multi jigsaw puzzle	Multi-lobate irregular pattern	
Leaf/culm indeterminate		
Leaf/culm long cells	Elongate psilate /– epidermal long cells	
Leaf/culm bilobe	Bilobate – epidermal short cells	
Leaf/culm saddle	Saddle - epidermal short cells	
Leaf/culm cross	Quadra-lobate – epidermal short cell	
Leaf/culm cf *Panicum*		
Leaf/culm *Oryza*		
Leaf/culm *Phragmites*		
Leaf/culm reed		Cf Phragmites but with fewer id criteria
Leaf/culm square cell		
Leaf/culm stomata		
	ICPN nomenclature	*Notes*
Multicells - husks		
Indeterminate husk		
Setaria husk		
Panicum husk		
Millet type 1 husk		Panicoid similar to *Setaria* without all id criteria
Millet type 2 husk		Panicoid similar to *Panicum* without all criteria
Multicells - husk		
Oryza husk	Bi-echinate - epidermal cells	Double peaked glume hairs – specific to *Oryza*
Cyperaceae		
Polyhedron		
Polyhedral hair base		
Multi tiered forms		
Mesophyll		
Indeterminate multicell		
Indeterminate phytolith		

	ICPN nomenclature	*Notes*
Others		
Diatoms		
Sponge spicules		
Starch		
Silica aggregate		

Appendix 5 Data

Data tables % per sample

BALIGANG														
	B17	B18	B19	B20	B21	B22	B23	B24	B25	B26	B27	B28	B29	B10
Long Smooth	35.60	27.44	32.09	19.63	25.00	19.39	28.33	19.88	22.44	22.88	23.78	24.32	17.83	10.77
Long Sinuate	3.60	2.89	3.52	1.37	3.39	1.90	2.54	4.29	1.46	3.39	2.97	3.86	2.17	0.00
Long Rods	4.00	3.61	6.15	3.65	7.20	9.51	8.46	10.53	21.46	1.69	2.12	1.93	4.35	4.62
Long Dendritic	2.40	0.72	0.00	0.46	0.00	0.38	0.85	0.00	1.46	0.00	0.42	1.93	0.43	0.00
Stomata	0.00	0.00	0.00	0.00	0.00	0.00	0.00	0.00	0.00	0.00	0.00	0.00	0.00	0.00
Total hairs	21.20	24.55	23.74	37.90	23.73	23.19	17.34	31.19	29.76	33.05	28.03	30.12	32.17	36.93
Bulliform	6.80	7.22	10.11	1.37	5.93	6.46	5.07	1.56	0.00	4.24	5.94	3.09	3.48	6.15
Cuneiform Bulliform	1.20	2.17	3.96	1.37	0.85	3.42	3.81	1.95	0.00	2.12	5.10	3.86	2.17	4.62
cf. Oryza Bulliform	0.40	0.00	0.00	0.00	0.00	0.38	0.85	0.00	0.00	0.00	0.00	0.00	0.00	0.00
Crenate	0.00	0.00	0.00	0.00	0.00	0.00	0.00	0.00	0.00	0.00	0.00	0.00	0.00	0.00
Bilobate	2.00	3.97	3.30	4.11	5.93	6.84	3.17	4.09	1.46	1.69	3.18	2.70	1.74	0.00
Polylobate	0.40	0.36	0.00	0.00	0.00	0.00	0.00	0.00	0.00	0.00	0.00	0.00	0.00	0.00
Cross	0.00	0.36	1.32	0.91	0.00	2.66	0.42	1.17	0.00	0.42	0.00	1.16	0.87	0.00
cf. Setaria bilobate	0.00	0.00	0.00	0.46	0.00	0.00	0.00	0.00	0.00	0.00	0.00	0.00	0.00	0.00
cf. Oryza bilobate	0.00	0.00	0.00	0.00	0.00	0.38	0.00	0.00	0.00	0.00	0.00	0.00	0.00	0.00
Rondel	4.80	6.86	5.27	8.22	6.78	4.56	4.65	4.29	4.39	6.78	6.37	5.79	10.43	13.85
cf. Stipa Rondel	0.40	0.00	0.00	0.00	0.00	0.38	0.42	0.00	0.00	0.00	0.00	0.00	0.87	0.00
Saddles	1.60	2.17	0.00	0.46	2.54	2.28	1.27	1.17	0.98	2.12	0.85	0.00	0.00	0.00
Collapsed saddle	0.80	0.36	0.00	0.00	0.42	0.38	0.42	0.00	0.00	0.00	0.00	0.00	0.00	0.00
Cone	1.20	1.44	0.88	2.74	1.69	2.66	0.85	1.17	2.93	2.97	1.27	0.77	3.04	0.00
Rugulose Spheroid	0.00	0.00	0.00	0.00	0.00	0.00	0.00	0.39	0.00	0.00	0.00	0.00	0.00	0.00
Smooth Spheroid	0.40	0.00	0.00	0.91	1.27	0.00	0.00	0.39	0.00	0.42	0.42	0.00	0.00	0.00
Verrucate	0.00	0.00	0.00	0.46	0.00	0.38	0.42	0.39	0.00	0.00	0.42	0.39	0.00	0.00
Crescent	0.00	0.00	0.00	0.00	1.27	0.00	0.00	0.00	0.00	0.00	0.85	0.39	0.00	0.00
Elongate	1.60	1.44	3.08	1.37	1.27	2.66	2.96	0.39	2.44	0.85	2.12	0.00	0.87	7.69
Tracheid	3.60	1.81	2.20	0.46	0.00	0.38	4.23	1.95	4.39	0.85	2.12	1.93	1.30	0.00
Two-Tiered	0.00	0.36	0.00	0.00	0.00	0.00	0.00	0.00	0.00	0.00	0.00	0.39	0.00	0.00
Block	0.40	0.72	0.00	1.37	1.27	1.90	2.54	0.00	0.00	1.27	2.97	0.39	1.30	3.08
Platey	6.40	9.03	2.64	11.87	8.47	6.84	6.77	12.87	5.85	11.44	6.79	11.20	13.48	12.31
Sheet	0.00	0.00	0.00	0.00	0.42	0.00	0.00	0.00	0.00	0.00	0.42	0.39	0.00	0.00
Single Polyhedron	1.20	2.17	0.88	0.91	2.54	3.04	4.23	2.34	0.98	3.39	3.82	5.02	3.48	0.00
Scalloped	0.00	0.00	0.88	0.00	0.00	0.00	0.42	0.00	0.00	0.42	0.00	0.39	0.00	0.00
Single Jigsaw puzzle	0.00	0.36	0.00	0.00	0.00	0.00	0.00	0.00	0.00	0.00	0.00	0.00	0.00	0.00
Leaf/culm indet	26.30	13.63	14.34	22.86	20.83	18.73	12.15	36.70	26.14	6.47	33.78	19.47	11.89	0.00
Leaf/culm jigsaw	0.00	0.00	0.00	0.00	0.00	0.00	0.00	0.53	0.00	0.00	0.00	0.00	0.21	0.00
Leaf/culm bilobate	0.00	0.00	0.35	0.00	0.00	0.53	0.22	0.53	0.00	0.00	0.00	0.00	0.00	0.00
Leaf/culm saddle	0.00	0.28	0.00	0.00	0.00	0.26	0.00	0.00	0.00	0.00	0.00	0.00	0.00	0.00
Leaf/culm cross	0.00	0.00	0.00	0.00	0.00	0.00	0.00	0.00	0.00	0.00	0.00	0.00	0.00	0.00
Leaf/culm cf Oryza	0.00	0.00	0.00	0.00	0.00	0.00	0.00	0.00	0.00	0.00	0.00	0.00	0.00	0.00
Leaf/culm long cells	14.13	11.41	10.84	3.14	24.40	4.22	6.85	15.96	28.41	2.55	15.54	4.20	3.13	0.00
Leaf culm stomata	0.00	0.00	0.00	0.00	0.00	0.00	0.00	0.00	0.00	0.00	0.00	0.00	0.00	0.00
Leaf/ culm Phragmites	0.00	0.28	0.00	0.00	0.00	0.53	0.00	0.53	0.00	0.00	0.00	0.00	0.00	0.00
BALIGANG														
	B17	B18	B19	B20	B21	B22	B23	B24	B25	B26	B27	B28	B29	B10
Leaf / culm reed	0.00	0.00	0.70	0.00	0.00	0.53	0.00	1.06	0.00	0.34	0.00	1.53	0.63	0.00
Leaf / culm Panicum	0.00	0.00	0.00	0.00	0.00	0.00	0.00	0.00	1.14	0.00	0.00	0.00	0.00	0.00
Square-cell leaf/stem	0.97	1.95	3.50	0.29	4.76	1.85	1.55	0.00	2.27	1.36	4.05	2.29	0.83	0.00
Indet husk	7.31	14.75	7.34	0.86	2.98	1.32	2.21	4.26	9.09	4.94	16.89	8.78	2.92	0.00
cf.Setaria husk	0.00	0.00	0.00	0.00	0.00	0.00	0.00	0.00	0.00	0.00	0.00	0.00	0.00	0.00
cf.Panicum husk	0.00	0.00	0.70	0.00	0.00	0.00	0.00	0.00	0.00	0.00	0.00	0.00	0.00	0.00
Millet type 1	0.00	0.00	0.35	0.00	0.00	0.00	0.00	0.00	0.00	0.00	0.00	0.00	0.00	0.00
Millet type 2	0.00	0.00	0.70	0.00	0.00	0.00	0.00	0.53	0.00	0.00	0.00	0.00	0.00	0.00
cf. Oryza husk	4.38	3.06	1.05	0.00	0.00	0.00	0.66	0.53	0.00	1.19	2.03	0.38	0.21	0.00
Cyperaceae	3.41	3.34	2.80	4.00	2.98	2.90	2.87	3.19	7.39	0.85	6.76	2.67	1.88	0.00

Polyhedron	0.97	0.56	1.05	4.86	5.36	0.00	2.87	4.26	0.57	0.51	1.35	0.38	0.42	0.00
Polyhedral hair base	0.00	0.00	0.00	0.00	0.60	1.85	0.00	0.00	0.57	0.00	0.68	0.76	0.00	0.00
Multi-Tiered forms	0.00	0.00	0.00	0.00	0.60	0.53	0.00	0.00	0.00	0.00	0.00	0.00	0.00	0.00
Mesophyll	0.00	0.00	0.00	0.00	0.00	0.00	0.00	0.00	0.00	0.00	0.00	0.00	0.00	0.00
Indet multicell	9.05	12.02	31.82	33.71	10.12	36.41	24.79	5.32	18.18	14.07	7.43	23.16	15.02	0.00
Indet phytolith	29.85	32.05	15.91	25.71	25.00	24.27	32.30	22.87	5.68	64.20	11.49	33.91	57.25	82.47
Diatom	0.90	0.00	0.00	0.00	0.60	0.00	0.00	1.60	0.00	0.00	0.00	0.00	0.00	0.00
Sponge Spicule	0.90	2.67	0.00	0.00	1.79	0.00	1.50	1.06	0.00	1.76	0.00	0.83	0.00	0.00

BALIGANG											
	B8	B9	B5	B6	B7	B4	B11	B12	B13	B14	B15
Long Smooth	18.59	6.25	29.74	18.48	21.31	14.65	213.84	23.47	17.49	19.65	18.02
Long Sinuate	0.50	0.39	2.14	0.00	1.43	3.37	3.85	2.53	1.39	3.49	1.94
Long Rods	2.48	0.39	1.75	1.66	1.02	0.79	34.68	4.69	3.78	1.27	2.30
Long Dendritic	0.74	0.19	1.36	0.24	0.00	0.20	13.49	1.08	0.99	1.27	0.71
Stomata	0.00	0.19	0.19	0.24	0.00	0.00	0.00	0.00	0.00	0.00	0.00
Total hairs	23.30	34.89	16.91	33.18	27.05	17.23	287.05	32.85	16.50	35.82	16.78
Bulliform	9.67	14.03	8.55	9.00	9.43	9.90	52.02	9.39	7.55	4.44	8.30
Cuneiform Bulliform	3.47	6.63	2.92	3.55	2.66	7.33	26.97	2.17	3.78	2.85	4.77
cf. Oryza Bulliform	0.50	0.78	0.39	0.71	0.61	1.19	3.85	2.17	1.19	1.58	0.53
Crenate	0.74	0.00	0.00	0.00	0.00	0.00	5.78	0.00	0.40	0.00	0.53
Bilobate	3.10	0.58	0.87	2.13	3.69	5.54	35.06	2.89	10.36	4.60	3.89
Polylobate	0.00	0.00	0.19	0.00	0.00	0.20	3.85	0.00	0.99	0.00	0.00
Cross	1.24	0.00	0.78	0.95	0.82	0.40	19.27	0.72	0.60	2.22	1.06
cf. Setaria bilobate	0.00	0.00	0.19	0.00	0.20	0.00	0.00	0.00	0.60	0.00	0.00
Cf. Oryza bilobate	0.50	0.00	0.58	0.95	0.20	0.40	28.90	0.36	8.15	0.63	1.77
Rondel	12.14	10.91	8.94	4.98	3.89	12.87	105.96	5.78	5.57	7.29	9.19
cf. Stipa Rondel	0.00	0.00	0.58	0.00	0.00	3.96	0.00	0.00	0.00	0.00	1.41
Saddles	2.23	2.73	1.94	1.66	5.94	3.96	30.82	0.72	3.18	1.27	3.18
Collapsed saddle	0.50	1.56	1.17	0.00	1.43	1.58	15.41	0.36	0.00	0.63	0.00
Cone	0.00	0.00	0.00	0.00	0.00	0.40	1.93	0.36	0.00	0.32	1.06
Rugulose Spheroid	0.00	0.00	0.00	0.00	0.61	0.00	3.85	0.00	0.00	0.00	0.18
Smooth Spheroid	0.25	0.19	0.19	0.47	0.20	0.00	1.93	0.36	0.40	0.32	0.18
Verrucate	0.00	0.19	0.00	0.00	0.00	0.00	0.00	0.00	0.00	0.00	0.00
Crescent	1.24	0.00	0.00	0.00	0.00	0.00	0.00	0.00	0.00	0.00	0.00
Elongate	5.70	4.29	2.53	3.32	3.89	4.16	52.02	2.17	6.76	1.58	2.83
Tracheid	2.23	0.39	3.30	0.47	2.05	0.59	17.34	2.17	1.79	2.22	1.41
Two-Tiered	0.00	0.00	0.78	0.00	0.00	0.00	0.00	0.00	0.00	0.00	0.00
Block	0.74	0.39	0.39	0.47	0.00	0.00	1.93	0.36	0.99	1.58	0.53
Platey	4.71	10.14	6.41	10.66	1.84	6.93	40.46	5.42	6.56	6.66	17.49
Sheet	0.25	0.58	0.19	0.24	0.82	0.79	7.71	0.00	0.20	0.00	0.88
Single Polyhedron	4.71	3.90	6.03	5.45	8.81	2.38	3.85	0.00	0.40	0.32	0.00
Scalloped	0.50	0.19	0.19	0.47	1.02	0.40	1.93	0.00	0.20	0.00	0.88
Single Jigsaw puzzle	0.00	0.19	0.78	0.71	1.02	0.79	0.00	0.00	0.20	0.00	0.18
Leaf/culm indet	12.02	6.19	39.22	30.67	22.86	42.65	40.00	17.37	9.15	22.96	39.83
Leaf/culm jigsaw	0.13	0.95	0.00	0.00	7.95	0.00	0.00	0.00	0.00	0.84	0.00
Leaf/culm bilobate	0.52	0.00	0.00	0.26	0.25	0.41	0.78	0.00	2.44	2.51	0.00
Leaf/culm saddle	0.26	0.00	0.49	1.28	0.00	0.00	0.39	0.00	1.22	0.00	0.00
Leaf/culm cross	0.13	0.00	0.00	0.77	0.75	0.00	0.00	0.00	0.00	0.00	0.00

BALIGANG											
	B8	B9	B5	B6	B7	B4	B11	B12	B13	B14	B15
Leaf/culm cf Oryza	0.00	0.00	0.00	0.26	0.00	0.00	0.00	0.00	0.00	1.67	3.20
Leaf/culm long cells	2.09	0.24	15.69	6.90	8.20	5.80	16.08	22.75	14.63	6.68	7.82
Leaf culm stomata	0.13	0.00	0.00	0.00	0.00	0.00	0.00	0.00	0.00	0.00	0.00
Leaf/ culm Phragmites	0.13	0.00	0.00	0.26	0.00	0.00	0.78	0.00	3.66	0.00	4.27
Leaf / culm reed	0.65	0.00	0.00	0.00	0.00	0.41	0.00	0.60	2.44	0.00	3.20
Leaf / culm Panicum	0.00	0.00	0.00	0.00	0.00	0.00	0.00	0.00	0.00	0.00	0.00
Square-cell leaf/stem	0.52	0.71	6.37	1.02	1.49	7.04	5.88	4.79	41.46	5.85	14.58
Indet husk	1.70	0.24	7.35	5.62	7.46	1.66	16.47	8.98	4.88	5.43	3.56

cf.Setaria husk	0.00	0.00	0.00	0.00	0.00	0.00	0.00	1.80	0.61	0.00	0.00
cf.Panicum husk	0.00	0.00	0.00	0.00	0.00	0.00	0.00	0.00	0.00	0.42	0.00
Millet type 1	0.00	0.00	0.49	0.00	0.99	0.83	0.39	0.00	0.61	0.00	0.00
Millet type 2	0.26	0.00	1.96	0.51	2.24	1.66	4.31	1.20	0.00	1.25	0.00
cf. Oryza husk	3.27	0.24	1.96	0.00	1.99	0.41	4.31	7.78	3.05	0.42	2.13
Cyperaceae	2.61	1.43	5.88	0.77	2.24	4.14	2.75	9.58	3.05	3.34	2.13
Polyhedron	0.78	0.24	0.98	1.28	0.99	0.41	0.39	0.00	3.05	1.25	1.07
Polyhedral hair base	0.78	1.43	1.96	1.53	1.74	2.07	1.96	0.60	0.00	0.42	0.71
Multi-Tiered forms	0.39	2.85	3.92	7.92	2.49	14.49	3.92	0.00	0.00	0.00	2.84
Mesophyll	0.00	0.00	0.00	0.00	0.00	0.00	0.00	0.00	0.00	0.00	0.00
Indet multicell	0.13	0.00	0.00	0.00	0.00	0.00	0.00	12.57	0.00	10.44	0.00
Indet phytolith	60.74	61.14	1.96	34.25	19.88	0.00	634.27	4.17	2.17	0.95	1.17
Diatom	5.88	0.00	8.33	0.84	2.11	0.00	0.00	0.00	0.00	1.04	0.00
Sponge Spicule	3.92	3.05	3.43	1.68	3.17	3.73	1.57	2.99	0.00	4.18	1.82
Starch	0.00	0.00	0.00	0.00	0.00	0.00	0.00	0.00	0.00	0.00	0.00
Silica aggregate	2.94	21.32	0.00	4.20	13.20	14.29	0.00	2.99	0.61	4.18	10.00

XIPO														
	X1	X2	X3	X4	X5	X6	X7	X8	X9	X10	X11	X12	X13	X14
Long Smooth	13.83	10.69	14.80	13.44	13.53	12.60	1.23	16.59	13.54	13.14	11.32	9.79	21.31	26.88
Long Sinuate	1.13	2.10	1.35	2.32	1.46	0.24	0.11	2.73	2.26	3.08	0.00	1.40	0.26	0.00
Long Rods	2.04	1.05	1.79	1.66	1.65	7.77	0.13	1.82	1.04	2.05	1.35	1.96	4.36	0.36
Long Dendritic	1.13	1.05	0.90	2.32	0.00	0.71	0.07	1.82	0.87	3.29	1.08	1.12	1.54	0.00
Stomata	0.00	0.00	0.00	0.00	0.00	0.00	0.02	0.23	0.00	0.21	0.00	0.00	0.00	0.00
Total hairs	20.86	31.55	28.48	26.22	23.22	26.15	2.72	29.77	32.81	31.42	33.42	39.72	22.08	36.92
Bulliform	8.39	1.75	2.02	10.29	6.76	3.53	0.72	9.09	7.12	4.93	6.74	2.52	5.39	7.89
Cuneiform Bulliform	2.72	1.05	0.22	1.66	4.39	1.18	0.71	0.91	1.91	1.44	3.50	0.84	1.80	2.51
cf. Oryza Bulliform	0.23	0.00	0.00	0.00	0.18	0.00	0.00	0.23	0.00	0.00	0.27	0.00	0.00	0.00
Crenate	0.00	0.35	0.45	0.00	0.18	0.47	0.02	0.23	0.00	0.21	0.00	0.00	0.00	0.00
Bilobate	8.39	9.90	14.57	10.37	7.50	2.83	0.42	12.61	6.25	5.54	3.77	14.13	11.94	2.15
Polylobate	0.68	0.18	0.22	0.17	0.00	0.00	0.00	0.45	0.35	0.21	0.27	0.00	0.00	0.00
Cross	1.81	2.45	3.36	0.66	1.10	1.18	0.04	2.27	1.39	2.05	0.27	1.68	0.51	0.36
cf. Setaria bilobate	0.00	0.18	0.90	0.00	0.00	0.00	0.00	0.00	0.00	0.21	0.27	0.00	0.00	0.00
cf. Oryza bilobate	2.49	0.53	0.00	0.33	0.00	0.00	0.04	0.68	0.35	0.62	0.00	0.00	0.77	0.36
Rondel	5.44	5.96	7.62	7.47	9.14	11.54	0.71	13.64	6.25	10.68	8.36	7.83	11.55	8.96
cf. Stipa Rondel	0.45	0.00	0.00	1.16	0.73	0.94	0.09	0.45	0.52	1.64	0.27	0.00	0.51	0.36
Saddles	3.17	4.38	3.59	0.66	2.19	0.94	0.00	3.86	1.74	2.67	2.43	1.96	4.62	0.00
Collapsed saddle	3.17	3.51	0.90	0.50	1.10	0.00	0.05	0.45	0.87	0.00	0.00	0.00	0.00	0.00
XIPO														
	X1	X2	X3	X4	X5	X6	X7	X8	X9	X10	X11	X12	X13	X14
Cone	0.00	0.35	0.90	0.00	0.37	0.00	0.00	0.23	0.00	0.00	0.27	1.12	0.00	0.00
Rugulose Spheroid	0.00	0.00	0.00	0.00	0.00	0.00	0.00	0.23	0.00	0.00	0.00	0.00	0.00	0.00
Smooth Spheroid	1.59	0.88	0.90	0.33	0.37	0.94	0.09	1.14	0.35	0.82	0.54	0.00	0.00	0.36
Verrucate	0.68	0.00	0.00	0.00	0.00	0.00	0.00	0.23	0.69	0.82	0.81	0.00	0.00	0.72
Crescent	0.00	0.70	0.90	0.17	0.00	0.94	0.00	0.68	0.17	0.21	0.27	0.28	0.77	0.72
Elongate	3.85	4.38	3.36	3.82	4.94	3.53	0.45	2.95	2.26	2.05	3.23	1.96	1.80	0.00
Tracheid	0.00	0.88	0.45	0.83	0.91	2.12	0.09	0.91	1.04	0.21	1.62	0.28	0.51	0.36
Two-Tiered	0.00	0.00	0.22	0.50	0.73	0.94	0.04	0.00	0.17	0.62	0.27	0.00	0.00	0.00
Block	0.91	1.23	1.12	0.50	1.10	1.18	0.09	0.00	0.35	0.21	0.00	0.56	0.77	1.08
Platey	6.12	7.54	6.05	10.29	14.26	16.49	1.56	5.68	7.99	2.87	18.33	10.35	7.19	6.09
Sheet	2.04	2.28	2.24	0.33	1.65	0.71	0.05	1.59	1.39	1.64	0.00	1.12	0.26	0.00
Single Polyhedron	8.16	4.03	2.24	3.15	1.28	3.06	0.56	4.77	7.99	6.57	1.35	1.40	2.05	3.94
Scalloped	0.00	0.35	0.00	0.17	0.55	0.00	0.00	0.00	0.17	0.00	0.00	0.00	0.00	0.00
Single Jigsaw puzzle	0.68	0.70	0.45	0.66	0.73	0.00	0.00	0.00	0.17	0.62	0.00	0.00	0.00	0.00
Leaf/culm indet	32.01	41.48	57.47	28.20	15.65	7.53	13.92	57.43	51.18	37.97	24.44	55.66	18.40	6.70
Leaf/culm jigsaw	3.96	14.64	10.75	5.78	2.61	3.76	4.41	6.05	7.25	3.80	0.00	0.59	0.26	0.00
Leaf/culm bilobate	1.08	1.22	3.27	0.72	0.00	0.00	0.55	4.03	2.56	1.17	1.29	6.45	1.84	0.00
Leaf/culm saddle	0.72	0.81	0.00	0.00	0.00	0.00	0.00	0.00	0.43	0.58	0.64	1.17	0.26	0.00
Leaf/culm cross	0.36	0.00	0.00	0.00	0.00	0.07	0.00	0.00	0.43	1.46	0.00	1.17	0.26	0.00
Leaf/culm cf Oryza	0.72	0.00	0.00	0.00	0.00	0.07	0.14	0.00	0.00	0.29	0.00	0.00	0.00	0.00

Leaf/culm long cells	3.24	8.95	8.88	7.59	7.61	3.35	2.21	2.02	0.00	9.64	6.43	8.79	2.89	0.00
Leaf culm stomata	0.00	0.00	0.00	0.00	0.00	0.00	0.14	1.01	3.41	0.00	0.00	0.00	0.00	0.00
Leaf/ culm Phragmites	0.00	0.00	0.00	1.45	0.00	0.00	0.69	2.02	0.00	0.00	0.00	0.00	0.53	0.00
Leaf / culm reed	0.00	0.00	0.00	0.00	0.00	0.00	0.00	0.00	0.00	0.00	1.29	0.59	0.00	0.00
Leaf / culm Panicum	0.00	0.00	0.00	0.00	0.00	0.00	0.00	0.00	0.00	0.00	0.00	0.59	0.00	0.00
Square-cell leaf/stem	3.24	1.63	0.00	0.72	1.74	0.21	2.07	2.02	1.28	0.88	4.50	0.00	3.94	1.24
Indet husk	3.24	14.23	8.88	3.25	3.04	1.44	4.69	8.56	14.50	12.85	6.43	6.45	4.21	0.50
cf.Setaria husk	0.00	0.41	0.00	0.00	0.22	0.00	0.00	0.00	0.00	0.00	0.00	0.00	0.00	0.00
cf.Panicum husk	0.00	1.22	0.00	0.00	0.00	0.00	0.00	1.01	0.00	0.00	1.29	0.00	0.00	0.00
Millet type 1	0.36	0.41	1.87	1.81	0.00	0.27	0.14	1.01	2.99	2.92	0.00	1.76	0.26	0.00
Millet type 2	0.72	2.85	0.93	0.00	0.00	0.55	0.55	2.52	0.85	3.50	0.00	2.34	0.26	0.00
cf. Oryza husk	0.36	0.00	0.00	0.00	0.00	0.21	0.28	0.50	0.00	0.00	0.00	0.00	0.00	0.00
Cyperaceae	0.00	0.00	0.00	0.00	0.00	0.14	0.00	1.51	0.85	0.00	0.00	1.17	0.00	0.00
Polyhedron	1.44	3.25	1.87	0.72	0.43	0.00	0.41	2.52	4.26	3.80	0.64	0.59	0.79	0.00
Polyhedral hair base	0.72	1.22	0.93	1.45	0.00	0.07	0.69	0.50	0.00	0.58	0.64	0.59	0.00	0.00
Multi-Tiered forms	1.80	4.47	1.40	0.36	2.39	0.27	0.96	1.01	1.28	2.92	0.00	0.00	0.26	0.00
Mesophyll	0.00	0.00	0.00	2.17	0.87	0.89	1.24	1.01	2.99	2.04	0.00	0.00	0.00	0.00
Indet multicell	0.00	0.37	0.01	20.83	55.87	79.19	64.86	2.26	3.19	12.68	27.98	4.49	49.31	88.08
Indet phytolith	20.86	1.63	3.74	22.77	9.35	1.30	1.79	2.52	2.13	2.04	22.51	6.45	16.52	3.48
Diatom	0.00	0.00	0.00	0.00	0.00	0.00	0.00	0.00	0.00	0.00	0.64	0.00	0.00	0.00
Sponge Spicule	0.00	1.22	0.00	0.36	0.00	0.00	0.14	0.50	0.00	0.00	0.64	1.17	0.00	0.00
Starch	0.36	0.00	0.00	1.45	0.22	0.62	0.14	0.00	0.00	0.00	0.64	0.00	0.00	0.00
Silica aggregate	24.82	15.05	17.29	9.04	14.35	0.00	0.00	8.56	0.00	0.00	23.15	3.52	33.04	0.00

XIPO														
	X15	X16	X20	X21	X22	X23	X24	X25	X26	X27	X28	X29	X30	X31
Long Smooth	16.71	16.21	17.39	17.69	9.37	20.90	6.75	17.79	12.56	8.41	13.42	195.27	15.41	16.69
Long Sinuate	2.63	2.15	2.90	0.00	0.30	3.39	0.00	1.55	1.02	0.00	2.10	14.29	5.60	1.00
Long Rods	3.34	1.37	4.71	2.00	2.42	2.26	1.93	1.55	3.06	0.88	0.42	14.29	0.56	2.34
Long Dendritic	0.72	1.37	3.26	0.86	1.81	1.13	0.32	0.39	1.36	0.00	0.42	0.00	1.12	0.00
Stomata	0.95	0.20	0.00	0.00	0.00	0.28	0.00	0.00	0.68	0.00	0.00	0.00	0.28	0.00
Total hairs	30.07	32.61	28.62	38.23	33.23	22.88	32.47	35.98	24.11	46.90	47.80	266.71	31.93	31.05
Bulliform	1.67	3.32	2.17	3.14	2.72	2.54	3.86	0.77	0.00	0.88	1.68	85.73	4.76	1.67
Cuneiform Bulliform	0.95	0.78	1.81	0.86	1.51	1.13	1.29	0.39	0.00	0.44	2.52	14.29	9.52	0.67
cf. Oryza Bulliform	0.00	0.00	0.00	0.29	0.00	0.00	0.00	0.00	0.00	0.00	0.00	0.00	0.00	0.00
Crenate	0.24	0.00	0.00	0.29	0.00	0.00	0.00	0.00	0.00	0.00	0.00	0.00	0.00	0.00
Bilobate	14.32	9.98	10.14	8.42	14.20	12.43	3.22	18.38	25.64	0.44	1.47	9.53	1.40	4.51
Polylobate	0.24	0.20	0.00	0.00	0.00	0.28	0.00	0.39	0.68	0.00	0.00	0.00	0.00	1.00
Cross	1.43	0.59	0.36	0.86	0.30	0.85	0.00	3.09	1.70	0.00	0.00	0.00	0.28	0.00
cf. Setaria bilobate	0.00	0.00	0.00	0.00	0.00	0.56	0.00	1.16	1.02	0.00	0.00	0.00	0.00	0.00
cf. Oryza bilobate	0.48	0.39	0.36	0.29	0.60	0.28	0.00	0.39	0.34	0.00	0.00	0.00	0.00	1.34
Rondel	9.07	8.79	4.35	6.85	9.37	9.60	6.11	7.35	4.41	8.85	7.55	85.73	9.52	12.02
cf. Stipa Rondel	0.24	0.00	0.72	0.00	1.21	0.56	0.32	1.55	1.02	0.00	0.00	0.00	0.00	1.00
Saddles	2.15	0.98	2.90	1.43	2.42	8.76	0.64	0.77	5.09	0.44	0.42	4.76	1.12	1.00
Collapsed saddle	0.24	0.98	0.72	0.00	0.00	1.69	0.00	0.39	1.02	0.00	0.00	0.00	0.28	0.33
Cone	0.95	0.78	1.45	0.00	0.00	0.00	0.00	0.00	1.70	0.88	0.00	9.53	0.28	0.67
Rugulose Spheroid	0.00	0.00	0.00	0.00	0.00	0.00	0.00	0.00	0.00	0.00	0.00	0.00	0.00	0.00
Smooth Spheroid	0.00	0.39	0.00	0.29	1.21	0.00	0.96	0.00	0.00	0.44	0.00	4.76	0.56	0.67
Verrucate	0.95	0.00	0.36	0.86	0.00	0.00	0.32	0.00	0.00	0.00	0.00	0.00	0.00	0.00
Crescent	0.00	0.39	0.00	0.00	0.30	0.00	0.00	0.00	0.34	0.00	0.00	0.00	0.00	0.00
Elongate	3.58	3.71	3.99	3.14	3.93	0.28	0.96	2.32	4.41	0.88	0.00	33.34	0.56	2.34
Tracheid	0.72	3.51	3.26	1.71	1.21	2.26	1.61	0.00	1.02	0.00	0.00	4.76	1.96	0.00
Two-Tiered	0.00	0.00	0.00	0.00	0.30	0.00	0.96	0.39	0.00	0.44	0.00	0.00	0.00	0.00
Block	0.24	1.76	0.36	1.43	0.60	1.13	0.00	3.87	1.36	3.10	2.10	9.53	3.08	0.67
Platey	5.25	7.23	8.33	7.42	7.55	4.24	31.51	0.77	5.77	23.01	15.93	180.98	10.08	17.03
Sheet	0.00	0.00	0.72	0.86	0.60	0.00	2.89	0.77	1.36	2.65	0.00	0.00	0.00	1.67
Single Polyhedron	2.86	2.34	1.09	3.14	2.42	2.54	3.86	0.00	0.34	1.33	4.19	66.68	1.68	2.34
Scalloped	0.00	0.00	0.00	0.00	2.42	0.00	0.00	0.00	0.00	0.00	0.00	0.00	0.00	0.00
Single Jigsaw	0.00	0.00	0.00	0.00	0.00	0.00	0.00	0.00	0.00	0.00	0.00	0.00	0.00	0.00

puzzle														
Leaf/culm indet	46.79	2.81	26.46	37.73	36.92	0.01	19.78	53.06	80.76	1.80	29.00	20.53	49.24	30.02
Leaf/culm jigsaw	0.44	0.04	1.20	1.22	0.00	0.00	0.00	0.00	0.34	0.00	0.00	0.00	0.00	0.44
Leaf/culm bilobate	7.06	0.33	3.01	1.22	4.71	0.00	2.20	4.86	2.05	0.09	1.66	0.33	2.80	0.88
Leaf/culm saddle	1.32	0.00	1.80	1.22	1.09	0.00	1.10	0.00	0.17	0.00	0.00	0.00	1.12	0.44
Leaf/culm cross	0.00	0.00	0.00	0.00	0.00	0.01	0.00	0.44	0.51	0.05	0.00	0.00	0.00	0.44
Leaf/culm cf Oryza	0.00	0.00	0.00	0.00	0.00	0.00	0.00	0.00	0.00	0.00	0.00	0.00	0.00	0.00
Leaf/culm long cells	14.57	0.62	8.42	6.69	10.50	0.00	3.30	12.82	4.27	0.43	2.90	5.96	7.27	3.53
Leaf culm stomata	1.32	0.00	0.00	0.00	0.00	0.00	0.00	0.88	0.00	0.00	0.41	0.00	0.00	0.00
Leaf/ culm Phragmites	0.00	0.08	0.00	0.00	0.00	0.00	0.00	0.00	0.00	0.00	0.00	0.00	1.68	0.00
XIPO														
	X15	X16	X20	X21	X22	X23	X24	X25	X26	X27	X28	X29	X30	X31
Leaf / culm reed	0.88	0.12	0.60	0.00	2.53	0.00	0.00	0.00	0.00	0.00	0.00	0.66	1.12	0.00
Leaf / culm Panicum	0.44	0.00	0.00	0.61	2.17	0.01	0.00	0.00	0.85	0.00	0.00	0.00	1.12	0.00
Square-cell leaf/stem	2.21	0.21	3.01	3.04	2.17	0.00	4.40	0.00	0.51	0.09	0.83	0.66	3.36	0.44
Indet husk	14.12	0.58	8.42	12.78	7.60	0.00	5.49	7.08	5.13	0.19	0.41	2.65	6.71	3.53
cf.Setaria husk	0.00	0.00	0.00	0.00	0.00	0.00	0.00	0.00	0.68	0.00	0.00	0.00	0.56	0.00
cf.Panicum husk	0.00	0.04	0.60	0.00	0.00	0.00	0.00	0.00	0.17	0.00	0.00	0.00	0.00	0.00
Millet type 1	0.88	0.04	1.20	0.00	1.09	0.00	0.00	0.00	0.51	0.00	0.00	0.00	0.00	0.88
Millet type 2	0.44	0.04	1.80	0.61	0.36	0.00	0.00	0.44	1.20	0.00	0.00	0.00	0.00	2.21
cf. Oryza husk	0.00	0.00	0.00	0.00	0.00	0.00	0.00	0.00	0.00	0.00	0.00	0.00	0.00	0.00
Cyperaceae	0.00	0.08	3.01	1.83	2.90	0.00	2.20	0.44	1.20	0.00	2.07	1.99	1.12	0.88
Polyhedron	0.44	0.04	1.80	1.83	0.36	0.00	5.49	0.00	0.00	0.05	0.00	0.00	0.56	0.00
Polyhedral hair base	0.00	0.00	0.00	0.00	0.72	0.00	0.00	0.88	0.00	0.05	2.90	0.00	1.12	0.00
Multi-Tiered forms	0.00	0.04	0.00	0.00	0.00	0.00	1.10	0.00	0.00	0.00	0.00	0.00	0.00	0.00
Mesophyll	0.00	0.00	0.00	0.00	0.00	0.03	0.00	0.00	0.00	0.00	0.00	0.00	0.00	0.00
Indet multicell	2.01	94.22	29.03	19.66	3.72	0.00	12.09	0.06	0.26	49.18	0.00	55.63	6.87	10.60
Indet phytolith	4.41	0.62	9.62	10.96	21.72	0.00	24.18	19.01	1.37	48.06	59.81	11.59	15.35	44.43
Diatom	0.44	0.00	0.00	0.00	0.00	0.00	0.00	0.00	0.00	0.00	0.00	0.00	0.00	0.00
Sponge Spicule	1.77	0.08	0.00	0.61	1.45	0.01	1.10	0.00	0.00	0.00	0.00	0.00	0.00	1.27
Starch	0.00	0.00	0.00	0.00	0.00	0.00	0.00	0.00	0.00	0.00	0.00	0.00	0.00	0.00
Silica aggregate	3.97	0.25	0.04	6.69	7.24	0.00	17.58	5.75	0.34	1.04	2.33	0.33	3.84	11.42

Xipo								
	X33	X34	X35	X36	X38	X39	X40	X41
Long Smooth	33.88	16.36	22.22	15.12	18.34	13.00	13.85	18.51
Long Sinuate	1.41	1.82	0.87	1.94	2.42	1.08	0.81	2.64
Long Rods	5.65	0.45	1.44	3.10	2.77	2.53	4.48	0.48
Long Dendritic	1.88	0.00	0.00	0.78	0.35	0.72	1.22	0.96
Stomata	0.00	0.00	0.00	0.39	0.00	0.00	0.00	0.24
Total hairs	0.00	37.27	34.34	29.07	34.95	36.82	32.18	31.73
Bulliform	3.29	3.64	5.19	3.10	2.42	3.97	0.81	5.05
Cuneiform Bulliform	0.94	5.00	0.58	0.00	2.42	2.17	0.41	1.68
	0.00	0.00	0.00	0.00	0.35	0.00	0.00	0.00
cf. Oryza Bulliform	0.00	0.00	0.00	0.00	0.00	0.00	0.00	0.00
Crenate	20.47	0.91	5.34	15.50	9.00	3.97	0.20	7.93
Bilobate	0.00	0.00	0.00	0.39	0.35	0.00	20.37	0.24
Polylobate	0.00	0.00	0.58	0.78	1.38	1.08	1.22	0.72
Cross	0.00	0.00	0.29	1.55	0.00	0.00	0.00	0.00
cf. Setaria bilobate	1.88	0.00	0.29	0.00	0.00	0.00	0.81	0.96
cf. Oryza bilobate	9.88	7.73	12.12	10.85	8.30	11.19	6.52	8.41
Rondel	0.00	0.00	0.29	0.00	0.35	1.81	1.22	1.20
cf. Stipa Rondel	5.18	0.45	1.44	4.65	1.38	0.72	4.07	0.72
Saddles	0.00	0.00	0.00	0.00	0.00	0.00	0.41	0.24
Collapsed saddle	0.47	0.00	0.58	0.00	2.42	0.36	0.00	0.00
Cone	0.00	0.00	0.00	0.00	0.00	0.00	0.00	0.00
Rugulose Spheroid	0.94	0.00	0.00	0.00	0.35	0.00	0.00	0.00

Smooth Spheroid	0.47	0.00	0.00	0.00	0.35	0.36	0.41	0.00
Verrucate	0.00	0.00	0.00	0.00	0.00	0.36	0.00	0.48
Crescent	2.35	5.00	2.89	5.04	3.11	3.25	2.04	2.16
Elongate	0.00	0.00	0.87	1.16	0.35	0.00	0.00	0.72
Tracheid	0.00	0.00	0.00	0.00	0.00	0.72	0.00	0.00
Two-Tiered	0.47	0.45	2.60	1.94	0.35	1.81	0.00	0.24
Block	9.41	20.91	7.79	4.26	5.54	12.64	6.11	9.86
Platey	0.47	0.00	0.00	0.39	1.38	0.72	0.00	0.00
Sheet	0.94	0.00	0.29	0.00	1.38	0.72	2.44	4.81
Single Polyhedron	0.00	0.00	0.00	0.00	0.00	0.00	0.00	0.00
Scalloped	0.00	0.00	0.00	0.00	0.00	0.00	0.41	0.00
Single Jigsaw puzzle	54.59	15.61	28.61	65.73	38.78	54.85	37.75	40.69
Leaf/culm indet	0.00	0.00	1.07	2.36	0.54	0.87	0.00	0.98
Leaf/culm jigsaw	9.73	0.00	1.43	3.55	3.23	1.74	4.41	2.94
Leaf/culm bilobate	1.08	0.00	0.00	0.39	0.00	0.00	0.00	0.49
Leaf/culm saddle	1.62	0.00	0.00	0.79	2.15	1.74	0.49	0.00
Leaf/culm cross	0.00	0.00	0.00	0.00	0.00	0.00	0.00	0.00
Leaf/culm cf Oryza	9.73	7.17	4.65	4.73	8.62	13.06	12.25	11.76
Leaf/culm long cells	2.16	0.00	0.00	0.00	0.00	0.00	0.00	0.49
Leaf culm stomata	0.00	0.00	0.00	0.00	0.00	0.87	0.00	0.98
Leaf/ culm Phragmites	1.08	0.42	0.00	0.79	0.54	0.00	2.45	0.49
Leaf / culm reed	3.24	0.00	0.00	0.00	0.00	0.87	3.43	1.96
Leaf / culm Panicum	2.70	0.00	0.36	0.00	2.15	4.35	0.49	2.94
Square-cell leaf/stem	5.95	2.95	6.79	0.00	6.46	9.58	10.29	10.29
Indet husk	0.00	0.00	0.00	0.00	0.00	0.00	0.49	0.00
cf.Setaria husk	0.00	0.00	0.00	0.00	0.00	0.87	0.98	0.00
cf.Panicum husk	0.54	0.00	0.00	0.39	0.00	0.00	0.98	1.47
Millet type 1	0.00	0.00	0.00	1.18	0.54	0.00	0.49	3.92
Millet type 2	0.00	0.00	0.00	0.00	0.00	1.74	0.00	0.49
cf. Oryza husk	5.95	0.00	0.00	0.39	3.23	4.35	0.98	1.96
Cyperaceae	0.00	0.84	2.50	0.39	3.23	3.48	0.98	2.94
Polyhedron	0.54	0.00	1.07	0.00	0.00	0.00	0.00	0.00
Polyhedral hair base	0.00	0.00	0.00	0.00	0.00	0.00	0.00	0.00
Multi-Tiered forms	0.00	0.00	0.00	0.00	0.00	0.00	0.00	0.00
Mesophyll	0.54	2.11	4.25	10.64	13.46	1.02	10.29	9.80
Indet multicell	0.00	70.89	48.41	8.27	16.16	0.58	12.75	4.90
Indet phytolith	0.00	0.00	0.00	0.00	0.00	0.00	0.00	0.00
Diatom	0.54	0.00	0.00	0.39	0.90	0.02	0.49	0.49
Sponge Spicule	0.00	0.00	0.85	0.00	0.00	0.00	0.00	0.00
Starch	3.24	1.48	18.69	0.39	1.80	0.36	4.41	3.43
Silica aggregate	33.88	16.36	22.22	15.12	18.34	13.00	13.85	18.51

Huizui														
	HE04	HE05	HE06	HE07	HE08	HE09	HE12	HE17	HE18	HE19	HE90	HE91	HE92	HE93
Long Smooth	23.77	22.22	19.75	23.33	28.87	7.58	20.55	23.58	18.29	21.43	26.69	31.09	19.72	20.99
Long Sinuate	2.16	3.98	1.44	2.08	0.42	2.04	2.77	1.22	1.80	1.34	2.56	2.24	4.43	4.94
Long Rods	2.47	2.32	2.15	15.42	5.44	1.75	4.74	0.41	1.20	0.89	3.66	5.45	6.84	3.29
Long Dendritic	0.62	2.65	0.00	0.00	0.42	0.87	3.16	1.22	0.90	0.89	1.10	6.09	1.21	0.41
Stomata	0.00	0.00	0.00	0.00	0.00	0.00	0.00	0.00	3.00	0.00	0.00	0.00	0.40	0.00
Total hairs	23.15	27.53	32.67	27.08	27.20	33.53	32.41	30.08	23.09	37.95	30.35	24.04	35.01	39.09
Bulliform	5.86	2.65	5.39	4.17	5.02	4.96	2.77	4.47	2.40	1.79	3.29	2.24	3.62	3.29
Cuneiform Bulliform	6.48	3.98	5.03	3.33	0.42	2.04	0.79	3.25	1.50	1.79	1.83	2.88	2.41	0.82
cf. Oryza Bulliform	0.00	0.00	0.72	0.00	0.00	0.29	0.00	0.00	0.00	0.00	0.00	0.00	0.00	0.00
Crenate	0.00	0.00	0.00	0.00	0.00	0.00	0.00	0.00	0.00	0.00	0.00	0.00	0.00	0.00
Bilobate	5.25	2.49	2.33	0.42	2.93	0.29	0.00	2.03	2.85	4.02	6.40	2.24	6.64	0.00
Polylobate	0.00	0.00	0.00	0.00	0.42	0.00	0.00	0.00	0.00	0.00	0.37	0.00	0.80	0.00
Cross	0.00	0.00	0.36	0.00	0.00	0.00	0.00	0.41	0.00	0.00	0.00	0.00	0.00	0.00
cf. Setaria bilobate	0.31	0.33	0.36	0.42	0.00	0.00	0.00	0.00	0.00	0.00	0.00	0.32	0.00	0.00
cf. Oryza bilobate	1.23	1.00	0.00	0.00	0.00	0.00	0.40	0.41	1.50	0.00	0.00	0.00	0.00	0.00
Rondel	10.49	11.28	12.21	6.67	8.79	7.87	3.16	4.88	6.30	1.34	7.31	5.13	7.65	4.53

Huizui														
	HE04	HE05	HE06	HE07	HE08	HE09	HE12	HE17	HE18	HE19	HE90	HE91	HE92	HE93
cf. Stipa Rondel	0.31	0.33	0.72	0.83	0.42	0.58	0.00	0.00	0.00	0.00	1.46	0.00	0.00	0.00
Saddles	0.93	0.66	0.72	0.83	0.42	3.50	0.40	0.00	0.30	0.45	0.00	0.32	2.41	1.23
Collapsed saddle	0.00	1.33	0.00	0.00	0.42	0.00	0.00	0.00	0.00	0.45	0.00	0.32	0.40	0.41
Cone	0.62	1.66	0.00	2.08	0.00	0.29	0.00	0.81	0.00	0.00	1.10	0.00	1.21	1.23
Rugulose Spheroid	0.00	1.00	0.00	0.00	0.00	0.00	0.00	0.00	0.30	0.00	0.00	0.00	0.00	0.00
Smooth Spheroid	0.00	0.66	0.00	0.00	0.00	0.00	0.00	0.00	0.00	0.89	0.37	0.00	0.00	0.00
Verrucate	0.00	0.00	0.00	0.00	0.00	0.00	0.00	0.00	0.00	0.00	0.00	0.00	0.00	0.00
Crescent	0.00	0.00	0.00	0.00	0.42	0.00	0.00	0.00	0.00	0.00	0.00	0.00	0.40	0.00
Elongate	2.16	1.66	5.74	0.00	2.09	0.87	1.98	2.03	1.20	1.34	0.00	2.56	0.40	5.35
Tracheid	1.54	1.00	0.36	1.25	2.09	0.00	3.16	2.03	0.00	1.34	1.46	4.49	1.61	0.00
Two-Tiered	0.00	1.00	1.08	0.00	0.42	1.75	0.00	0.00	0.00	0.45	0.00	0.00	0.00	0.00
Block	1.54	4.31	1.08	4.17	0.42	1.17	0.00	0.81	0.60	1.79	1.46	0.00	0.40	1.23
Platey	7.41	5.64	6.46	7.08	11.30	26.82	22.92	21.55	30.88	19.20	9.87	9.62	3.22	7.41
Sheet	3.70	0.00	0.36	0.42	0.42	0.58	0.79	0.41	0.90	0.45	0.00	0.00	0.00	0.82
Single Polyhedron	0.00	0.33	1.08	0.42	1.67	2.92	0.00	0.41	3.30	2.23	0.00	0.00	1.21	4.94
Scalloped	0.00	0.00	0.00	0.00	0.00	0.29	0.00	0.00	0.00	0.00	0.37	0.96	0.00	0.00
Single Jigsaw puzzle	0.00	0.00	0.00	0.00	0.00	0.00	0.00	0.00	0.00	0.00	0.37	0.00	0.00	0.00
Leaf/culm indet	29.51	8.31	34.26	21.78	41.38	48.93	21.27	12.22	13.50	23.48	16.18	20.55	19.20	3.22
Leaf/culm jigsaw	0.00	0.50	0.55	0.00	0.99	3.26	0.00	0.00	0.36	0.00	0.00	0.00	0.00	0.06
Leaf/culm bilobate	2.42	0.33	0.00	0.57	0.00	0.00	0.00	0.18	0.00	0.59	1.03	0.00	0.36	0.06
Leaf/culm saddle	0.00	0.17	0.00	0.00	0.49	0.00	0.00	0.00	0.00	0.00	0.00	0.00	0.00	0.00
Leaf/culm cross	0.00	0.00	0.00	0.00	0.00	0.00	0.00	0.00	0.00	0.00	0.00	0.00	0.00	0.00
Leaf/culm cf Oryza	0.00	0.00	0.00	0.00	0.00	0.00	0.00	0.00	0.00	0.00	0.00	0.00	0.00	0.00
Leaf/culm long cells	13.06	7.97	5.53	19.48	9.85	5.44	3.77	2.19	3.10	4.11	4.88	8.30	9.42	0.94
Leaf culm stomata	0.00	0.00	0.00	0.00	0.00	0.00	0.00	0.18	0.00	0.00	0.00	0.00	0.00	0.00
Leaf/ culm Phragmites	0.00	0.00	0.55	0.00	0.00	0.00	0.00	0.00	0.73	0.29	0.00	0.00	0.00	0.00
Leaf / culm reed	0.97	0.00	0.00	1.15	0.99	0.00	0.00	0.00	0.00	0.00	0.00	0.00	1.09	0.06
Leaf / culm Panicum	0.00	0.00	0.00	0.00	0.00	2.17	0.00	0.00	0.00	0.29	0.00	0.79	0.00	0.00
Square-cell leaf/stem	2.42	2.16	3.32	5.16	1.97	6.52	0.34	1.46	0.18	0.59	1.03	1.19	2.17	0.22
Indet husk	11.61	2.49	13.26	9.17	20.69	4.35	11.32	0.36	3.47	0.59	1.28	14.62	3.62	0.50
cf.Setaria husk	0.00	0.00	0.00	0.00	0.00	0.00	0.34	0.00	0.55	0.00	0.00	0.00	0.00	0.00
cf.Panicum husk	0.00	0.00	0.00	0.00	0.00	0.00	0.34	0.00	0.00	0.00	0.00	0.00	0.00	0.00
Millet type 1	0.97	0.00	0.00	0.00	2.46	0.00	0.34	0.00	0.00	0.00	0.00	0.00	0.00	0.17
Millet type 2	3.39	0.17	0.00	0.00	1.97	0.00	2.40	0.36	0.00	0.00	0.00	2.77	0.00	0.06
cf. Oryza husk	1.94	0.00	0.00	0.00	5.42	0.00	0.00	0.00	0.00	0.00	0.00	0.00	0.00	0.00
Cyperaceae	2.90	1.33	0.00	5.73	0.99	0.00	0.69	0.18	0.00	0.29	0.00	0.40	1.45	0.83
Polyhedron	1.94	0.00	0.00	1.72	0.99	4.35	0.00	0.18	0.00	0.00	0.51	0.00	1.81	0.00
Polyhedral hair base	0.00	0.00	0.00	0.57	0.99	0.00	0.00	0.18	0.00	0.29	0.00	0.00	0.00	0.50
Huizui														
	HE04	HE05	HE06	HE07	HE08	HE09	HE12	HE17	HE18	HE19	HE90	HE91	HE92	HE93
Multi-Tiered forms	0.00	0.00	0.00	0.00	0.00	0.00	0.00	0.00	0.00	0.00	0.00	0.00	0.00	0.00
Mesophyll	0.00	0.17	0.00	0.00	0.00	2.17	0.34	0.18	0.00	0.00	0.00	0.00	0.00	0.00
Indet multicell	18.25	1.91	2.76	6.30	4.43	8.70	25.73	4.95	5.11	6.32	2.53	27.67	25.36	6.20
Indet phytolith	0.00	63.04	36.47	12.61	3.45	8.70	20.58	47.65	17.52	26.67	61.59	15.81	28.74	86.80
Diatom	0.00	0.00	0.00	0.00	0.99	0.00	0.00	0.00	1.46	0.00	1.69	1.32	0.85	0.00
Sponge Spicule	0.00	5.73	1.11	12.61	0.99	1.09	0.00	0.00	4.38	0.00	0.00	1.32	0.85	0.20
Starch	0.00	0.00	0.00	0.00	0.00	0.00	0.00	0.00	0.00	0.00	0.84	0.00	0.00	0.00
Silica aggregate	10.64	5.73	2.21	3.15	0.99	4.35	11.84	29.70	49.63	36.49	8.44	5.27	5.07	0.20

Huizui														
	HE04	HE05	HE06	HE07	HE08	HE09	HE12	HE17	HE18	HE19	HE90	HE91	HE92	HE93
Long Smooth	23.77	22.22	19.75	23.33	28.87	7.58	20.55	23.58	18.29	21.43	26.69	31.09	19.72	20.99
Long Sinuate	2.16	3.98	1.44	2.08	0.42	2.04	2.77	1.22	1.80	1.34	2.56	2.24	4.43	4.94
Long Rods	2.47	2.32	2.15	15.42	5.44	1.75	4.74	0.41	1.20	0.89	3.66	5.45	6.84	3.29
Long Dendritic	0.62	2.65	0.00	0.00	0.42	0.87	3.16	1.22	0.90	0.89	1.10	6.09	1.21	0.41

Stomata	0.00	0.00	0.00	0.00	0.00	0.00	0.00	0.00	3.00	0.00	0.00	0.00	0.40	0.00
Total hairs	23.15	27.53	32.67	27.08	27.20	33.53	32.41	30.08	23.09	37.95	30.35	24.04	35.01	39.09
Bulliform	5.86	2.65	5.39	4.17	5.02	4.96	2.77	4.47	2.40	1.79	3.29	2.24	3.62	3.29
Cuneiform Bulliform	6.48	3.98	5.03	3.33	0.42	2.04	0.79	3.25	1.50	1.79	1.83	2.88	2.41	0.82
cf. Oryza Bulliform	0.00	0.00	0.72	0.00	0.00	0.29	0.00	0.00	0.00	0.00	0.00	0.00	0.00	0.00
Crenate	0.00	0.00	0.00	0.00	0.00	0.00	0.00	0.00	0.00	0.00	0.00	0.00	0.00	0.00
Bilobate	5.25	2.49	2.33	0.42	2.93	0.29	0.00	2.03	2.85	4.02	6.40	2.24	6.64	0.00
Polylobate	0.00	0.00	0.00	0.00	0.42	0.00	0.00	0.00	0.00	0.00	0.37	0.00	0.80	0.00
Cross	0.00	0.00	0.36	0.00	0.00	0.00	0.00	0.41	0.00	0.00	0.00	0.00	0.00	0.00
cf. Setaria bilobate	0.31	0.33	0.36	0.42	0.00	0.00	0.00	0.00	0.00	0.00	0.00	0.32	0.00	0.00
cf. Oryza bilobate	1.23	1.00	0.00	0.00	0.00	0.00	0.40	0.41	1.50	0.00	0.00	0.00	0.00	0.00
Rondel	10.49	11.28	12.21	6.67	8.79	7.87	3.16	4.88	6.30	1.34	7.31	5.13	7.65	4.53
cf. Stipa Rondel	0.31	0.33	0.72	0.83	0.42	0.58	0.00	0.00	0.00	0.00	1.46	0.00	0.00	0.00
Saddles	0.93	0.66	0.72	0.83	0.42	3.50	0.40	0.00	0.30	0.45	0.00	0.32	2.41	1.23
Collapsed saddle	0.00	1.33	0.00	0.00	0.42	0.00	0.00	0.00	0.00	0.45	0.00	0.32	0.40	0.41
Cone	0.62	1.66	0.00	2.08	0.00	0.29	0.00	0.81	0.00	0.00	1.10	0.00	1.21	1.23
Rugulose Spheroid	0.00	1.00	0.00	0.00	0.00	0.00	0.00	0.00	0.30	0.00	0.00	0.00	0.00	0.00
Smooth Spheroid	0.00	0.66	0.00	0.00	0.00	0.00	0.00	0.00	0.00	0.89	0.37	0.00	0.00	0.00
Verrucate	0.00	0.00	0.00	0.00	0.00	0.00	0.00	0.00	0.00	0.00	0.00	0.00	0.00	0.00
Crescent	0.00	0.00	0.00	0.00	0.42	0.00	0.00	0.00	0.00	0.00	0.00	0.00	0.40	0.00
Elongate	2.16	1.66	5.74	0.00	2.09	0.87	1.98	2.03	1.20	1.34	0.00	2.56	0.40	5.35
Tracheid	1.54	1.00	0.36	1.25	2.09	0.00	3.16	2.03	0.00	1.34	1.46	4.49	1.61	0.00
Two-Tiered	0.00	1.00	1.08	0.00	0.42	1.75	0.00	0.00	0.00	0.45	0.00	0.00	0.00	0.00
Block	1.54	4.31	1.08	4.17	0.42	1.17	0.00	0.81	0.60	1.79	1.46	0.00	0.40	1.23
Platey	7.41	5.64	6.46	7.08	11.30	26.82	22.92	21.55	30.88	19.20	9.87	9.62	3.22	7.41
Sheet	3.70	0.00	0.36	0.42	0.42	0.58	0.79	0.41	0.90	0.45	0.00	0.00	0.00	0.82
Single Polyhedron	0.00	0.33	1.08	0.42	1.67	2.92	0.00	0.41	3.30	2.23	0.00	0.00	1.21	4.94
Scalloped	0.00	0.00	0.00	0.00	0.00	0.29	0.00	0.00	0.00	0.00	0.37	0.96	0.00	0.00
Huizui														
	HE04	HE05	HE06	HE07	HE08	HE09	HE12	HE17	HE18	HE19	HE90	HE91	HE92	HE93
Single Jigsaw puzzle	0.00	0.00	0.00	0.00	0.00	0.00	0.00	0.00	0.00	0.00	0.37	0.00	0.00	0.00
Leaf/culm indet	29.51	8.31	34.26	21.78	41.38	48.93	21.27	12.22	13.50	23.48	16.18	20.55	19.20	3.22
Leaf/culm jigsaw	0.00	0.50	0.55	0.00	0.99	3.26	0.00	0.00	0.36	0.00	0.00	0.00	0.00	0.06
Leaf/culm bilobate	2.42	0.33	0.00	0.57	0.00	0.00	0.00	0.18	0.00	0.59	1.03	0.00	0.36	0.06
Leaf/culm saddle	0.00	0.17	0.00	0.00	0.49	0.00	0.00	0.00	0.00	0.00	0.00	0.00	0.00	0.00
Leaf/culm cross	0.00	0.00	0.00	0.00	0.00	0.00	0.00	0.00	0.00	0.00	0.00	0.00	0.00	0.00
Leaf/culm cf Oryza	0.00	0.00	0.00	0.00	0.00	0.00	0.00	0.00	0.00	0.00	0.00	0.00	0.00	0.00
Leaf/culm long cells	13.06	7.97	5.53	19.48	9.85	5.44	3.77	2.19	3.10	4.11	4.88	8.30	9.42	0.94
Leaf culm stomata	0.00	0.00	0.00	0.00	0.00	0.00	0.00	0.18	0.00	0.00	0.00	0.00	0.00	0.00
Leaf/ culm Phragmites	0.00	0.00	0.55	0.00	0.00	0.00	0.00	0.00	0.73	0.29	0.00	0.00	0.00	0.00
Leaf / culm reed	0.97	0.00	0.00	1.15	0.99	0.00	0.00	0.00	0.00	0.00	0.00	0.00	1.09	0.06
Leaf / culm Panicum	0.00	0.00	0.00	0.00	0.00	2.17	0.00	0.00	0.00	0.29	0.00	0.79	0.00	0.00
Square-cell leaf/stem	2.42	2.16	3.32	5.16	1.97	6.52	0.34	1.46	0.18	0.59	1.03	1.19	2.17	0.22
Indet husk	11.61	2.49	13.26	9.17	20.69	4.35	11.32	0.36	3.47	0.59	1.28	14.62	3.62	0.50
cf.Setaria husk	0.00	0.00	0.00	0.00	0.00	0.00	0.34	0.00	0.55	0.00	0.00	0.00	0.00	0.00
cf.Panicum husk	0.00	0.00	0.00	0.00	0.00	0.00	0.34	0.00	0.00	0.00	0.00	0.00	0.00	0.00
Millet type 1	0.97	0.00	0.00	0.00	2.46	0.00	0.34	0.00	0.00	0.00	0.00	0.00	0.00	0.17
Millet type 2	3.39	0.17	0.00	0.00	1.97	0.00	2.40	0.36	0.00	0.00	0.00	2.77	0.00	0.06
cf. Oryza husk	1.94	0.00	0.00	0.00	5.42	0.00	0.00	0.00	0.00	0.00	0.00	0.00	0.00	0.00
Cyperaceae	2.90	1.33	0.00	5.73	0.99	0.00	0.69	0.18	0.00	0.29	0.00	0.40	1.45	0.83
Polyhedron	1.94	0.00	0.00	1.72	0.99	4.35	0.00	0.18	0.00	0.00	0.51	0.00	1.81	0.00
Polyhedral hair base	0.00	0.00	0.00	0.57	0.99	0.00	0.00	0.18	0.00	0.29	0.00	0.00	0.00	0.50
Multi-Tiered forms	0.00	0.00	0.00	0.00	0.00	0.00	0.00	0.00	0.00	0.00	0.00	0.00	0.00	0.00
Mesophyll	0.00	0.17	0.00	0.00	0.00	2.17	0.34	0.18	0.00	0.00	0.00	0.00	0.00	0.00
Indet multicell	18.25	1.91	2.76	6.30	4.43	8.70	25.73	4.95	5.11	6.32	2.53	27.67	25.36	6.20
Indet phytolith	0.00	63.04	36.47	12.61	3.45	8.70	20.58	47.65	17.52	26.67	61.59	15.81	28.74	86.80
Diatom	0.00	0.00	0.00	0.00	0.99	0.00	0.00	0.00	1.46	0.00	1.69	1.32	0.85	0.00
Sponge Spicule	0.00	5.73	1.11	12.61	0.99	1.09	0.00	0.00	4.38	0.00	0.00	1.32	0.85	0.20
Starch	0.00	0.00	0.00	0.00	0.00	0.00	0.00	0.00	0.00	0.00	0.84	0.00	0.00	0.00
Silica aggregate	10.64	5.73	2.21	3.15	0.99	4.35	11.84	29.70	49.63	36.49	8.44	5.27	5.07	0.20

Huizui												
	HE94	HE95	HE96	HE97	HE98	HE99	HE100	HE101	HE102	HE103	HE105	HE107
Long Smooth	26.80	18.71	26.37	24.95	10.96	28.25	9.59	16.35	23.57	21.29	10.41	29.48
Long Sinuate	0.34	3.24	0.91	1.25	0.69	0.45	0.00	0.00	0.76	0.80	0.00	0.00
Long Rods	8.93	7.91	2.27	0.42	0.00	4.48	1.37	3.85	5.32	1.20	0.00	9.16
Long Dendritic	0.69	3.96	0.91	0.83	0.69	0.90	0.68	0.00	1.52	0.00	0.00	1.20
Huizui												
	HE94	HE95	HE96	HE97	HE98	HE99	HE100	HE101	HE102	HE103	HE105	HE107
Stomata	0.00	0.00	0.00	0.00	0.00	0.00	0.00	0.00	0.00	0.00	0.00	0.00
Total hairs	28.87	39.57	51.82	29.52	37.68	17.04	44.52	46.16	28.90	27.31	27.60	22.31
Bulliform	3.44	1.44	4.09	4.16	10.28	9.42	1.37	2.88	5.70	3.21	5.88	3.19
Cuneiform Bulliform	0.69	0.00	0.00	6.65	1.37	5.38	0.68	0.00	4.18	1.20	0.45	2.79
cf. Oryza Bulliform	0.00	0.00	0.00	0.83	0.00	0.00	0.00	0.00	0.00	0.00	0.45	0.00
Crenate	0.00	0.00	0.00	0.42	0.00	0.00	0.00	0.00	0.00	0.00	0.00	0.00
Bilobate	4.81	8.27	0.00	5.20	0.00	1.79	0.00	0.00	0.76	2.01	0.00	19.92
Polylobate	1.37	2.52	0.00	0.83	0.00	0.00	0.00	0.00	0.00	0.00	0.00	0.00
Cross	0.00	0.00	0.00	0.00	0.00	0.00	0.00	0.00	0.00	0.00	0.00	1.59
cf. Setaria bilobate	0.00	0.00	0.00	0.42	0.00	0.00	0.00	0.00	0.00	0.00	0.00	0.40
cf. Oryza bilobate	0.34	0.00	0.00	0.83	0.00	0.00	0.00	0.00	0.00	0.00	0.00	3.19
Rondel	5.50	6.47	6.36	5.82	4.80	4.93	2.74	2.40	6.08	4.82	0.45	0.80
cf. Stipa Rondel	0.34	0.00	0.00	0.00	0.00	0.00	0.00	0.48	0.00	0.00	0.00	0.00
Saddles	0.34	0.00	0.00	0.42	0.00	0.90	0.00	0.00	0.38	0.00	0.00	0.00
Collapsed saddle	0.00	1.08	0.45	0.42	0.00	0.00	0.00	0.00	0.00	0.00	0.00	0.00
Cone	1.03	0.00	0.00	0.00	0.00	0.00	0.00	0.00	0.76	0.40	0.00	0.40
Rugulose Spheroid	0.00	0.00	0.00	0.00	0.00	0.00	0.00	0.00	0.00	0.00	0.00	0.00
Smooth Spheroid	0.00	0.00	0.00	0.00	0.00	0.00	0.00	0.48	0.38	0.00	0.00	0.00
Verrucate	0.00	0.00	0.00	0.00	0.00	0.00	0.00	0.00	0.00	0.00	0.00	0.00
Crescent	0.00	0.00	0.00	0.42	0.69	0.00	0.00	0.00	0.00	0.40	0.00	0.00
Elongate	2.41	4.32	1.82	1.66	0.69	2.24	1.37	0.00	2.66	1.20	1.36	1.20
Tracheid	0.69	2.16	0.00	2.49	0.69	0.00	0.00	0.48	0.76	0.00	0.00	0.00
Two-Tiered	0.69	0.00	0.00	0.42	0.69	0.90	1.37	0.00	0.38	0.00	0.00	0.00
Block	0.34	0.00	0.00	1.66	1.37	0.45	0.00	0.00	0.00	0.00	0.00	0.00
Platey	9.28	0.36	5.00	8.73	27.40	21.08	35.62	25.96	17.87	34.94	52.94	3.59
Sheet	0.00	0.00	0.00	1.25	0.00	0.00	0.68	0.96	0.00	0.80	0.00	0.40
Single Polyhedron	2.41	0.00	0.00	0.00	0.00	0.00	0.00	0.00	0.00	0.40	0.45	0.40
Scalloped	0.69	0.00	0.00	0.00	2.06	1.79	0.00	0.00	0.00	0.00	0.00	0.00
Single Jigsaw puzzle	0.00	0.00	0.00	0.42	0.00	0.00	0.00	0.00	0.00	0.00	0.00	0.00
Leaf/culm indet	58.15	38.16	21.75	32.98	15.72	34.35	37.07	27.95	5.55	13.25	35.71	64.53
Leaf/culm jigsaw	0.00	0.00	0.00	1.00	15.72	0.59	0.00	0.00	0.29	0.00	0.00	0.00
Leaf/culm bilobate	1.63	2.42	0.00	1.00	0.00	0.59	3.53	4.41	0.00	0.49	0.00	8.14
Leaf/culm saddle	0.00	0.00	0.00	0.50	0.00	0.00	0.00	0.00	0.00	0.00	0.00	0.00
Leaf/culm	0.54	0.00	0.00	1.00	0.00	0.59	5.30	0.00	0.00	0.00	0.00	1.16

cross												
Leaf/culm cf Oryza	0.00	0.00	0.00	1.00	0.00	0.00	0.00	0.00	0.00	0.00	0.00	0.00
Leaf/culm long cells	14.13	10.14	0.00	13.99	4.49	8.88	0.00	2.94	9.35	6.56	14.29	11.63
Leaf culm stomata	0.54	0.00	0.00	0.00	0.00	0.00	0.00	0.00	0.00	0.00	0.00	0.00
Huizui												
	HE94	HE95	HE96	HE97	HE98	HE99	HE100	HE101	HE102	HE103	HE105	HE107
Leaf/ culm Phragmites	0.00	0.00	0.00	0.00	2.25	0.00	0.00	0.00	0.29	0.49	0.00	0.00
Leaf / culm reed	0.54	0.00	0.00	2.50	0.00	0.00	0.00	0.00	0.00	0.00	0.00	0.00
Leaf / culm Panicum	0.00	0.00	0.00	0.00	0.00	0.00	0.00	0.00	0.00	0.24	0.00	0.00
Square-cell leaf/stem	1.63	0.00	0.33	1.00	1.12	1.18	0.00	1.47	1.46	1.22	7.14	1.74
Indet husk	5.98	27.05	0.67	20.48	1.12	14.81	17.65	29.42	3.21	2.67	0.00	6.40
cf.Setaria husk	0.54	5.31	0.00	2.00	0.00	0.59	0.00	0.00	0.00	0.00	0.00	0.00
cf.Panicum husk	0.00	0.48	0.00	1.00	0.00	0.00	0.00	0.00	0.00	0.00	0.00	0.00
Millet type 1	0.00	5.31	0.00	1.00	0.00	0.59	0.00	0.00	0.29	0.00	0.00	0.00
Millet type 2	0.54	6.76	0.00	3.00	0.00	4.15	0.00	0.00	0.00	0.00	0.00	0.00
cf. Oryza husk	0.00	0.00	0.00	0.50	0.00	0.00	0.00	1.47	0.00	0.00	0.00	0.00
Cyperaceae	1.09	0.00	0.00	0.00	0.00	0.00	0.00	0.00	1.75	0.49	0.00	0.00
Polyhedron	4.89	0.00	0.00	1.00	0.00	0.59	0.00	0.00	0.58	0.49	0.00	0.58
Polyhedral hair base	0.00	1.45	0.00	0.00	0.00	0.59	1.77	0.00	0.00	0.00	0.00	0.00
Multi-Tiered forms	0.00	0.00	0.00	0.00	0.00	0.00	0.00	0.00	0.00	0.00	0.00	0.00
Mesophyll	0.00	1.45	0.00	0.00	0.00	0.00	0.00	0.00	4.27	0.00	0.00	0.00
Indet multicell	5.43	1.45	1.08	8.92	5.61	5.08	11.56	23.54	24.21	0.00	32.14	2.91
Indet phytolith	2.72	0.00	74.01	0.00	52.78	26.40	13.87	0.00	37.03	54.68	0.00	2.33
Diatom	0.00	0.00	0.00	1.78	0.00	0.00	0.00	0.00	0.00	1.22	0.00	0.00
Sponge Spicule	0.00	0.00	0.00	0.00	0.00	1.02	0.00	0.00	7.12	3.65	0.00	0.58
Starch	0.00	0.00	1.08	0.00	1.12	0.00	2.31	8.83	0.00	0.00	0.00	0.00
Silica aggregate	1.63	0.00	1.08	5.35	0.00	0.00	6.94	0.00	4.27	14.58	10.71	0.00

HUIZUI												
	HE108	HE110	HGS4	HGS5	HE1	HE2	HE52	HE53	HE54	HE55	HE56	HE57
Long Smooth	26.03	18.60	15.53	18.78	26.78	12.71	18.95	20.60	19.68	26.12	21.45	20.57
Long Sinuate	1.65	1.55	2.82	2.71	1.67	1.94	2.46	1.87	1.94	2.45	1.07	0.00
Long Rods	2.48	3.88	1.41	0.90	4.60	5.17	0.70	1.87	1.94	6.12	5.90	1.71
Long Dendritic	5.79	2.71	1.41	1.81	0.84	1.51	1.05	0.75	0.97	2.04	0.54	2.29
Stomata	0.41	0.00	0.00	0.00	0.00	0.00	0.35	0.00	0.00	0.00	0.00	0.00
Total hairs	28.10	28.68	21.18	24.21	34.31	22.40	38.95	38.58	28.71	24.90	31.64	46.29
Bulliform	5.37	5.81	16.71	9.95	3.35	1.29	4.91	3.00	7.10	4.90	1.07	0.00
Cuneiform Bulliform	0.41	1.55	9.88	5.43	0.84	2.15	2.46	0.75	2.90	1.22	0.54	0.57

cf. Oryza Bulliform	0.00	0.39	1.18	0.45	0.00	0.22	0.00	0.00	0.00	0.00	0.00	0.00
Crenate	0.00	0.00	0.00	0.45	0.00	0.00	0.00	0.00	0.00	0.00	0.00	0.00
Bilobate	3.72	6.20	1.18	0.90	7.11	11.03	5.26	2.62	7.10	5.31	16.89	5.14
Polylobate	0.41	0.78	0.24	0.23	0.84	0.00	0.70	0.00	2.90	0.00	0.00	2.29
Cross	0.00	0.78	0.71	0.68	0.00	1.29	0.00	0.00	0.00	0.41	1.07	0.00
HUIZUI												
	HE108	HE110	HGS4	HGS5	HE1	HE2	HE52	HE53	HE54	HE55	HE56	HE57
cf. Setaria bilobate	0.00	1.55	0.00	0.00	0.00	0.22	0.00	0.00	0.00	0.00	0.00	0.00
cf. Oryza bilobate	0.00	0.00	0.24	0.68	0.42	2.59	0.00	0.75	0.00	2.04	0.00	1.71
Rondel	10.74	13.18	0.94	1.36	2.93	10.12	5.26	4.12	6.13	5.31	5.90	1.71
cf. Stipa Rondel	0.00	0.00	0.24	0.90	0.00	0.86	0.00	0.00	0.65	0.82	0.00	4.57
Saddles	0.41	2.71	0.71	1.13	0.84	3.02	0.35	0.37	0.32	0.41	0.54	0.00
Collapsed saddle	0.00	0.00	0.00	1.13	0.00	1.51	0.35	0.00	0.00	0.41	0.00	0.57
Cone	1.24	1.55	1.41	4.75	0.42	0.22	1.05	0.37	0.00	0.00	0.00	1.71
Rugulose Spheroid	0.00	0.00	0.00	0.00	0.00	0.22	0.00	0.00	0.00	0.00	0.00	0.00
Smooth Spheroid	0.41	0.39	0.24	1.36	0.00	0.43	0.00	0.00	0.00	0.00	0.00	0.00
Verrucate	0.00	0.00	0.47	1.13	0.00	0.22	0.35	0.00	0.00	0.00	0.00	0.00
Crescent	0.00	0.00	0.71	0.23	0.00	0.86	0.00	0.00	0.00	0.00	0.00	0.00
Elongate	0.00	0.78	3.76	3.85	3.35	3.45	1.75	1.12	2.58	9.39	11.26	4.00
Tracheid	6.20	4.26	0.24	0.68	1.67	2.37	2.46	0.37	0.32	2.86	1.07	2.29
Two-Tiered	0.00	0.39	0.00	0.90	0.42	0.22	0.35	0.37	0.00	0.41	0.00	0.00
Block	0.83	0.39	0.71	0.23	0.42	1.94	2.11	1.50	0.00	0.00	0.00	1.71
Platey	4.13	2.33	8.47	6.56	8.79	5.17	8.42	20.60	15.48	4.49	1.07	2.29
Sheet	0.41	0.00	0.24	0.45	0.00	0.43	0.70	0.00	1.29	0.41	0.00	0.57
Single Polyhedron	1.24	1.55	9.18	8.14	0.42	6.03	1.05	0.00	0.00	0.00	0.00	0.00
Scalloped	0.00	0.00	0.00	0.00	0.00	0.00	0.00	0.37	0.00	0.00	0.00	0.00
Single Jigsaw puzzle	0.00	0.00	0.24	0.00	0.00	0.43	0.00	0.00	0.00	0.00	0.00	0.00
Leaf/culm indet	20.10	15.98	30.13	43.55	27.89	28.18	22.78	14.37	8.90	31.40	32.89	37.23
Leaf/culm jigsaw	0.00	0.24	2.28	3.76	0.40	3.78	0.00	0.21	0.31	0.41	0.00	0.00
Leaf/culm bilobate	0.00	0.00	0.91	1.08	0.80	2.75	0.00	0.21	0.00	0.41	0.48	0.87
Leaf/culm saddle	0.00	0.00	0.00	0.00	0.00	0.34	0.00	0.00	0.00	0.00	0.00	0.00
Leaf/culm cross	0.00	0.00	0.00	0.00	0.00	0.69	0.56	0.00	0.00	0.00	0.00	0.00
Leaf/culm cf Oryza	0.00	0.00	0.46	0.00	0.00	0.00	0.00	0.00	0.00	0.00	0.00	0.00
Leaf/culm long cells	4.64	3.39	23.29	22.58	7.17	18.21	6.11	2.57	5.31	2.07	4.77	4.33
Leaf culm stomata	0.00	0.00	0.00	0.00	0.00	0.00	0.00	0.00	0.00	0.00	0.00	0.43
Leaf/ culm Phragmites	0.00	0.24	2.28	2.15	0.00	0.69	0.00	0.21	0.00	0.00	0.00	0.00
Leaf / culm reed	0.00	0.00	0.00	0.00	0.00	0.00	0.00	0.00	0.00	0.00	0.00	0.00
Leaf / culm Panicum	0.00	0.00	0.00	0.00	0.00	0.00	0.00	0.00	0.00	0.41	0.95	0.87
Square-cell leaf/stem	0.00	0.24	6.39	3.23	0.40	1.72	1.67	0.64	1.56	0.83	0.00	0.00
Indet husk	22.16	5.57	4.57	4.30	6.37	12.03	11.67	1.29	1.56	18.60	31.94	27.27

cf.Setaria husk	2.58	1.69	0.00	0.00	0.80	1.03	7.22	0.00	0.00	3.72	4.77	3.90
HUIZUI												
	HE108	HE110	HGS4	HGS5	HE1	HE2	HE52	HE53	HE54	HE55	HE56	HE57
cf.Panicum husk	1.55	0.97	0.00	0.00	0.80	0.69	1.11	0.00	0.00	1.24	2.38	3.03
Millet type 1	1.03	2.42	0.46	0.00	0.00	2.75	2.22	0.00	0.31	8.68	4.77	3.90
Millet type 2	12.89	2.42	0.00	0.00	1.99	0.34	7.78	0.64	0.62	15.70	12.87	9.09
cf. Oryza husk	0.00	0.00	0.00	0.00	0.00	1.03	0.00	0.64	0.00	0.00	1.43	0.00
Cyperaceae	1.03	0.00	0.00	0.00	0.00	0.69	0.56	0.43	0.31	0.00	1.43	0.87
Polyhedron	0.00	0.00	7.31	4.84	0.00	1.03	2.22	0.00	0.00	0.00	0.00	0.00
Polyhedral hair base	0.52	0.00	1.37	1.08	0.00	3.09	1.11	0.64	0.00	0.00	0.00	0.87
Multi-Tiered forms	0.00	0.00	4.57	1.08	0.00	0.34	0.56	0.43	0.00	0.00	0.00	0.43
Mesophyll	0.00	0.00	0.00	2.15	0.00	2.06	0.00	0.00	0.00	0.83	0.95	0.43
Indet multicell	10.82	23.24	5.02	3.76	32.27	3.09	0.00	5.18	7.00	0.83	0.19	1.73
Indet phytolith	11.34	37.77	5.02	3.76	14.74	6.87	27.22	60.68	29.09	13.22	0.00	2.60
Diatom	0.52	2.91	0.00	0.00	0.40	0.00	0.00	0.74	1.08	0.00	0.00	0.00
Sponge Spicule	2.06	2.91	1.49	2.69	0.80	0.69	0.56	1.48	0.00	0.00	0.00	0.43
Starch	0.00	0.00	1.49	0.00	0.00	0.00	0.00	1.48	13.47	0.00	0.00	0.00
Silica aggregate	8.76	0.00	2.98	0.00	5.18	7.90	6.67	8.14	30.16	1.24	0.19	1.73

	HE81	HE82	HE83	HE84	HE85	HE86	HW1	HW2	HW3	HW4	HW5
Long Smooth	18.33	17.72	28.04	18.18	29.98	18.11	14.69	11.61	13.05	6.56	10.48
Long Sinuate	0.42	2.76	2.21	1.44	0.00	0.75	3.51	1.04	0.22	1.19	0.97
Long Rods	2.08	3.54	1.11	0.96	5.31	3.40	3.73	5.37	2.88	4.57	2.92
Long Dendritic	0.42	0.00	0.37	0.48	1.90	1.51	1.54	0.69	0.00	0.60	0.73
Stomata	0.00	0.00	0.00	0.48	0.38	0.00	0.22	0.00	0.22	0.60	0.00
Total hairs	30.83	31.10	30.63	44.02	25.81	32.83	28.51	21.84	23.01	32.01	24.60
Bulliform	6.67	5.51	3.32	7.66	7.21	3.40	3.07	9.53	5.09	2.78	0.97
Cuneiform Bulliform	2.92	1.57	0.74	4.78	2.28	1.51	1.10	2.43	0.22	0.40	0.97
cf. Oryza Bulliform	0.00	0.00	0.00	0.00	0.00	0.38	0.22	0.00	0.00	0.00	0.00
Crenate	0.00	0.00	0.00	0.00	0.00	0.00	0.00	0.00	0.00	0.00	0.24
Bilobate	0.83	3.54	0.37	0.48	5.88	8.68	12.06	9.19	16.15	16.70	25.21
Polylobate	0.00	0.39	0.00	0.00	0.00	0.75	0.00	0.00	0.00	1.19	0.00
Cross	0.00	0.00	0.00	0.00	0.00	0.00	1.75	1.56	1.99	1.39	0.00
cf. Setaria bilobate	0.00	0.39	0.00	0.00	0.38	0.00	0.00	0.35	0.00	0.40	0.24
cf. Oryza bilobate	0.00	0.39	0.00	0.00	1.52	1.13	6.36	3.99	4.87	3.58	5.36
Rondel	0.83	5.91	3.32	6.70	5.69	6.79	5.70	6.76	8.63	7.75	7.55
cf. Stipa Rondel	0.00	0.00	0.00	0.00	0.00	0.00	0.88	0.87	1.77	0.60	1.46
Saddles	0.42	0.39	0.74	0.00	1.14	1.13	0.88	1.21	0.66	2.58	2.68
Collapsed	0.00	0.00	0.00	0.00	0.00	0.00	1.54	1.39	1.11	0.60	0.97

saddle											
HUIZUI											
	HE81	HE82	HE83	HE84	HE85	HE86	HW1	HW2	HW3	HW4	HW5
Cone	0.00	0.39	0.37	0.48	1.14	1.89	1.32	1.04	1.77	0.60	1.95
Rugulose Spheroid	0.00	0.39	0.00	0.00	0.00	0.00	0.00	0.00	0.00	0.20	0.49
Smooth Spheroid	0.00	0.00	0.00	0.00	0.38	0.00	0.88	0.17	0.22	0.99	0.00
Verrucate	0.00	0.00	0.00	0.00	0.00	0.00	0.00	0.35	1.33	0.60	0.00
Crescent	0.00	0.39	0.00	0.48	0.00	0.00	0.00	1.21	0.44	1.99	0.97
Elongate	0.83	0.39	2.21	0.96	0.76	0.38	4.39	4.68	4.42	6.36	3.90
Tracheid	0.00	0.39	0.74	0.00	3.04	0.38	0.22	1.04	0.88	0.60	0.24
Two-Tiered	0.00	0.00	0.00	0.00	0.00	0.00	0.00	0.17	0.88	0.20	0.24
Block	0.42	0.39	2.95	2.39	1.52	0.38	0.00	0.17	0.22	0.20	0.73
Platey	33.75	22.05	22.51	10.05	5.31	14.72	2.19	7.97	5.75	2.19	0.97
Sheet	0.83	0.39	0.37	0.48	0.00	1.51	0.44	1.04	2.21	0.80	0.49
Single Polyhedron	0.42	1.97	0.00	0.00	0.38	0.38	4.82	3.81	1.99	1.79	4.14
Scalloped	0.00	0.00	0.00	0.00	0.00	0.00	0.00	0.17	0.00	0.00	0.00
Single Jigsaw puzzle	0.00	0.00	0.00	0.00	0.00	0.00	0.00	0.35	0.00	0.00	0.49
Leaf/culm indet	4.84	4.05	1.41	4.05	23.74	46.19	45.95	37.46	44.93	40.10	45.17
Leaf/culm jigsaw	0.00	0.00	0.00	0.00	0.00	0.51	0.00	5.40	4.64	1.21	2.41
Leaf/culm bilobate	0.00	0.08	0.00	0.00	0.79	2.03	10.81	4.13	2.61	9.18	11.72
Leaf/culm saddle	0.00	0.00	0.00	0.00	0.00	0.00	0.39	0.95	0.87	0.48	0.69
Leaf/culm cross	0.00	0.08	0.00	0.00	0.00	0.00	0.77	0.00	0.87	1.93	0.69
Leaf/culm cf Oryza	0.00	0.00	0.00	0.00	0.00	0.00	0.00	0.00	0.58	0.00	0.00
Leaf/culm long cells	5.06	2.53	1.41	0.90	15.04	5.58	23.17	9.84	10.43	10.14	0.00
Leaf culm stomata	0.00	0.00	0.00	0.00	0.00	0.00	0.00	0.00	0.29	0.00	0.00
Leaf/ culm Phragmites	0.00	0.17	0.00	0.00	0.00	0.51	0.00	0.95	0.58	0.00	0.00
Leaf / culm reed	0.00	0.00	0.00	0.00	0.00	0.00	0.00	0.00	0.29	0.97	0.00
Leaf / culm Panicum	0.00	0.00	0.00	0.00	0.00	0.00	0.00	0.00	0.00	0.00	0.00
Square-cell leaf/stem	0.45	0.67	0.32	0.00	0.79	5.08	0.77	1.59	0.00	10.14	1.03
Indet husk	1.69	0.25	0.32	0.00	1.58	4.57	5.02	8.57	10.14	9.42	18.62
cf.Setaria husk	0.00	0.00	0.00	0.00	0.00	0.00	0.00	0.32	0.00	0.00	0.00
cf.Panicum husk	0.11	0.00	0.00	0.00	0.00	0.51	0.00	0.63	0.00	0.48	0.00
Millet type 1	0.11	0.00	0.00	0.00	0.00	0.00	0.00	1.90	0.00	0.97	2.07
Millet type 2	0.00	0.00	0.00	0.00	0.00	0.00	0.00	0.95	0.00	0.24	2.41
cf. Oryza husk	0.00	0.08	0.00	0.00	0.00	0.00	0.00	0.00	0.00	0.24	0.00
Cyperaceae	0.00	0.00	0.00	0.00	1.19	0.51	0.00	2.22	2.32	0.00	4.48
Polyhedron	0.00	0.00	0.00	0.00	0.79	2.54	1.93	3.81	0.58	2.90	2.07
Polyhedral hair base	0.00	0.00	0.00	0.00	0.00	1.02	0.77	0.63	0.00	0.72	0.34
Multi-Tiered	0.00	0.25	0.00	0.00	0.00	0.51	1.54	0.00	0.00	0.00	0.69

forms											
Mesophyll	0.00	0.00	0.00	0.00	0.00	0.00	1.93	2.86	1.16	1.21	1.38
Indet multicell	10.12	5.56	2.35	1.15	0.90	4.57	1.54	1.90	3.48	3.38	0.69
Indet phytolith	0.00	73.27	94.18	82.46	29.85	23.35	2.70	0.63	2.90	0.97	0.34
Diatom	0.00	0.93	0.00	0.00	1.81	0.51	0.00	0.00	0.00	0.24	0.69
Sponge Spicule	6.75	0.00	0.00	1.15	3.62	0.00	0.39	0.42	0.00	0.24	0.34
Starch	0.00	0.00	0.00	1.15	0.00	0.00	0.00	0.00	0.00	0.00	0.00
Silica aggregate	70.87	12.06	0.00	9.16	19.90	2.03	2.32	14.81	13.33	4.83	3.45

Survey											
	S1	S4	S6	S7	S8	S9	S10	S12	S13	S14	S15
Long Smooth	33.20	20.38	18.70	21.05	3.36	14.73	13.04	19.62	24.90	22.17	18.23
Long Sinuate	1.23	2.26	0.87	0.44	0.59	1.55	0.87	2.09	0.00	1.89	2.88
Long Rods	6.15	3.77	1.74	4.82	8.79	1.94	2.17	2.92	9.73	3.30	4.80
Long Dendritic	1.23	1.51	1.30	1.75	1.17	0.39	1.74	0.84	0.39	0.94	0.96
Stomata	0.00	0.75	0.00	1.32	0.00	0.00	0.43	0.00	0.00	0.47	0.00
Total hairs	20.08	36.23	34.78	31.58	48.03	33.33	41.30	35.07	39.30	38.21	38.85
Bulliform	2.05	1.89	4.35	2.19	1.76	1.55	4.78	2.09	3.11	4.25	0.48
Cuneiform Bulliform	4.92	4.53	0.87	0.88	0.59	0.78	1.30	0.42	0.39	1.42	0.48
cf. Oryza Bulliform	0.00	0.00	0.00	0.00	0.00	0.00	0.00	0.00	0.00	0.00	0.00
Crenate	0.00	0.00	0.00	0.00	0.00	0.78	0.00	0.00	0.00	0.00	0.00
Bilobate	9.02	13.96	14.35	18.42	15.23	18.22	15.65	19.00	10.12	7.55	9.35
Polylobate	0.00	0.75	0.87	1.32	0.59	0.39	0.87	0.00	0.00	0.94	0.00
Cross	0.00	0.75	0.00	0.00	0.59	0.78	0.00	0.42	0.00	0.00	0.00
cf. Setaria bilobate	0.00	0.00	0.00	0.44	0.59	0.39	0.00	0.00	0.39	0.00	0.48
cf. Oryza bilobate	2.05	2.26	5.22	3.07	1.76	3.10	1.74	1.67	1.17	3.30	1.44
Rondel	2.05	3.02	5.22	3.51	2.93	4.26	1.74	2.09	1.56	3.30	1.92
cf. Stipa Rondel	0.00	0.00	0.00	0.00	1.17	0.39	0.00	0.00	0.00	0.00	0.00
Saddles	3.28	2.26	2.17	0.88	2.93	1.16	0.87	0.00	0.78	0.94	2.40
Collapsed saddle	0.00	0.00	0.00	0.00	0.59	0.39	0.00	0.00	0.00	0.00	0.00
Cone	0.41	0.38	2.61	0.88	0.00	1.55	0.87	4.18	3.11	1.89	0.96
Rugulose Spheroid	0.00	0.00	0.00	0.00	0.00	0.00	0.00	0.00	0.00	0.00	0.00
Smooth Spheroid	0.00	0.00	0.43	0.00	0.59	0.00	0.00	0.00	0.78	0.47	0.00
Verrucate	0.00	0.00	0.00	0.44	0.59	0.39	0.00	0.00	0.00	0.00	0.48
Crescent	0.00	0.00	0.00	0.00	0.00	0.39	0.87	0.84	0.00	0.00	0.00
Elongate	0.41	1.51	2.61	1.32	1.76	5.04	4.78	2.51	2.72	0.47	4.80
Tracheid	2.46	0.38	0.43	1.75	0.00	1.16	0.87	1.25	0.39	2.83	4.32
Two-Tiered	0.00	0.00	0.00	0.00	1.17	0.00	0.00	0.00	0.00	0.00	0.00
Block	0.82	0.38	0.00	0.00	0.59	0.78	0.00	0.42	1.17	0.00	0.48
Platey	10.25	1.89	2.61	3.51	2.93	6.59	6.09	4.18	0.00	5.19	5.76
Sheet	0.00	0.00	0.00	0.00	0.59	0.00	0.00	0.00	0.00	0.00	0.00
Single Polyhedron	0.41	1.13	0.87	0.44	1.17	0.00	0.00	0.42	0.00	0.47	0.96
Scalloped	0.00	0.00	0.00	0.00	0.00	0.00	0.00	0.00	0.00	0.00	0.00
Survey											
	S1	S4	S6	S7	S8	S9	S10	S12	S13	S14	S15
Single Jigsaw puzzle	0.00	0.00	0.00	0.00	0.00	0.00	0.00	0.00	0.00	0.00	0.00

Leaf/culm indet	14.50	40.57	41.29	38.39	38.63	38.46	55.14	36.14	32.00	58.68	50.00
Leaf/culm jigsaw	0.00	0.00	0.00	0.00	0.00	0.00	0.00	0.00	0.00	0.00	0.00
Leaf/culm bilobate	1.18	3.56	7.10	14.69	2.15	3.62	7.00	3.61	4.50	2.61	5.07
	0.00	0.00	0.97	0.00	0.00	0.45	0.00	0.00	0.00	0.00	0.36
Leaf/culm saddle	0.00	0.00	1.61	0.47	0.43	0.00	0.00	0.00	0.50	0.65	0.00
Leaf/culm cross	0.00	0.00	0.00	0.00	0.00	0.00	0.00	0.00	0.00	0.00	0.00
Leaf/culm cf Oryza	5.03	9.61	13.55	8.06	9.87	4.52	5.76	13.25	10.00	11.08	10.87
Leaf/culm long cells	0.89	0.00	0.32	0.00	0.00	0.00	0.00	0.60	0.00	0.00	0.36
Leaf culm stomata	0.00	0.00	0.00	0.00	0.00	0.00	0.00	0.00	0.00	0.00	0.00
Leaf/ culm Phragmites	1.48	0.00	0.65	0.47	1.29	0.90	0.00	0.60	0.00	0.00	0.72
Leaf / culm reed	0.00	0.71	0.65	0.00	0.00	0.00	0.41	0.00	0.00	0.65	0.00
	S1	S4	S6	S7	S8	S9	S10	S12	S13	S14	S15
Leaf / culm Panicum	2.37	2.49	0.32	0.00	0.86	0.90	0.82	2.41	0.50	0.00	0.00
Square-cell leaf/stem	3.85	7.47	9.68	6.16	9.87	8.14	9.05	3.61	2.00	14.34	3.62
Indet husk	0.00	1.42	1.29	0.00	0.00	0.45	0.41	0.00	0.00	0.00	0.72
cf.Setaria husk	0.00	0.71	0.00	0.00	0.43	0.00	0.00	0.00	0.00	0.00	1.09
cf.Panicum husk	0.00	0.36	0.00	0.00	0.43	0.90	0.41	0.00	0.00	1.30	0.36
Millet type 1	1.18	1.07	0.65	0.00	1.29	0.45	0.82	0.00	0.00	1.30	4.35
Millet type 2	0.00	0.00	0.00	0.00	0.86	0.00	0.00	0.00	0.00	0.00	0.00
cf. Oryza husk	0.00	1.78	2.58	0.95	0.86	2.71	1.23	3.61	0.50	5.22	0.00
Cyperaceae	0.00	0.36	0.97	0.00	0.43	0.45	0.82	0.00	0.00	0.65	0.00
Polyhedron	0.00	0.36	0.32	2.84	0.86	0.00	0.82	2.41	2.50	3.26	2.17
Polyhedral hair base	0.30	0.71	0.65	0.00	0.00	0.90	0.41	1.20	1.00	0.00	0.00
Multi-Tiered forms	0.00	1.42	0.65	0.00	0.86	0.00	0.82	1.81	0.00	0.00	0.00
Mesophyll	10.21	16.01	10.32	18.48	18.03	20.81	9.47	9.04	18.00	0.16	10.14
Indet multicell	40.85	7.12	3.87	8.53	9.01	10.41	3.70	16.87	25.50	0.05	5.80
Indet phytolith	1.13	0.00	0.00	0.00	0.00	0.00	0.00	0.00	0.00	0.00	0.00
Diatom	0.00	0.00	0.00	0.00	0.00	0.45	0.00	0.00	0.00	0.00	0.00
Sponge Spicule	0.00	0.00	0.00	0.00	0.00	0.00	0.00	0.00	0.00	0.00	0.00
Starch	17.02	3.91	2.58	0.95	3.86	5.43	2.88	4.82	3.00	0.02	4.35
Silica aggregate	14.50	40.57	41.29	38.39	38.63	38.46	55.14	36.14	32.00	58.68	50.00

Appendix 6 Phytolith reference collection

Ref Number	Taxon	Plant part	Abundance	Diagnostic	Provenience	Common name
59	Aegilops crassa	awn	common	x	Teaching collection, UCL	Goatgrass
58	Aegilops crassa	glume	common	x	Teaching collection, UCL	Goatgrass
203	Aeluropus littoralis	husk & floret			Herbarium, UCL	
205	Agrostis gigantea	husk	common		Herbarium, UCL	Black bent
284	Aruebsia decumbrens	floret	present		Herbarium, UCL	Black bent
210	Aruebsia decumbens	floret			Herbarium, UCL	Black bent
28	Arundo donax	culm	common		Seville, riverbank Feb 2006	Giant reed
31	Arundo donax	inflorescence	abundant	x	Seville, riverbank Feb 2006 - infl has ergot	Giant reed
30	Arundo donax	leaf	abundant	x	Seville, riverbank Feb 2006	Giant reed
29	Arundo donax	node	present		Seville, riverbank Feb 2006	Giant reed
60	Avena barbata	awn	abundant		Teaching collection, UCL	Oat
61	Avena barbata	glume	abundant	x	Teaching collection, UCL	Oat
62	Avena barbata	glume hairs & seed coat	abundant	possibly	Teaching collection, UCL	Oat
163	Avena fatua	husk & awn	abundant	x	Teaching collection, UCL	Spring or common wild oat
162	Avena sterilis	husk & awn	abundant	x	Teaching collection, UCL	Wild winter oat
292	Avena sterilis	husk	abundant	x	Teaching collection, UCL	Wild winter oat
285	Bambusa	culm	abundant	x	Teaching collection, UCL	Bamboo
233	Bambusa sindhudurg	culm	common		Maharastra	Bamboo
231	Bambusa sindhudurg	infl	abundant	x	Maharastra	Bamboo
230	Bambusa sindhudurg	leaf	abundant	x	Maharastra	Bamboo
219	Brachiaria ramosa	culm	abundant	x	Grown 2007 London	Browntop millet
218	Brachiaria ramosa	leaf	abundant	x	Grown 2007 London	Browntop millet
90	Brachiaria ramosa	seed			D6	Browntop millet
220	Brachiaria ramosa	seed coat	abundant	x	Grown 2007 London	Browntop millet
189	Brachypodium pinnatum	seeds	abundant	x	Kew, collected from Shelly Cleve July 2015	Tor grass

Ref Number	Taxon	Plant part	Abundance	Diagnostic	Provenience	Common name
181	Bromus sterilis	infl & awn	common	x	Herbarium, UCL	Barren Brome
293	Bromus catharticus	leaves	common	x	Herbarium, UCL	Rescuegrass
294	Bromus catharticus	culm	common	x	Herbarium, UCL	Rescuegrass
161	Broussonetta papyrifera	culm	present	possibly	Col. F. Itzstein Davey 2007	Paper mulberry
159	Broussonetta papyrifera	leaf	present		Col. F. Itzstein Davey 2007	Paper mulberry
160	Broussonetta papyrifera	seed	present		Col. F. Itzstein Davey 2007	Paper mulberry
208	Buglossoides arvensis	floret	common	possibly	Herbarium, UCL	Lithospermum - stoney gromwell
206	Buglossoides tenniflora	culm			Herbarium, UCL	Lithospermum - stoney gromwell
184	Buglossoides tenniflora	floret	rare	x	Herbarium, UCL	Lithospermum - stoney gromwell
207	Buglossoides tenniflora	leaf			Herbarium, UCL	Lithospermum - stoney gromwell
126	Capsella Bursa Pastoris	leaf	common	x	Brassicaceae, DQF 2006 11 April Tian Luo Shan Zhejiang arable weed flowers & fruit cooked as green vegetable	Shepherd's Purse
127	Capsella Bursa Pastoris	seed	common	possibly	Col. UK 2006	Shepherd's Purse
281	Carex divisia	culm	common	possibly	Herbarium, UCL	Divided sedge
280	Carex divisia	infl.	common	x	Herbarium, UCL	Divided sedge
222	Carex limosa	leaf	abundant	x	Herbarium, UCL	Bog sedge
223	Carex limosa	seed	abundant	x	Herbarium, UCL	Bog sedge
234	Carex remota	leaf	common		Herbarium, UCL	Remote sedge
235	Carex remota	seed	abundant	x	Herbarium, UCL	Remote sedge
286	Carex riparia	leaf	abundant	possibly	Herbarium, UCL	Great marsh sedge
283	Carex stenophylla	infl	common	x	Herbarium, UCL	Narrow leafed sedge
108	Cenchrus pennisetiformis	culm	abundant		Herbarium, UCL	Conclurry
39	Cenchrus pennisetiformis	culm	rare		Grown in London 2007	Conclurry

Ref Number	Taxon	Plant part	Abundance	Diagnostic	Provenience	Common name
110	Cenchrus pennisetiformis	inflorescence	abundant	x	Grown in London 2007	Conclurry
41	Cenchrus pennisetiformis	inflorescence	common	x	Grown in London 2007	Conclurry
107	Cenchrus pennisetiformis	leaf	abundant	x	Grown in London 2007	Conclurry
109	Cenchrus pennisetiformis	node	common		Grown in London 2007	Conclurry
40	Cenchrus pennisetiformis	rachis			Grown in London 2007	Conclurry
255	Chloris virgata	culm	abundant	x	Baligang, China 2006 DQF	Feather fingergrass
255	Chloris virgata	culm	abundant	x	Baligang, China 2006 DQF	Feather fingergrass
256	Chloris virgata	husk	abundant	x	Baligang, China 2006 DQF	Feather fingergrass
254	Chloris virgata	leaf	abundant	x	Baligang, China 2006 DQF	Feather fingergrass
133	Citrillus colocynthis	rind	present		El Terief 4th Cataract Sudan Jan 2004	Bitter apple, bitter cucumber
116	Cocos nucifera	culm				Coconut
115	Cocos nucifera	leaf	abundant	x	Palmar,Koanmanal Tamil Nadu	Coconut
226	Coix lachryma jobi	culm	abundant	x	Kyoto st nr Rihn vac lot, also on nearby stream banks DQF (2007)	Job's tears, adlay
229	Coix lachryma jobi	infl	abundant	x	Kyoto st nr Rihn vac lot, also on nearby stream banks DQF (1-9-2007)	Job's tears, adlay
227	Coix lachryma jobi	leaf	abundant	x	Kyoto st nr Rihn vac lot, also on nearby stream banks DQF (1-9-2007)	Job's tears, adlay
266	Coix lachyrma jobi	culm	abundant	x	Kew 2006 ELH ARW Shelly Cleeve	Job's tears, adlay
267	Coix lachyrma jobi	husk			Kew 2006 ELH ARW Shelly Cleeve	Job's tears, adlay
265	Coix lachyrma jobi	leaf	abundant	x	Kew 2006 ELH ARW Shelly Cleeve	Job's tears, adlay
128	Curcubita pepo	rind	rare	x		Squash

Ref Number	Taxon	Plant part	Abundance	Diagnostic	Provenience	Common name
154	Cyperus papyrus	awn	common		Cyperus papyrus Burunga Uganda M. McClatchie	Papyrus
155	Cyperus papyrus	culm			Cyperus papyrus Burunga Uganda M. McClatchie	Papyrus
287	Cyperus papyrus	floret	abundant	x	Cyperus papyrus Burunga Uganda M. McClatchie	Papyrus
156	Cyperus papyrus	leaf	common	x	Cyperus papyrus Burunga Uganda M. McClatchie	Papyrus
232	Cyperus papyrus (immature)	floret	abundant	x	Cyperus papyrus Burunga Uganda M. McClatchie	Papyrus
288	Cyperus rotundus	infl	abundant	x	Herbarium, UCL	
158	Dactylis	culm	common	x	Herbarium, UCL	Orchard grass,
157	Dactylis	seed	abundant	x	Herbarium, UCL	Orchard grass
193	Dactylis glomerata	culm			Kew, collected from Shelly Cleve July 2019	Cocksfoot
195	Dactylis glomerata	husk	abundant	x	Kew, collected from Shelly Cleve July 2021	Cocksfoot
194	Dactylis glomerata	seed coat	abundant	x	Kew, collected from Shelly Cleve July 2020	Cocksfoot
26	Dactyloctenium aegytium	culm	common	x	Kurnool dist 8.9.04	Egyptian crowfoot
27	Dactyloctenium aegytium	inflorescence	abundant	x	Kurnool dist 8.9.04	Egyptian crowfoot
10	Dennstaediaceae	culm			Richmond Park, London Nov 2004	Bracken
9	Dennstaediaceae	leaf	abundant	x	Richmond Park, London Nov 2005	Bracken
23	Dioscorea rotunda	leaf	rare	possibly	grown in lab winter 1998/9	Yam
24	Dioscorea rotunda	leaf	rare		Herbarium, UCL	Yam
25	Dioscorea rotunda	nodes/culm	rare		Herbarium, UCL	Yam
20	Echinochloa crus-gali	leaf			Fuller 2026	Barnyard millet, Cockspur
249	Echinochloa crus-gali var. mitis	culm	common	x	Beijing PKU, China DQF 2010	Barnyard millet, Cockspur

Ref Number	Taxon	Plant part	Abundance	Diagnostic	Provenience	Common name
249	Echinochloa crus-gali var. mitis	culm	common	x	Beijing PKU, China DQF 2010	Barnyard millet, Cockspur
250	Echinochloa crus-gali var. mitis	husk	abundant	x	Beijing PKU, China DQF 2011	Barnyard millet, Cockspur
248	Echinochloa crus-gali var. mitis	leaf	abundant	x	Beijing PKU, China DQF 2009	Barnyard millet, Cockspur
253	Echinochloa FOC	husk			Baligang, China 2006 DQF	Barnyard millet
251	Echinochloa FOC	leaf			Baligang, China 2006 DQF	Barnyard millet
96	Eleusine coracana	infl	abundant	x	H4	Finger millet
89	Eleusine coracana	seed & husk	abundant	x	C5S2	Finger millet
13	Eleusine indica	culm	common			Goose Grass
270	Eleusine indica	culm	common		USDA-ARS PI 217600 grown autumn/ winter 1998/9 DQF	Goose Grass
268	Eleusine indica	floret	common		USDA-ARS PI 217600 w/ unfertilsed florets grown autumn/ winter 1998/9 DQF	Goose Grass
14	Eleusine indica	inflorescence	common	possibly	USDA-ARS PI 217600 w/ unfertilsed florets grown autumn/ winter 1998/9 DQF	Goose Grass, Yard grass,
15	Eleusine indica	leaf	common	possibly	USDA-ARS PI 217600 w/ unfertilsed florets grown autumn/ winter 1998/9 DQF	Goose Grass
269	Eleusine indica	leaf	common		USDA-ARS PI 217600 grown autumn/ winter 1998/9 DQF	Goose Grass
182	Festuca glauca Elija blue	culm + infl	abundant	x	Kew, collected from Shelly Cleve July 2008	Blue grass
183	Festuca glauca Elija blue	seeds and seed coat	abundant	x	Kew, collected from Shelly Cleve July 2009	Blue grass
1	Equisetum varigatum	culm	common	x	London Nov 2005	Horsetail
2	Equisetum varigatum	culm	common		London Nov 2005	Horsetail
7	Equisetum varigatum	culm	common		London Nov 2005	Horsetail
3	Equisetum varigatum	culm/node	common		London Nov 2005	Horsetail
4	Equisetum varigatum	culm/top	common		London Nov 2005	Horsetail

Ref Number	Taxon	Plant part	Abundance	Diagnostic	Provenience	Common name
8	Equisetum varigatum	inflorescence	common		London Nov 2005	Horsetail
6	Equisetum varigatum	node	common		London Nov 2005	Horsetail
151	Eragrostis minor	culm	abundant	x	Col F. Itzstein Davey 2007	Little lovegrass
152	Eragrostis minor	leaf	abundant	x	Col F. Itzstein Davey 2007	Little lovegrass
153	Eragrostis minor	seed	abundant	x	Col F. Itzstein Davey 2007	Little lovegrass
185	Eragrostis tef	culm	abundant	x	Kew, collected from Shelly Cleve July 2011	Tef
184	Eragrostis tef	leaf	abundant	x	Kew, collected from Shelly Cleve July 2010	Tef
187	Eragrostis tef	nodes	abundant	x	Kew, collected from Shelly Cleve July 2013	Tef
186	Eragrostis tef	seeds	abundant	x	Kew, collected from Shelly Cleve July 2012	Tef
238	Eremopyrum	culm			UCL Herbarium	
236	Eremopyrum	husk	common	x	UCL Herbarium	
237	Eremopyrum	leaf	common	possibly	UCL Herbarium	
42	Festuceae	culm	abundant	x	Syrian collection Festuceaea Loc 20 1158	Fescue
139	Festuceae	rachis	common		Syrian collection Festuceaea Loc 20 1158	Fescue
183	Helianthemum salicfolium	floret	rare		UCL Herbarium	
140	Hordeum vulgare	husk	abundant	x	UCL Herbarium	Six-rowed barley
181	Imperata cylindrica rubens	culm	abundant		Kew, collected from Shelly Cleve July 2007	Japanese blood grass
180	Imperata cylindrica rubens	leaf	abundant	x	Kew, collected from Shelly Cleve July 2006	Japanese blood grass
137	Lagenaria Ciseria	rind	present	possibly	UCL Herbarium	Bottle gourd, calabash
182	Lamarchia aurea	husk	common	x	UCL Herbarium	Bottle gourd, calabash
44	Lolium	leaf	common		Syrian collection Gram/17 Lolium 29	Ryegrass
211	Lolium temulentum	husk & awn	abundant	x	Syrian collection Gram/17 Lolium 29	Darnel
241	Lollium	culm			Syrian collection Gram/17 Lolium 29	Ryegrass
239	Lollium	husk			Syrian collection Gram/17 Lolium	Ryegrass
240	Lollium	leaf			Syrian collection Gram/17 Lolium	Ryegrass
201	Lollium Persicarium	husk & awn	abundant	x	Syrian collection Gram/17Lolium	Ryegrass

Ref Number	Taxon	Plant part	Abundance	Diagnostic	Provenience	Common name
202	Lollium Rigidum	awn & rachis			Syrian collection Gram/17 Lolium	Ryegrass
131	Luffa acutangula	rind	rare		purchased London July 03, feels waxy	Sinkwa towel gourd, dishcloth gourd
130	Momordian dioecia	rind	present		purchased London July 03	Spiny gourd
136	Musa bajoo	leaf	abundant		Japanese hardy banana purchased as pot plant Eden project Cornwall 2003	Japanese banana
112	Musa phradisica	culm	common		Kodumanal(14.9.02)	Banana
111	Musa phradisica	leaf	abundant		Kodumanal (14.9.02)	Banana
216	Oryza (sativa)	straw	abundant		Tian Luo Shan collected DQF 11 November 2007	Rice
217	Oryza (sativa)	straw ash	abundant		Tian Luo Shan collected DQF 11 November 2008	
72	Oryza japonica	husk	abundant	x	Tian Luo Shan col. DQF 2008	
71	Oryza japonica	leaf	abundant	x	Tran Luo Shan, Zhejiang	
214	Oryza nivara	culm	abundant		Tian Luo Shan col. DQF 2008	
212	Oryza nivara	husk	abundant		Tian Luo Shan col. DQF 2008	
213	Oryza punct ata	husk	abundant			
73	Oryza rufipogon/spontanea	leaf	abundant	x	Lhavrador 22 jan 2006	
215	Oryza sativa	culm	abundant	x		Rice
180	Palmyra Borrassus	leaf	common	x	UCL Herbarium	Palmyra palm
32	Panicum miliaceum	culm	common		Grown London 2006	Proso millet
35	Panicum miliaceum	inflorescence	abundant	x	Grown London 2006	Proso millet
34	Panicum miliaceum	leaf	abundant	x	Grown London 2006	Proso millet
33	Panicum miliaceum	nodes	common		Grown London 2006	Proso millet
49	Panicum turgidum	culm	common	x	Grown London 2006	
51	Panicum turgidum	inflorescence	abundant		Grown London 2006	
48	Panicum turgidum	leaf	abundant	x	Forsk. Fuller 97-112 21-10>1997 Wadi Muqqadam, Bayuda West Sudan	
50	Panicum turgidum	node	common		Wadi Muqqadam, West Sudan	
17	Panicum typheron	inflorescence	abundant		Fuller 2029	

Ref Number	Taxon	Plant part	Abundance	Diagnostic	Provenience	Common name
16	Panicum typheron	leaf	abundant		Fuller 2029	
97	Pennisetum glaucum	chaff			I5	Pearl millet
100	Pennisetum glaucum	culm			K7	Pearl millet
272	Pennisetum glaucum	culm			seeds Karnakata India grown summer / winter 1998/9 DQF	Pearl millet
271	Pennisetum glaucum	leaf			seeds Karnakata India grown summer / winter 1998/9 DQF	Pearl millet
185	Phalaris arundinaceae	husk	common	x	Col. F Itzstein Davey	Reed Canary grass
204	Phalaris minor	husk	common		UCL Herbarium	Lesser Canary grass
56	Phragmites australis	culm			Col. F Itzstein Davey	Common reed
191	Phragmites australis	culm	present		Kew, collected from Shelly Cleve July 2017	Common reed
36	Phragmites australis	inflorescence	common	x	Emma Jenkins Catal Huyuk July 2004	Common reed
37	Phragmites australis	leaf	abundant	x	E Jenkins Catal Huyuk 2004	Common reed
190	Phragmites australis	leaf	abundant	x	Kew, collected from Shelly Cleve July 2016	Common reed
275	Polypogon monspeliensis	culm	present		Argolid Peloponnese South Greece	Rabbit's foot grass, beard grass
276	Polypogon monspeliensis	husk	common	possibly	Argolid Peloponnese South Greece	Rabbit's foot grass, beard grass
274	Polypogon monspeliensis	leaf	common	x	Argolid Peloponnese South Greece	Rabbit's foot grass, beard grass
22	Saccharum officinarum L.	culm/node	common			Sugar cane
77	Saccharum officinarum L.	leaf	abundant	x	Hulaskhem 22.1.06 Fuller	Sugar cane
21	Saccharum officinarum L.	nodes	common		Hainan island DQF	Sugar cane
113	Saccharum spontaneum	leaf	abundant	x	Tamil Nadu DQF	Wild sugarcane
146	Scirpus lacustris	inflorescence	abundant	x	UK	Common club rush, bulrush

Ref Number	Taxon	Plant part	Abundance	Diagnostic	Provenience	Common name
261	Setaria (wild sp)	culm	present		MiluYuan (Deer Park) China 2006 DQF	Bristle grass
262	Setaria (wild sp)	husk	present		MiluYuan (Deer Park) China 2006 DQF	Bristle grass
260	Setaria (wild sp)	leaf	common	possibly	MiluYuan (Deer Park) China 2006 DQF	Bristle grass
118	Setaria Faberii	culm	abundant	x		Giant foxtail
119	Setaria Faberii	inflorescence	abundant	x		Giant foxtail
117	Setaria Faberii	leaf	abundant	x	Her.coll.D Fuller 9 ov 2004 Zheijang Prov PRC Shimadou near Hangzhou	Giant foxtail
120	Setaria Faberii	node	abundant	x	Zheijiang Prov. Shimadou Nr Honngzhou, China 9th Nov 2004 DQF	Giant foxtail
247	Setaria faberii	culm			Beijing PKU, China DQF 2008	Giant foxtail
245	Setaria faberii	husk			Beijing PKU, China DQF 2006	Giant foxtail
246	Setaria faberii	leaf			Beijing PKU, China DQF 2007	Giant foxtail
244	Setaria faberii (Herm)	culm	abundant	x	Zheijiang Prov. Shimadou Nr Honngzhou, China 9th Nov 2004 DQF	Giant foxtail
242	Setaria faberii (Herm)	husk	abundant	x	Zheijiang Prov. Shimadou Nr Honngzhou, China 9th Nov 2004 DQF	Giant foxtail
243	Setaria faberii (Herm)	leaf	abundant	x	Zheijiang Prov. Shimadou Nr Honngzhou, China 9th Nov 2004 DQF	Giant foxtail
11	Setaria italica	culm			grown summer - winter 1998/9 collected Sanganakallu DQF 12.2.98	Foxtail millet
258	Setaria italica	culm	common	x	Shandong, China 2006 DQF	Foxtail millet
263	Setaria italica	culm	abundant		Kurnool Dist. A.P. India 8th Sept 2004 DQF	Foxtail millet

Ref Number	Taxon	Plant part	Abundance	Diagnostic	Provenience	Common name
278	Setaria italica	culm	abundant	x	Morocco	Foxtail millet
91	Setaria italica	husk	abundant	x	E1	Foxtail millet
259	Setaria italica	husk	abundant	x	Shandong, China 2006 DQF	Foxtail millet
92	Setaria italica	husk			E1S3	Foxtail millet
279	Setaria italica	husk	common		Morocco	Foxtail millet
12	Setaria italica	leaf	common	possibly		Foxtail millet
257	Setaria italica	leaf	common	x	Shandong, China 2006 DQF	Foxtail millet
264	Setaria italica	leaf	abundant	x	Kurnool Dist. A.P. India 8th Sept 2004 DQF	Foxtail millet
277	Setaria italica	leaf	abundant	x	Morocco	Foxtail millet
174	Setaria macrostachya	culm	common		UCL Herbarium	Large-spike bristlegrass
173	Setaria macrostachya	husk	abundant	x	UCL Herbarium	Large-spike bristlegrass
290	Setaria palmifolia	husk	abundant	x	UCL Herbarium	Palmgrass, pit pit
291	Setaria plicata	husk	abundant	x	UCL Herbarium	Slender pigeongrass
19	Setaria pumila	inflorescence	abundant	x	syn glauca Hubskhera 221.1.06	Yellow foxtail millet
82	Setaria pumila	leaf	abundant	x	syn glauca Hubskhera 221.1.06	Yellow foxtail millet
74	Setaria pumila	panicle	abundant	x	syn glauca Hulskhem 22.1 06	Yellow foxtail millet
81	Setaria pumila	panicle	common		(prob) Hainan Jian Fengling forest	Yellow foxtail millet
79	Setaria verticillata	culm	abundant	x	Atarauli Dorian Fuller	Bristly foxtail millet, Rough bristle grass
80	Setaria verticillata	leaf	abundant	x	Atarauli Dorian Fuller	Bristly foxtail millet, Rough bristle grass
289	Setaria verticillata	husk	abundant	x	Atarauli Dorian Fuller	Bristly foxtail millet, Rough bristle grass
78	Setaria verticillata	panicle and infl	abundant	x	Atarauli Dorian Fuller	Bristly foxtail millet, Rough bristle grass
147	Setaria viridis	awn & rachis	abundant	x	Shanghai street millet Aug 2006 DQF	Green bristle grass

Ref Number	Taxon	Plant part	Abundance	Diagnostic	Provenience	Common name
122	Setaria viridis	culm	abundant		Shanghai street millet Aug 2006 DQF	Green bristle grass
123	Setaria viridis	inflorescence	abundant	x	Shanghai street millet Aug 2006 DQF	Green bristle grass
121	Setaria viridis	leaf	abundant	x	Shanghai street millet Aug 2006 DQF	Green bristle grass
125	Setaria viridis	node	abundant	x	Shanghai street millet Aug 2006 DQF	Green bristle grass
124	Setaria viridis	awn and rachis	abundant	x	Shanghai street millet Aug 2006 DQF	Green bristle grass
122	Setaria viridis	culm	abundant		Shanghai street millet Aug 2006 DQF	Green bristle grass
123	Setaria viridis	inflorescence	abundant	x	Shanghai street millet Aug 2006 DQF	Green bristle grass
121	Setaria viridis	leaf	abundant	x	Shanghai street millet Aug 2006 DQF	Green bristle grass
125	Setaria viridis	node	abundant	x	Shanghai street millet Aug 2006 DQF	Green bristle grass
124	Setaria viridis	awn and rachis	abundant	x	Shanghai street millet Aug 2006 DQF	Green bristle grass
132	Siraitia grosvenorii	rind	present	possibly	Luoshan fruit, Swingles C Jeffery ex Lu et Z Y Zhang	*Luohan guo*
86	Sorghum bicolor	inflorescence	common	possibly	from collection 313 A3	Sorghum, jowar
149	Sorghum bicolor	leaf	abundant	x	UCL Herbarium	Sorghum, jowar
84	Sorghum bicolor	seed			UCL Herbarium A1S2	Sorghum, jowar
85	Sorghum bicolor	culm	common		UCL Herbarium A2	Sorghum, jowar
148	Sorghum bicolor	glume	abundant	x	UCL Herbarium	Sorghum, jowar
55	Sorghum sudani	inflorescence	abundant	possibly	UCL Herbarium	
52	Sorghum sudani	leaf	abundant	x	UCL Herbarium	
54	Sorghum sudani	node	common		UCL Herbarium	
150	Stipa capilata	inflorescence	abundant	x	UCL Herbarium	Hair like feather grass
199	Stipa tirsa	awn	present		Kew, collected from Shelly Cleve July 2025	Bristle leaved feather grass
200	Stipa tirsa	awn			Kew, collected from Shelly Cleve July 2025	Bristle leaved feather grass

Ref Number	Taxon	Plant part	Abundance	Diagnostic	Provenience	Common name
198	Stipa tirsa	leaf	common	x	Kew, collected from Shelly Cleve July 2024	Bristle leaved feather grass
197	Stipa tirsa	seed coat	abundant	x	Kew, collected from Shelly Cleve July 2023	Bristle leaved feather grass
129	Trichosanthe cucmerina	rind	present	possibly	Snake gourd DF BII 67	Snake gourd
105	Triticum aestivum	husk	abundant	x	IoA teaching collection	Bread wheat
106	Triticum aestivum	spikelets	abundant	x	IoA teaching collection	Bread wheat
168	Triticum compactum	culm	present		IoA teaching collection	Club wheat
170	Triticum compactum	husk & awn	present		IoA teaching collection	Club wheat
169	Triticum compactum	leaf	present		IoA teaching collection	Club wheat
104	Triticum diccoccoides	husk	abundant	x	IoA teaching collection	Wild Emmer
164	Triticum dicoccum	husk & awn	abundant	x	IoA teaching collection	Emmer
64	Triticum durum	awn	abundant	possibly	IoA teaching collection	Durum wheat
65	Triticum durum	husk	abundant	x	IoA teaching collection	Durum wheat
103	Triticum durum	husk	abundant	x	IoA teaching collection	Durum wheat
66	Triticum monococcum	husk	abundant	x	IoA teaching collection	Einkorn
172	Triticum monococum	culm	abundant	x	IoA teaching collection	Einkorn
67	Triticum spelta	husk	abundant	x	IoA teaching collection	Spelt
69	Triticum timopheevi	awn	abundant	x	IoA teaching collection	Zanduri wheat
70	Triticum timopheevi	husk	common	x	IoA teaching collection	Zanduri wheat
166	Triticum turgidum	culm	present		IoA teaching collection	Rivet wheat, Poulard wheat
165	Triticum turgidum	husk & awn	present		IoA teaching collection	Rivet wheat, Poulard wheat
68	Triticum vavilovii	husk	abundant	x	IoA teaching collection	
171	Triticum vulgare	husk	abundant	x	IoA teaching collection	
175	Typha	leaf	none		London, pond	Cat Tail
221	Zingiber officinale	leaf	common	x	Grown 2007 London	Ginger
179	Zizania latifolia	culm	abundant	x	DQF	Wild rice, Indian rice
178	Zizania latifolia	leaf	abundant	x	DQF	Wild rice, Indian rice

REFERENCES CITED

Adler (1996). Land tenure, archaeology, and the ancestral Pueblo social landscape. *Journal of Anthropological* Archaeology 15: 337-371.

Ahn, S. M. (1993). *Origin and differentiation of domesticated rice in Asia.* PhD dissertation. Institute of Archaeology. London, University College London.

Albert, R. M., Lavi, O., Estrof, L., Weiner, S., Tsatskin, A., Ronen, O., Lev-Yadun, S. (1999). Mode of occupation of Tabun Cave, Mt Carmel, Israel during the Mousterian period: a study of sediments and phytoliths. *Journal of Archaeological Science* 26: 1249-1260.

Albert, R. M., Weiner, S. (2001). Study of phytoliths in prehistoric ash layers from Kebara and Tabun caves using a quantative approach. *Phytoliths: Applications in Earth Sciences and Human History*. F. C. J.D. Meunier. Lisse, Balkema: 251-266.

Allard, J.-L., Kon, K-F., Morishima, Y., Kotzian, R. (2005) *The crop protection industry's view on trends in rice crop establishment in Asia and their impact on weed management techniques. Rice is life: scientific perspective for the 21st century.* Copyright International Rice Research Institute 2005: 205.

Allen, B. J., Ballard, C. (2001). Beyond intensification. *Asia Pacific Viewpont* 42(2/3): 157-162.

Alexandre, A., Meunier, J-D., Lezine, A-M, Vincens, A., Schwartz (1997). Phytoliths: indicators of grassland dynamics during the late Holocene in intertropical Africa. *Palaeogeography, Palaeoclimatology, Palaeoecology* 136: 213-229.

Ames, O. (1939). *Economic annuals and human cultures.* Cambridge, Massachusetts.

An, C., Feng, Z. and Barton, L. (2006). Dry or humid? Mid Holocene humidity changes in arid and semi-arid China. *Quaternary Science Reviews* 25: 351-361.

An, C. B., Feng, Z.D., Tang, L. (2004). Environmental change and cultural response between 8000 and 4000 cal. yr BP in the western Loess Plateau, northwest China. Journal of *Quaternary Science* 19(6): 529-535.

An, Z. (1989). Prehistoric agriculture in China. Foraging and farming. *The evolution of plant exploitation.* D. Harris, Hillman, G. Boston, Unwin Hyman: 643-649.

An, Z. (1988) a. Archaeological research on Neolithic China. *Current Anthropology* 29: 753-759.

Andersson, J. G. (1929). *Children of the Yellow Earth.* London, Kegan Paul, Trench, Trubner & Co.

Archibold, O. W. (1995). *Ecology of world vegetation.* London, Chapman and Hall.

Arora, R. K. (1977). Jobs Tears *Coix lachryma-jobi* a minor food and fodder crop of north eastern India. *Economic Botany* 31(3): 358-366.

Atahan, P., Itzstein Davey, F, Taylor, D., Dodson, J., Qin, J., Zheng, H., Brooks, A. (2008). Holocene-aged sedimentary records of environmental changes and early agriculture in the lower Yangtze, China. *Quaternary Science Reviews* 27: 556-570.

Bagley, R. (1999). Shang Archaeology. *The Cambridge History of Ancient China.* M. Loewe, Shaughnessy, E. Cambridge, Cambridge University Press: 124-231.

Barker, G. (1985). *Prehistoric Farming in Europe.* Cambridge, Cambridge University Press.

Barnes, G. (1990). Paddy soils now and then. *World Archaeology* 22(1): 1-17.

Barnes, G. (1999). *China Korea and Japan: The rise of civilisation in East Asia.* New York, Thames and Hudson.

Barton, L., Newsome, S.D., Chen, F-A., Wang, H., Guilderson, T.P., Bettinger, R.L. (2009). Agricultural origins and the isotopic identity of domestication in northern China. *Proceedings of the National Academy of Sciences* 106(14): 5523-8.

Behrendt, S., Hanf, M. (1979). *Grass Weeds in World Agriculture.* Ludwigshafen am Rhein, BASF Aktiengeselleschaft.

Bellwood, P. (2005). Examining the farming/ language hypothesis in the East Asian context. *The peopling of East Asia: Putting together archaeology, linguistics and genetics.* L. Sagart, Blench, R., Sanchez-Mazes, A. London, Routledge-Carson: 17-30.

Bellwood, P. (2005). First *farmers. Origins of agricultural societies.* Oxford, Blackwell.

Bettinger, R. L., Barton, L., Richerson, R.J., Boyd, R., Wang, H., Won, C. (2007). *The transition to agriculture in northwestern China. Late Quaternary Climate Change and Human Adaption.* D. B. Madsen, Gao, X. Chen, F-H. Amsterdam, Elsevier: 83-101.

Bettinger, R. L., Barton, L.W, Elston, R.G. (2009). *Report on 2004 archaeological fieldwork at the Dadiwan site, Shao Dian Village, Gansu Province, PRC.* Davis California, Reno, Nevada, Dept Anthropology, University of California-Davis, University of Nevada-Reno. Field Report http://www.anthro.ucdavis.edu/card/usprc/pubs/Dadiwan%202004%20Report.pdf

Bogaard, A. (2002). Questioning the relevance of shifting cultivation to Neolithic farming in the loess belt of Europe: evidence from the Hambach Forest experiment. *Vegetation History and Archaeobotany* 11(1-2): 155-168.

Bogaard, A., Palmer, C., Jones, G., Charles, M. (1999). A FIBS approach to the use of weed ecology for the archaeobotanical recognition of crop rotation regimes. *Journal of Archaeological Science* 26: 1211-1224.

Bowdery, D. (2007). Phytolith analysis, sheep, diet and fecal material at Ambathala pastoral station (Queensland, Australia). *Plants, people and places: recent studies in phytolith analysis.* M. Madella, Zurro, D. Oxford, Oxbow Books: 134-150.

Bozarth, S. R. (1992). Classification of Opal phytoliths formed in selected dicotyledons native to the Great Plains. *Phytolith Systematics*. G. R. Rapp, Mulholland, S.C. New York, Plenum Press: 193-214.

Bozarth, S. R. (1987). Opal phytolith analysis of edible fruits and nuts naive to the Central Plains. *Phytolitharien newsletter* 4(3): 9-10.

Bozarth, S. R. (1985). Distinctive phytoliths from various dicot species. *Phytolitharien Newsletter* 3(3): 7-8.

Bray, F. (1984). *Agriculture. Science and civilization in China*. Cambridge, Cambridge University Press. Volume 6 part 2.

Bremond, L., Alexandre, A., Vela, E., Guiot, J. (2004). Advantages and disadvantages of phytolith analysis for the reconstruction of Mediterranean vegetation: an assessment based on modern phytolith, pollen and botanical data (Luberon, France). *Review of Palaeobotany and Palynology* 129: 213-228.

Brothwell, D., Brothwell, P.B. (1969). *Food in Antiquity.* London, Thames and Hudson.

Bush, M. B., Piperno, D.R., Colinvaux, De Oliveira, P.E., Krissek, L.A., Miller, M.C., Rowe, W.E. (1992). A 14,300-Yr Palaeoecological profile of a lowland tropical lake in Panama. *Ecology* 62: 251-275.

Canti, M. G. (1999). The production and preservation of faecal spherulites: animals, environment and taphonomy. *Journal of Archaeological Science* 26(3): 251 - 258.

Canti, M. G. (1998). The micromorphological identification of faecal spherulites from archaeological and modern materials. *Journal of Archaeological Science* 25(5): 435-444.

Cao, Z. H., Ding, J.L., Hu, Z.H., Knicker, H., Kogel-Knabner, I., Yang, L.Z., Yin, R., Lin, X.G., Dong, Y.H. (2006). Ancient paddy soils from the Neolithic age in China's Yangtze River delta. *Naturwissenschaften* 93: 232-236.

Carnelli, A. L., Madella, M., Theurillat, J.P. (2001). Biogenic silica production in selected alpine plant species and plant communities. *Annals of Botany* 87: 425-434.

Carnerio, R. L. (1981). Chiefdom: precursor of state. *The transition to statehood in the New World.* G. D. Jones, Kautz, R.R. Cambridge, Cambridge University Press: 37-75.

Chang, K. C. (1999). China on the eve of the historical period. *The Cambridge History of Ancient China: From the Origins of Civilisation to 221BC.* M. Loewe, E. Shaughnessy. Cambridge, Cambridge University Press: 37-73.

Chang, K. C. (1989). Ancient China and its anthropological significance. *Archaeological thought in America.* C. C. Lamberg-Karlovsky. New Haven, Cambridge University Press.

Chang, K. C. (1986). *The archaeology of Ancient China.* New Haven & London, Yale University Press.

Chang, K. C. (1984). Ancient China and its Anthropological significance. *Symbols* 2-4: 20-22.

Chang, K. C. (1983). *Art Myth and Ritual.* Cambridge MA, Harvard University Press.

Chang, K. C. (1980). *Shang civilisation.* New Haven, Yale University Press.

Chang, K. C. (1976). *Early Chinese Civilization: Anthropological Perspectives.* Cambridge M.A. Harvard University Press.

Chang, K. C. (1970). The beginnings of agriculture in the Far East. *Antiquity* 44: 175-185.

Chang, T. T. (1995). Rice. *Evolution of Crop Plants.* J. Smartt, Simmonds, N.W. Essex, Longman Scientific.

Charles, M. G. (1998). Fodder from dung: the recognition and interpretation of dung derived plant material from archaeological sites. *Environmental Archaeology* 1: 111-122.

Charles, M., Hoppe, C., Jones, G., Bogaard, A., Hodgson, J.G., (2003). Using weed functional attributes for the identification of irrigation regimes in Jordan. *Journal of Archaeological Science* 30: 1429-1441.

Charles, M., Jones, G. (1997). FIBS in archaeobotany: functional interpretation of weeds floras in relation to husbandry practices. *Journal of Archaeological Science* 24: 1151-1161.

Chen, B., Jiang, Q. (1997). Antiquity of the earliest cultivated rice in central China and its implications. *Economic Botany* 51(3): 307-310.

Chen, W. -B., Nakamura, I., Sato, Y-I, Nakai, H. (1993). Distribution of deletion type in cp DNA of cultivated and wild rice. *Japanese Journal of Genetics* 68: 597-603.

Cheng, C., Motohashi, R., Tchuchimoto, S., Fukuta, Y., Ohtsubo, H., Ohtsubo, E. (2003). Polyphyletic Origin of Cultivated Rice: Based on the Interspersion patterns of SINEs. *Molecular Biology and Evolution* 20: 67-75.

Clarke, J. E., Blake, M. (1994). Factional Competition and Political Development in the New World. *Factional Competition and Political Development in the New World.* E. M. F. Brumfiel, J.W., Cambridge, Cambridge University Press: 17-30.

Cohen, D. J. (1998). The origins of domesticated cereals and the Pleistocene-Holocene transition in East Asia. *The Review of Archaeology* 19: 22-29.

Coil, J. K., M.A., Archer, S., Hastorf, C.A. (2003). Laboratory goals and considerations for multiple microfossil extraction in archaeology. *Journal of Archaeological Science* 30: 991-1008.

Coley, P., Barone, J. (1996). Herbivory and plant defences in tropical forests. *Annual Review of Ecology and Systematics* 27: 305-335.

Colledge, S. (2001). *Plant exploitation on Epipalaeolithic and Early Neolithic sites in the Levant.* Oxford, BAR International Series 986.

Colledge, S. (1998). Identifying pre-domestication cultivation using multivariate analysis. *The origins of agriculture and crop domestication.* A. B. Damania, Valkoun, J., Wilcox, G., Qualset, C.O. Aleppo, Syria, ICARDA: 121-131.

Collins, R. P., Jones, M.B. (1985). The influence of climatic factors on the distribution of C_4 species in Europe. *Vegetatio* 64: 121-129.

Cowgill, G. L. (1988). Onward and upward with collapse. *The collapse of ancient states and civilisations.* N. Yoffee, Cowgill, G.L. Tuscon, The University of Arizona Press: 244-276.

Crawford, G. (2009). Agricultural origins in North China pushed back to the Pleistocene-Holocene boundary. *Proceedings of the National Academy of Sciences* 106(18): 7271-7272.

Crawford, G. (2006). East Asian plant domestication. *Archaeology of Asia.* M. Stark. Oxford, Blackwell: 77-96.

Crawford, G. (1992). Prehistoric plant domestication in East Asia. *The origins of agriculture: An international perspective.* C. W. Cowan, P.J. Watson, Washington, Smithsonian Institution Press: 7-38.

Crawford, G. W., Chen, S. (1998). The origins of rice agriculture: recent progress in East Asia. *Antiquity* 72: 858-66.

Crawford, G., Takamiya, H. (1990). The origins and implications of late prehistoric plant husbandry in northern Japan. *Antiquity* 64: 889-911.

Crawford, G., Underhill, A., Zhao, Z., Lee, G-A, Feinman, G., Nicholas, L., Luan, F., Yu, H., Fang, H., Cai, F. (2005). Late Neolithic plant remains from Northern China: Preliminary results from Liangchengzhen, Shandong. *Current anthropology* 46(2): 309-315.

Cribb, A. B., Crib, J.W. (1974). *Wild food in Australia.* Brisbane.

Crüger, H. (1857). Westindische Fragmente, 9. *El Cauto Botanische Zeitung* 15: 281-292, 297-308.

D'Andrea, C. A. (1999). The dispersal of domesticated plants into north eastern Japan. *The Prehistory of Food: Appetites for change.* C. Gosden, Hather, J. London, Routledge: 166-183.

Davy, H. (1814). Elements of agricultural chemistry, *A course of lectures for the Board of Agriculture.* Ed 2. London, JG Barnard.

De Garine, I. (1994). The diet and nutrition of human populations. *Companion of Anthropology: Humanity, culture and social life.* T. Ingold. London, Routledge: 226-264.

Delhon, C., Alexandre, A., Berger, J-F., Thiebault, S., Brochier, J-L., Meunier, J-D. (2003). Phytolith assemblages as a promising tool for reconstructing Mediterranean Holocene vegetation. *Quaternary Research* 59: 48-60.

Dekker, J. (2003) a. Evolutionary biology of the foxtail (Setaria) species-group. *Weed biology and management.* Inderjit. The Netherlands, Kluwer Academic Publishers: 65-114.

Dekker, J. (2003) b. The foxtail (*Setaria*) species-group. *Weed Science* 51: 641-656.

Dematte, P. (1999). Longshan-Era urbanism: The role of cities in pre-dynastic China. *Asian Perspectives*: (Honolulu) 38(2): 119-153.

De Montmollin, O. (1989). Land tenure and politics in the late/terminal classic Rosario valley, Chiapas, Mexico. *Journal of Anthropological Research* 45(3): 293-314.

Dennell, R. W. (1976). The economic importance of plant resources represented on archaeological sites. *Journal of Archaeological Science* 3: 339-47.

Dennell, R. W. (1974). Botanical evidence for crop processing activities. *Journal of Archaeological Science* 3: 275-284.

Dennell, R. W. (1972). The interpretation of plant remains: Bulgaria. *Papers in Economic Prehistory*. E. S. Higgs. Cambridge: 149-159.

Dietler, M., Hayden, B. (2001). Digesting the feast: good to eat, good to drink, good to think: an introduction. *Feasts*. M. Dietler, Hayden, B. Washington, Smithsonian Institution.

Doust, A. N., Kellog, E.A., Devos, K.M., Bennetzen, J.L. (2009). Foxtail Millet: A sequence-driven grass model system. *Plant Physiology* 149: 137-141.

Doust, A. N., Kellog, E.A. (2006). Effect of genotype and environment on branching in weedy green millet (*Setaria viridis*) and domesticated foxtail millet (*Setaria italica*) Poaceae. *Molecular Ecology* 15: 1335-1349.

Dove, M. R. (1985). *Swidden Agriculture in Indonesia. The subsistence strategies of the Kaliman Kantu.* Berlin. De Gruyter

Drennan, R. D. (2000). Games, Players, Rules, and circumstances: Looking for understanding of social change at different levels. *Cultural Evolution: Contemporary Viewpoints*. G. M. Feinman, L. Manzanilla. New York, Kluwer Academic Publishers: 177-196.

Drennan, R. D., Uribe, C.A. (editors) (1987). *Chiefdoms in the Americas.* University Press of America, Lanham.

Dykoski, C. A., Lawrence Edwards, R., Hai Cheng, Daoxian Yuan, Yanjun Cai, Meiliang Zhang, Yushi, Lin, Jiaming Qing, Zhisheng An, Justin Revenaugh (2005). A high-resolution, absolute-dated Holocene and deglacial Asian monsoon record from Donge Cave, China. *Earth and Planetary Science Letters* (233): 71-86.

Earle, T. K. (1991). The evolution of kingdoms. *Chiefdoms: Power, economy, and ideology.* T. K. Earle. Cambridge, Cambridge University Press: 1-15.

Earle, T. K. (1987). Chiefdoms in archaeological and ethno historical perspective. *Annual Review of Anthropology* 16: 279-308.

Ehrenberg, C. G. (1854). *Mikrogeologie*. Leipzig, Leopold Voss.

Ehrenberg, C. G. (1841). *Über Verbreitung und Einfluss des Mikroskopischen* Lebens in Sud- und Nordamerika. Monatsberichte der Königlich Preussischen Akademie der Wissenschaften zu Berlin: 139-44.

Englemark, R. (1989). Weed-seeds in archeological deposits. Models, experiments and interpretations. Approaches to Swedish Prehistory. *A spectrum of problems and perspectives in contemporary research.* T. B. Larsson, Lundmark, H. Oxford, BAR International Series No. 500: 179-87.

Epstein, E. (1999). Silicon. *Annual Review of Plant Physiology and Plant Molecular Biology* 50: 641-664.

Ehleringer, J. R., Monson, R.K. (1993). Evolutionary and ecological aspects of photosynthetic pathway variation. *Annual Review of Ecology and Systematics* 24: 411-439.

von Faulkenhausen, L. (2008). Stages in the development of 'Cities' in Pre-Imperial China. *The Ancient City: New Perspectives on Urbanism in the Old and New World.* J. Marcus, Sabloff, J.A. Santa Fe, School for Advanced Research.

von Falkenhausen, L. (1993). On the Historiographical Orientation of Chinese anthropology. *Antiquity* 67: 839-849.

Feng, Z.-D., An, C.B., Wang. H.B., (2006). Holocene climatic and environment changes in the arid and semi-arid areas of China - a review. *Holocene* 1(16): 119-130.

Feng, Z.-D., Chen, F.H. (1999). Problems of the magnetic susceptibility signature as the proxy of the summer monsoon intensity in the Chinese Loess Plateau. *Chinese Science Bulletin* 44(Supplement 1): 97-104.

Ficken, K. J., Wooller, M.J., Swain, D.L., Street-Perrot, F.A., Eglinton, G. (2002). Reconstruction of a sub alpine grass-dominated ecosystem, Lake Rutundu, mount Kenya: a novel multi approach. *Palaeogeography, Palaeoclimatology, Palaeoecology* 177: 137-149.

Flad, R. K., Yuan, J., Li, S. (2007). Zooarchaeological evidence for animal domestication. *Late Quaternary climate change and human adaptation in arid China.* D. B. Madsen, Chen, F-H., Gao, X. Amsterdam, Elsevier. 167-204.

Flannery, K. V. (1969). The Origins and Ecological Effects of Early Domestication in Iran and Near East. *The Domestication and Exploitation of Animals and Plants.* P. Ucko, Dimbleby, G.W. London, Duckworth: 73-100.

Ford, A. (2004). Ground stone tool production at Huizui, China: an analysis of a manufacturing site in the Yilou River Basin. *Bulletin of Indo-Pacific Prehistory Association* 24: 71-77.

Ford, A. (2001). *States and stones: Ground stone tool production at Huizui, China.* Melbourne, La Trobe. ejournal.anu.edu.au/index.php/bippa/article/view/110/100

Fredlund, G. G., Tieszen, L.L. (1997). Calibrating grass phytolith assemblages in climatic terms: Applications to late Pleistocene assemblages from Kansas and Nebraska. *Palaeogeography, Palaeoclimatology, Palaeoecology* 136: 199-211.

Fried, M. F. (1967). *The evolution of political society.* New York, Random House.

Fujio, S. I. (2004). The beginnings of agriculture in the Japanese islands. Cultural diversity in the Archaeology of the 21st Century. The Society of Archaeological Studies. Okayama, *The Society of Archaeological Studies*: 62-73.

Fujiwara, H. (1993). Research in the history of rice cultivation using plant opal analysis. *Current Research in Phytolith Analysis: applications in archaeology and palaeoecology.* D. Pearsall, Piperno, D.R., MASCA Research Papers in Science and Archaeology. Volume 10. Philadelphia, University of Pennsylvania.

Fuller, D. Q. (2006). *A Millet Atlas: Some identification Guidance*, University College London. Unpublished: 1-18.

Fuller, D. Q., Qin, L.,Zheng, Y., Zhao, Z., Hosoya, L.A., Sun, G-P (2009). The Domestication Process and Domestication Rate in Rice: Spikelet Bases from the Lower Yangtze. *Science* 323.5921: 1607-1610.

Fuller, D. Q., Madella, M. (2009). Banana cultivation in South Asia and East Asia: a review of the evidence from archaeology and Linguistics. *Ethnobotany Research and Applications* 7: 333-351.

Fuller, D. Q., Qin, L. (2009). Water management and labour in the origins dispersal of Asian rice. *World Archaeology* 41(1): 88-111.

Fuller, D. Q., Qin, L., Harvey, E. (2008). Evidence for a late onset of agriculture in the Lower Yangtze region and challenges for an archaeobotany of rice. *Past Human Migrations in East Asia: matching archaeology, linguistics and genetics.* A. Sanchez-Mazas, R. Blench, M. D. Ross, I. Peiros, M. Lin. London and New York, Routledge.

Fuller, D. Q., Qin, L. Harvey, E. (2007). A critical assessment of early agriculture in East Asia, with the emphasis in Lower Yangtze rice domestication. *Pradghara* (Journal of the Uttar Pradesh State Archaeology Department (special issue): 1-26.

Fuller, D. Q., Stevens, C. (2009). Agriculture and the development of complex societies: An archaeobotanical Agenda. *Ethnobotanist of distant pasts: Papers in honour of Gordon Hillman.* A. F. E. Weiss. Oxford, Oxbow: 37-57.

Fuller, D. Q., Zhang, H., (2007). A preliminary report of the survey archaeobotany of the upper Ying Valley (Henan Province). *Archaeological discovery and research at Wanchenggang site in Denfeng (2002-2005).* Peking University and Henan Provincial Institute of Cultural Relics and Archaeology. School of Archaeology and Museology. Zhengzhou, Great Elephant Publisher: 916-958.

Fung, C. (2000). The drinks are on us: Ritual, social status and practice in Dawenkou burials, North China. *Journal of East Asian Archaeology* 2(1-2): 67-92.

Gao, Q., Lee, Y.K. (1993). A biological perspective on Yangshao kinship. *Journal of Anthropological Archaeology* 12: 266-298.

Geis, J. W. (1973). Biogenic silica in selected species of deciduous angiosperms. *Soil Science* 116(113-130).

Glover, I. C., Higham, C.F.W. (1996). New evidence for early rice cultivation in South, Sotheast, and East Asia. *The origins and spread of agriculture and pastoralism in Eurasia.* D. Harris. London, University College London Press: 413-441

Gong, Z. T., Chen, H.Z., Yuan, D.G, Zhao, Y.G., Wu, Y.J., Zhang, G.L (2007). The temporal and spatial distribution of ancient rice in China and its implications. *Chinese Science Bulletin* 52(8): 1071-1079.

Graham, D. J., Midgley, N.G. (2000). Graphical representation of particle shape using triangular diagrams: an Excel spreadsheet method. *Earth Processes and Landforms* 25(13): 1473-1477.

Greiss, E. A. M. (1957). *Anatomical identification of some Egyptian plant materials.* Caire, Le C. Tsoumas.

Grob, A. (1896). Beitrage zur Anatomie der Epidermis der Gramineenblätter. *Bibliotheca Botanica.* 7 (36): 1-64.

Guo, X., Lin, K., Lu, X., Ma, H., Li, K., Yuan, J., Yuan, S., Wu, X., Liu, X. (2001). AMS radiocarbon dating of Tianma-Qucun site Shanxi, China. *Radiocarbon* 43: 1109-1114.

Halstead, P., O'Shea, J. (1989). Introduction: cultural responses to risk and uncertainty. *Bad year economics.* P. Halstead, J. O'Shea. Cambridge, Cambridge University Press: 1-8.

Harlan, J. (1995). *The living fields. Our agricultural heritage.* Cambridge, Cambridge University Press.

Harlan, J. R., Pasquereau, J. (1969). Decrue Agriculture in Mali. *Economic Botany* 23(1): 70-74.

Harvey, E. L. (2006). *Early agricultural communities in Northern and Eastern India: and archaeobotanical investigation.* Unpublished PhD thesis. Institute of Archaeology. London, University College London.

Harvey, E., Fuller, D. Q (2005). Investigating crop processing using phytolith analysis. *Journal of Archaeological Science* 32: 739-752.

Hastorf, C. A. (1988). The use of palaeoethnobotanic data in studies of crop production, processing and consumption. *Current Palaeoethnobotany: Analytical methods and cultural interpretations of archaeological plant remains.* C. A. Hastorf, V.S. Popper, Chicago, Chicago University Press: 119-144.

Hayden, B. (2001). Fabulous feasts: a prolegomenon to the importance of feasting. Feasts. B. Hayden, M. Dietler. Washington, *Smithsonian Institution Press*: 23-64.

Herzschuh, U., Tarasov, P., Wunnemann, B. & Hartmann, K. (2004). Holocene vegetation and climate of the Alashan Plateau, NW China, reconstructed from pollen data. *Palaeogeography, Palaeoclimatology, Palaeoecology* 211(1-2): 1-7.

Higham, C. F. W. (2005). East Asian Agriculture and its Impact. *The Human Past. World Prehistory and the Development of Human Societies.* C. Scarre. London, Thames and Hudson: 234-263.

Higham, C. (1995). The transition to rice cultivation in Southeast Asia. *Last Hunters-First Farmers. New Perspectives on the Prehistoric Transition to Agriculture.* T. D. Price, Gebauer, A.B. Santa Fe, New Mexico, School of American Research: 127-156.

Higham, C., Lu, T. (1998). The origins and dispersal of rice cultivation. *Antiquity* 72: 867-877.

Hillman, G. (1984). Interpretation of archaeological plant remains: the application of ethnographic models from Turkey. *Plants and ancient man - studies in palaeoethnobotany.* W. Van Zeist, W.A. Casparie. Rotterdam, A A Balkema.

Hillman, G. (1981). Reconstructing crop husbandry and food production: modern models for the interpretation of plant remains. *Farming practice in British Prehistory.* R. J. Mercer. Edinburgh, Edinburgh University Press: 123-162.

Hillman, G. C. (1973). Crop husbandry and food products: a modern basis for the interpretation of plant remains. *Anatolian studies* 23: 241 -244.

Hillman, G. C., Hedges, R., Moore, A.M.T., Colledge, S., Pettit, P. (2001). New evidence of Late Glacial cereal cultivation at Abu Hureyra on the Euphrates. *The Holocene* 11: 383-393.

Hillman, G., Legge, A.J., Rowley-Conwy, P.A. (1997). On the charred seeds from Epipalaeolithic Abu Hureya: food or fuel? *Current Anthropology* 38(4): 651-655.

Ho, C. K., Royaume-Uni, University of Southampton Department of Archaeology. (1986). *A New Perspective on the Origins of Chinese Agriculture. Recent Advances in the Understanding of Plant Domestication and Early Agriculture,* London: Allen and Unwin, 1986: 8.

Ho, P.-T. (1987). Review of Science and Civilisation in China, Vol 6: Biology and Biological Technology, Part II: Agriculture by Francesca Bray. *Journal of the American Oriental Society* 107(2): 347-348.

Ho, P.-T. (1977). The indigenous origins of Chinese agriculture. *The Origins of agriculture.* C. Reed. The Hague, Mouton: 413-484.

Ho, P.-T. (1956). Early-ripening rice in Chinese history. *The Economic History Review* 9: 200-218.

Hodson, M. J., Evans, D.E. (1995). Aluminum/silicon interactions in higher plants. *Journal of Experimental Botany* 46: 161-171.

Hodson, M. J., Williams, S.E., Sangster, A.G. (1997). Silica deposition in the needles of the Gymnosperms. I. Chemical analysis and light microscopy. *State of the Art of Phytoltihs in Soils and Plants.* A. Pinilla, Juan-Tresserras, J., Machado, M.J. Madrid, Monografias del Centro de Ciencias Medioambientales, Consejo Superior de Investigaciones Cientificas: 135-146

Hoilund-Neilson, K. (1988). Correspondence analysis applied to hoards and graves of the Germanic Iron Age. *Multivariate Archaeology.* T. Madsen. Jutland Archaeological Society Publications XXI, Aarhus University Press: 37-54.

Horrocks, M., Deng, Y. Ogden, J. Sutton, D.G. (2000). A reconstruction of the history of Holocene sand dune on Great Barrier Island, northern New Zealand, using pollen and phytolith analyses. *Journal of Biogeography* 27: 1269-1277.

Hsu, V. Y., Hsu, F.L.K. (1977). Modern China, North. *Food in Chinese culture: anthropological and historical perspectives.* K. C. Chang. New Haven and London, Yale University Press: 295-316.

Hu, Y., Ambrose, S.H., Wang, C. (2006). Stable isotopic analysis of human bones from Jiahu site, Henan, China: implications for the transition to agriculture. *Journal of Archaeological Science* 33: 1319-1330.

Hubbard, R. N. L. B. (1980). Development of agriculture in Europe and the Near East: Evidence from quantative studies. *Economic Botany* 34(1): 51-67.

Huber, L. (1981).The traditions of Chinese Neolithic pottery *Bulletin of the Museum of Far Eastern Antiquities* 53: 256-101.

Huang, F., Zhang, M. (2000). Pollen and phytolith evidence for rice cultivation during the Neolithic at Longquizhuang, eastern Jianghuai, China. *Vegetation History and Archaeobotany* 9: 161-168.

Hunt, H., Vander Linden, M., Liu, X., Motuzaite-Matuzeviciute, G., Colledge, S., Jones, M.K. (2008). Millets across Eurasia: chronology and context of early records of the genera *Panicum* and *Setaria* from archaeological in the Old World. *Vegetation History and Archaeobotany* 17(Supplement 1): S5-S18.

Imamura, K. (1996). The transition to agriculture in Japanese Prehistory. *The origins and spread of agriculture and pastoralism in Eurasia.* D. Harris. London, Institute of Archaeology, University College London: 442-464.

Iriarte, J. (2006). Vegetation and climate change since 148, 10 ^{14}C yr B.P. in south eastern Uruguay and implications for the rise of early Formative societies. *Quaternary Research* 65: 20-32.

Iriate, J., Paz, E.A. (2009). Phytolith analysis of selected native plants and modern soils from southeastern Uruguay and its implications for palaeoenvironmental and archaeological reconstruction. *Quaternary International* 193: 99-123.

Iskandar, J., Ellen, R.F. (2000). The contribution of *Paraserianthes (Albizia) falcataria* to sustainable swidden management practices among the Baduy of West Java. *Human Ecology* 28(1).

Isaacs, J. (1987). *Bush food. Aboriginal food and herbal medicine*. Sydney, Lansdowne Publishing

Itzstein-Davey, F., Atahan, P., Dodson, J., Taylor, D., Zheng, H. (2007). Environmental and cultural changes during the terminal Neolithic: Qingpu, Yangtze delta, eastern China. *Holocene* 17(7): 875-887.

Itzstein-Davey, F., Taylor, D., Dodson, J., Atahan, P., Zheng, H. (2007). Wild and domesticated forms of rice (*Oryza* sp.) in early agriculture at Qinpu, lower Yangtze, China: evidence from phytoliths. *Journal of Archaeological Science* 34: 2101-2108.

Jacobs, B. F., Kingston, J.D., Jacobs, L.J. (1999). The origin of grass dominated ecosystems. *Annals of Missouri Botanical Garden* 86: 590-643.

Jiang, Q. (1997). Phytolith assemblages and their implications from cal soils of Yingyang site in Denfeng County, Henan Province. *Chinese Science Bulletin* 42(1): 57-61.

Jiang, Q. (1995). Searching for evidence of early rice agriculture at prehistoric sites in China through phytolith analysis: an example from central China. *Review of Palaeobotany and Palynology* 89: 481-485.

Jiang, Q., Piperno, D. (1994). Phytolith analysis of an archaeological site (Longshan period) Zhumadian City, Henan Province China: Palaeoenvironmental and cultural implications. *Geoarchaeology* 9 (5): 409-417.

Jin, G.-Y. Y., S-D, Udatsu, T., Lan, Y-F, Wang, C-Y., Tong, P-H (2007). Neolithic rice paddy from the Zhaojizhuang site. *Chinese Science Bulletin* 52(24): 3376-3384.

Johannessen, S. (1988). Plant remains and culture change: Are palaeoethnobotanical data better than we think? *Current Palaeoethnobotany: Analytical methods and cultural interpretations of Archaeological plant remains*. C. A. Hastorf, V.S. Popper, Chicago, Chicago University Press: 145-166.

Jones, G. E. M. (1987). The statistical approach to the archaeological identification of crop processing. *Journal of Archaeological Science* 14: 311-323.

Jones, G. E. M. (1984). Interpretation of archaeological plant remains: ethnographic models from Greece. *Plants and ancient man,* Proceedings of the 6th Symposium of the International Workgroup for Palaeoethnobotany; 30th May - 3rd June, 1983, Groningen, Rotterdam and Boston.

Jones, G., Bogaard, A., Charles, M., Hodgson, J.G. (2000). Distinguishing the effects of agricultural practices relating to fertility and disturbance: a functional ecological approach in archaeobotany. *Journal of Archaeological Science* 27(11): 1073-1084.

Jones, M. K. (1985). Archaeobotany beyond subsistence reconstruction. *Beyond domestication in Prehistoric Europe, investigations in subsistence archaeology and social complexity*. G. Barker, Gamble, C. London, Academic Press: 107-128.

Jones, R. L. (1964). Note on the occurrence of opal phytoliths in some Cenozoic sedimentary rocks. *Journal of Palaeontology.*

Jones, R. L., Beavers, A.H. (1963). Some mineralogical and chemical properties of plan opal. *Soil Science* 96: 375-379.

Kaufman, P. B., Bigelow, W.C., Schmid, R.,Ghosheh, N.S. (1971). Electron microprobe analysis of silica in epidermal cells of Equisetum. *American Journal of Botany* 58: 309-316.

Keenan, D. J. (2002). Astro-Historiographic chronologies of early China are unfounded. *East Asian History* 23: 61-68.

Keenan, D. J. (2007). Defence of planetary conjunctions for early Chinese chronology is unmerited. *Journal of Astronomical History and Heritage* 10(2): 142-147.

Keightley, D. (1983). *The origins of Chinese civilisation.* Berkeley, C.A, University of California Press.

Kawaguchi, K., Kyuma, K. (1977). *Paddy soils in tropical Asia: their material nature and fertility*. Honolulu, the University Press of Hawaii.

Khush, G. S. (1997). Origin, dispersal, cultivation and variation of rice. *Plant Molecular Biology* 35: 24-34.

Kim, S.-O., (1994). Burials, pigs, and political prestige in Neolithic China. *Current Anthropology* 35 (2): 119-141.

Kittipong, P. (1983). *Weed control in farmer's fields in Thailand.* Proceedings of the Conference on weed control in rice. International Rice Research Institute, Los Banos: 193-200.

Klien, R. L., Geis, J.W. (1978). Biogenic silica in the Pinaceae. *Soil Science* 126: 145-155.

Leach, H. M. (1999). Intensification in the Pacific: a critique of the archaeological criteria and their application. *Current Anthropology* 40(3): 311-339.

Le Thierry d'Ennequin, M., Penaud, O., Toupance, B., Sarr, A. (2000). Assessment of genetic relationships between *Setaria italica* and its wild relative *S. viridis* using AFLP markers. *Theoretical and Applied Genetics* 100: 1061-1066.

Lee, G.-A., Bestel, S. (2007). Contextual analysis of plant remains at the Erlitou-period Huizui site, Henan, China. *Indo-Pacific Prehistory Association Bulletin* 27: 49-60.

Lee, G.-A., Crawford, G., Liu, L. Chen, X. (2007). Plants and people from the early Neolithic to the Shang periods in North China. *Proceedings of the National Academy of Sciences* 104(3): 1087-1092.

Lee, Y. K. (2004). Control strategies and polity competition in the lower Yi-Luo Valley, North China. *Journal of Anthropological Archaeology* 23: 172-195.

Lee, Y. K. (2002). Building the chronology of Early Chinese History. *Asian Perspectives* 41(1): 15-42.

Lee, Y-K. (1994). Burials, pigs, and political prestige in Neolithic China. Comments in Kim, S.-O., Burials, pigs, and political prestige in Neolithic China. *Current Anthropology* 35(2): 119-141.

Lee, Y. K. and N. Zhu. (2002). Social integration of religion and ritual in prehistoric China. *Antiquity* 76(293): 715-723.

Lentfer, C. J., Boyd, W.E. (1998). A comparison of three methods for the extraction of phytoliths from sediments. *Journal of Archaeological Science* 26: 31-44.

Li, H. W., Li, H., Pao, W.K. (1945). Cytological and genetical studies of the interspecific cross of the cultivated foxtail millet (*Setaria italica*) and the green foxtail (*S. viridis*). *Journal of the American Society of Agronomy* 9: 32-54.

Li, X. (2002). The Xia-Shang-Zhou Chronology Project: methodology and results. *Journal of East Asian Archaeology* 4: 321-333.

Li, X., Dodson, J., Zhou, X., Zhang, X., Masutomoto, R. (2007) c. Early cultivated wheat and broadening of agriculture in Neolithic China. *The Holocene* 17(5): 555-560.

Li, X., Zhou, X-Y, Zhang, H-B, Zhou, J., Shang, X, Dodson, J. (2007) a. The record of cultivated rice from archaeobiological evidence in northwestern China 5000 years ago. *Chinese Science Bulletin* 52(10): 1372-1378.

Li, X., Zhou, X., Zhou, J., Dodson, J., Zhang, H., Shang, X. (2007) b. The earliest archaeobiological evidence of the broadening agriculture in China recorded at Xishanping site in Gansu Province. Science in China Series D: *Earth Sciences* 50(11): 1707-1714.

Linduff, K. M., Drennan, D.D., Shelach, G. (2004). Early complex societies in NE China: The Chifeng International Collaborative Archaeological Research Project. *Journal of Field Archaeology* 29(1/2): 45-73.

Liu, J. (1985). Some observations on the archaeological site of Hemudu, Zheijiang Province, China. *Bulletin of Indo-Pacific Prehistory Association* 6: 40-45.

Liu, L. (2004). *The Chinese Neolithic: trajectories to early states*. Cambridge, Cambridge University Press.

Liu, L. (2000). The development and decline of social complexity in China: Some environmental and social factors. *Indo-Pacific Prehistory Association Bulletin* (Maleka Papers) 20(4): 14-33.

Liu, L. (1999). Who were the ancestors? The origins of Chinese ancestral cult and racial myths. *Antiquity* 73(281): 602-613.

Liu, L. (1996) a. Settlement patterns, chiefdom, variability and the development of early states in north China. *Journal of Anthropological Archaeology* 15: 237-88.

Liu, L. (1996) b. Mortuary ritual and social hierarchy in the Longshan Culture. *Early China* 21: 1-46.

Liu, L., Chen, X., (2007). Multidisciplinary research in the Yilou project: after 10 years. *Indo-Pacific Prehistory Association Bulletin* 27: 37-38.

Liu, L., Chen, X., (2006). Sociopolitical change from Neolithic to Bronze Age China. *Archaeology of Asia*. M. Stark. Oxford, Blackwell: 149-176.

Liu, L., Chen, X. (2003). *State formation in early China*. London, Duckworth.

Liu, L., Chen, X., Lee, Y.K., Wright, H. & Rosen, A. (2005). Settlement patterns and development of social complexity in the Yilou region, North China. *Journal of Field Archaeology* 29: 75-100.

Liu, L., Chen, X., Li, B. (2007). Non-state crafts in the early Chinese state: an archeological view from the Erlitou Hinterland. *Indo-Pacific Prehistory Association Bulletin* 27: 93-102.

Liu, L., Lee, G-A., Jiang, L., Zhang, J. (2007). Evidence for the early beginning (c. 9000 cal. BP) of rice domestication in China: a response. *Holocene* 17(8): 1059-1068.

Liu, L., Xu, H (2007). Rethinking Erlitou: legend, history and Chinese archaeology. *Antiquity* 81: 886-901.

Liu, X., Hunt, H.V., Jones, M.K. (2009). River Valleys and foothills: changing archeological perceptions of North China. *Antiquity* 83: 82-95.

Londo, J. P., Chiang, Y-C., Hung, K-H., Chiang, T-Y., Schaal, B., (2006). Phylogeography of Asian wild rice, *Oryza rufipogon*, reveals multiple independent domestications of cultivated rice, *Oryza sativa*. *Proceedings of the National Academy of Sciences* 103: 9578-9583.

Lu, H., Liu, K. (2000). *Illustrations of phytoliths from grasses in China and the USA*. Beijing, Institute of Geology and Geophysics, Chinese Academy of Sciences.

Lu, H., Liu, X., Wu, N., Berne, S., Saito, Y., Liu, B., Luo, W. (2002). Rice Domestication and climate change: phytoliths evidence from East China. *Boreas* 31: 378-385.

Lu, H., Wu, N-Q., Liu, K-B., Jiang, H., Liu, T-S. (2007). Phytoliths as quantative indicators for the reconstruction of past environmental conditions in China II: palaeoenvironmental reconstruction in the Loess Plateau. *Quaternary Science Reviews*. 26:759-772

Lu, H., Wu, N-Q., Yang, X-D., Liu, K-B., Jiang, H., Liu, T-S. (2006). Phytoliths as quantative indicators for the reconstruction of past environmental conditions in China I: phytolith-based transfer functions. *Quaternary Science Reviews* 25: 945-959.

Lu, H., Yang, X ., Ye, M., Liu, K-B, Xia, Z., Ren, X., Cai, L., Wu, N., Liu, T-S. (2005). Millet Noodles in Late Neolithic China. *Nature* 4737: 967-968.

Lu, H., Zhang, J., Liu, K., Wu, N., Li, Y., Zhou, K., Ye, M., Zhang, T., Zhang, H., Yang, X., Shen, L., Xu, D., Li, Q. (2009). Earliest domestication of common millet (*Panicum miliaceum*) in East Asia extended to 10,000 years ago. *Proceedings of the National Academy of Sciences*. 106:7367-7372

Lu, H., Zhang, J., Liu, K., Wu, N., Li, Y., Zhou, K., Ye, M., Zhang, T., Zhang, H., Yang, X., Shen, L., Xu, D., Li, Q. (2009) supplementary material www.pnas.org/cgi/content/short/0900158106

Lu, H., Zhang, J., Wu, N., Liu, K., Xu, D., Li, Q. (2009). Phytolith Analysis for the discrimination of Foxtail millet (*Setaria italica*) and Common millet (*Panicum miliaceum*). *Public Library of Science One* 4(2): e4448.

Lu, T. (2006). The occurrence of cereal cultivation in China. *Asian Perspectives* 45(2): 129-158.

Lu, T. (1999). *The transition from foraging to farming and origin of agriculture in China*. Oxford, British Archaeology Reports International Series 774.

Lundstrom-Bandais, K., Rachoud-Schneider, A.M., Baudais, D., Poissonnier, B., (2002). Le broyage dans la chaine de transformation du millet (*Panicum miliaceum*): outils, gestes et ecofacts. *Moudre Et Broyer, I, Methodes*. H. Procopion, Treuil, R. Paris, Comité des Travaux Historiques et Scientifiques: 155-180.

Jing, Y., Campbell, R. (2009). Recent archaeometric research on 'the origins of Chinese civilisation. *Antiquity* 83: 96-109.

Ma, X. (2005). *Emergent Social complexity in the Yangshao Culture: Analyses of settlement patterns and faunal remains from Lingbao, Western Henan, China*. Oxford, BAR International Series 1453.

Mabberley, D. J. (1987). *The plant book: a portable dictionary of the higher plants*. Cambridge, Cambridge University Press.

Macphail, R. I., Crowther, J. (2007). Soil micromorphology, chemistry and magnetic susceptibility studies at Huizui (Yilou Region, Henan Province, Northern China), with special focus on a typical Yangshao floor sequence. *Indo-Pacific Prehistory Association Bulletin* 27: 103-113.

Madella, M. (2000). Understanding archaeological structures by means of phytolith analysis: a test from the Iron Age site of Kilise Tepe - Turkey. *Phytoliths: Applications in earth sciences and human history*. J. D. Meunier, Colin, F. Lisse, Balkema: 173-182.

Madella, M. (1997). Phytolith analysis from the Indus Valley site of Kot Diji Sind, Pakistan. Archaeological Sciences 1995. *Proceedings of a conference on the application of scientific techniques to the study of archaeology. Liverpool, July 1995*. E. S. A. Sinclair, J. Gowlet, Oxbow Monograph 64.

Madella, M., Alexandre, A., Ball, T. (2005). International code for phytolith Nomenclature 1.0. *Annals of Botany* 96: 253-260.

Madella, M., Jones, M.K., Goldberg, P., Goren, Y., Hovers, E. (2002). The exploitation of plant resources by Neanderthals in Amud Cave (Israel): the evidence from phytolith studies. *Journal of Archaeological Science* 29: 703-719.

Madella, M., Powers-Jones, A.H., Jones, M.J. (1998). A simple method of extraction of opal phytoliths from sediments using a non-toxic heavy liquid. *Journal of Archaeological Science* 25: 801-803.

Madella, M., Powers-Jones, A.H. (1998). Phytoliths from Tagliente shelter: a discussion on deposition and taphonomy. Proceedings of XIII international congress of prehistoric and protohistoric sciences, Forli, Italia 8/14 September 1996: 501-507

Maher, B. A., Hu, M. (2006). A high resolution record of Holocene rainfall variations from the western Chinese Loess Plateau: antiphase behaviour of the African/ Indian and East Asian summer monsoons. *The Holocene*, 16(3): 309-319.

Maloney, B. K. (1991). Rice Origins: Recent advances. *Geojournal* 23: 121-124.

Mathews, R. B., Horie, T. Kropff, M.J., Bachelet, D., Centeno, H.G., Shin, J.C., Mohandass, S., Singh, S., Zhu, D., Lee, M-H (1995). Evaluation of effects of climate change. *Modeling the impact of climate change on rice production in Asia.* R. B. Mathews, Kropff, M.J., Bachelet, D., van Laar. Wallingford, Centre for Agriculture and Biosciences International in association with the International Rice Research Institute: 95-139.

Marshner, H. (1995). *Mineral Nutrition of Higher Plants.* London Academic Press, London.

McNaughton, S. J., Tarrants, J.L. (1983). Grass leaf silicification: natural selection for an inducible defense against herbivores. *Proceedings of the National Academy of Sciences* 80: 790-91.

Metcalfe, C. R., Chalk, L. (1983). *Anatomy of the dicotyledons. II Wood structure and conclusion of the general introduction.* Oxford, Clarendon Press.

Metcalfe, C. R. (1960). *Anatomy of the Monocotyledons I. Gramineae.* London, Oxford University Press.

McGovern, P. E., Underhill, A.P., Fang, H., Luan, F., Hall, G.R., Yu, H., Wang, C-S., Cai, F., Zhao, Z., Feinman, G.M. (2005). Chemical identification and cultural implications of a mixed fermented beverage from late prehistoric China. *Asian Perspectives* 44 (2): 249-275.

McGovern, P. E., Zhang, J., Tang, T., Zhang, Z., Hall, G.R., Moreau, R.,A., Nunez, A., Butrym, E.D., Richards, M.P., Wang, C-S., Cheng, C., Zhao, Z.,Wang, C. (2004). Fermented Beverages of pre- and proto-historic China. *Proceedings of the National Academy of Sciences* 101: 17593-17598.

Miller, N. F. (1988). Ratios in palaeoethnobotanical analysis. *Current Palaeoethnobotany: analytical methods and cultural interpretation of archaeological plant remains.* C. A. Hastorf, Popper, V.S. Chicago, University of Chicago Press: 72-85.

Miller, N. F. (1984). The use of dung as fuel: an ethnographic example and an archaeological application. *Palaeorient* 10: 71-79.

Moody, K. (1989). *Weeds reported in rice in South and Southeast Asia.* Manila, International Rice Research Institute.

Morrison, K. D., Feinman, G.M. Nicholas, L.M., Ladefoged, T.N., Myrdal-Runebjer, E., Stone, G.D., Wilk, R. (1996). Typological schemes and agricultural change: Beyond Boserup in South India [and comments and reply]. *Current Anthropology* 37: 583-608.

Morrison, K. D. (1994). The intensification of production: archaeological approaches. *Journal of Archaeological Method and Theory* 1(2): 111-59.

Mullholland, S. C., Rapp, G. Jr. (1992). Morphological classification of grass silica bodies. *Phytolith Systematics.* G. R. J. Rapp, Mulholland S.C. New York, Plenum Press: 68-90.

Murray, M. A. (2009). Questions of continuity: fodder and fuel use in Bronze Age Egypt. *Ethnobotanist of distant pasts: Papers in honour of Gordon Hillman.* A. Fairbairn, E. Weiss

Murrowchick, R., Ed. (1994). *Cradles of civilisation. China ancient culture, modern land.* Norman, University of Oklahoma Press.

Nasu, H., Gu, H. B., Momohara, A., & Yasuda, Y. (2012). Land-use change for rice and foxtail millet cultivation in the Chengtoushan site, central China, reconstructed from weed seed assemblages. *Archaeological and Anthropological Sciences*, 4(1), 1-14.

Nasu, H., Momohara, A., Yasuda, Y., He, J., (2007). The occurrence and identification of *Setaria italica* (L.) P. Beauv. (Foxtail millet) grains from the Chengtoushan site (ca. 5800 cal B.P.) in central China, with reference to the domestication centre in Asia. *Vegetation history and archaeobotany* 16: 481-494.

Nelson, D. M., Hu, F.S., Tian, J., Stefanova, I., Brown, T.A. (2004). Response of C3 and C4 plants to middle-Holocene climatic variation near the prairie- forest ecotone of Minnesota. *Proceedings of the National Academy of Sciences* 101: 562-567.

Netolitzky, F., Ed. (1929). *Die Kieselkorper. Linsbauer's Handbuch der Planzenanatomie Vol. III/1a.* Berlin. 1-19

Netolitzky, F. (1914). Die Hirse aus Antiken Funden. Sitzbuch der Keisekiche Akademie fur Wissenschat der Mathamatisch. *Naturwissenschaften* (123): 725-59.

Netolitzky, F. (1900). Mikroskopische Untersuchung Ganzlich verkholter Vorgeschlichter Nahrungsmittel aus Tirol. *Zeitschrift fur Unsuchung der Nahrungs- und Genusmittel* (3): 401-7.

Niu, S., Yuan, Z., Zhang, Y., Liu, W., Zhang, l., Huang, J., Wan, S. (2005). Photosynthetic responses of C_3 and C_4 species to seasonal water variability and competition. *Journal of Experimental Botany* 56(421): 2867-2876.

Oka, H.-I. (1988). *Origins for cultivated rice.* Amsterdam, Elsevier.

Ollendorf, A. L. (1992). Towards a classification scheme of sedge (Cyperaceae) phytoliths. *Phytolith Systematics: Emerging Issues.* Rapp, Jr. G., Mulholland, S.C. New York, Plenum Press: 91-111.

Olsen, J. W. (1987). The Practice of Archaeology in China. *Antiquity* 61: 282-290.

Osterrieth, M. L., Madella, M., Zurro, D., Alvarez, M.F. (2009). Taphonomical aspects of silica phytoliths in the loess sediments of the Argentinean Pampas. *Quaternary International* 193: 70-79.

Parker, A. G., Eckersley, L., Smith, M.M., Goudie, A.S., Stokes, S., Ward, S. White, K., Hodson, M.J, (2004). Holocene vegetation dynamics in the northeastern Rub' al-Khali desert, Arabian Peninsula: a phytolith, pollen and carbon isotope study. *Journal of Quaternary Science* 19 (7): 665-676.

Parr, J. F., Lentfer, C.J., Boyd, W.E. (2001). A comparative analysis of wet and dry ashing techniques for the extraction of phytoliths from plant material. *Journal of Archaeological Science* 28: 875-886.

Parry, D. W., Smithson, F. (1964). Types of Opaline Silica depositions in the leaves of British Grasses. *Annals of Botany* 30: 525-38.

Pearsall, D. M. (2000). *Palaeoethnobotany: a handbook of procedures.* San Diego, Academic Press.

Pearsall, D., Piperno, D.R. (1993). *Current research in Phytolith Analysis: Applications in Archaeology and Palaeoecology.* Philadelphia, MASCA, University of Pennsylvania.

Pearsall, D. M., Piperno, D.R., Dinan, E.H., Umlauf, M., Zhao, Z., Benfer, R.A., (1995). Distinguishing rice (*Oryza sativa* Poaceae) from wild *Oryza* species through phytolith analysis: results of preliminary research. *Economic Botany* 49: 183-196.

Pearson, R. (1981). Social complexity in Chinese Coastal Neolithic sites. *Science* 213: 1078-1086.

Pearson, R., Underhill, A. P. (1987). The Chinese Neolithic: Recent trends in research. *American Anthropologist* 89 (4): 807-822.

Pechenkina, E. A., Stanley, H.A., Ma, X. Benfer, R.A. (2005). Reconstructing northern Chinese Neolithic subsistence practices by isotopic analysis. *Journal of Archaeological Science* 32: 1176-1189.

Pechenkina, E. A., Benfer, R.A., Wang, Z. (2002). Diet and health changes at the end of the Chinese Neolithic: The Yangshao/Longshan transition in Shaanxi province. *American journal of physical anthropology* 117(1): 15-36.

Pei, A. (1998) b. Notes on new advancements and revelations in the agricultural archaeology of early rice domestication in the Dongting lake region. *Antiquity* 72: 878-885.

Peletz, M. G. (1988). *A share of the harvest: kinship, property, and social history among the Malays of Rembu.* Berkeley, University of California Press.

Pfeiffer, J. M., Dun, S., Mulawarman, B., Rice, K. (2006). Biocultural diversity in traditional rice-based agro systems: indigenous research and conservation of *mavo* (*Oryza sativa* L.) Upland land races of eastern Indonesia. *Environmental Development Sustainability*: 1-17.

Piperno, D. (2006). *Phytoliths: A comprehensive guide for archaeologists and palaeoecologists.* Oxford, Altimira Press.

Piperno, D. (1988). *Phytolith analysis, an archaeological and geological perspective.* San Diego, Academic Press.

Piperno, D. (1985). Phytolith Taphonomy and distributions in Archaeological sediments from Panama. *Journal of Archaeological Science* 12: 247-67.

Piperno, D., Pearsall, D. (1993). The nature and status of phytolith analysis. *Current research in phytolith analysis: applications in archaeology and palaeoecology.* D. R. Piperno. D.M. Pearsall. Philadelphia, MASCA, The University Museum of archaeology and anthropology, University of Pennsylvania. 10: 9-18.

Popper, V. S. (1988). Selecting quantative measurements in Paleoethnobotany. *Current Paleoethnobotany: Analytical methods and cultural interpretations of archaeological plant remains.* V. S. Popper. C.A. Hastorf. Chicago and London, University of Chicago Press: 36-52

Postek, M. T. (1981). The Occurence of Silica in the leaves of *Magnolia grandiflora* L. *Botanical Gazette* 142(1): 124-134.

Postgate, N., Wang, T., Wilkinson, T. (1995). The evidence for early writing: Utilitarian or ceremonial? *Antiquity* 69: 459-480.

Powers-Jones, A. H., Padmore, J. (1993). The use of quantitative methods and statistical analyses in the study of opal phytoliths. *Current research in phytolith analysis: applications in archaeology and palaeoecology.* D. Pearsall, D.R. Piperno, Philadelphia, University of Pennsylvania. 10: 47-56.

Price, B. (1977). Shifts in production and organisation. *Current Anthropology* 18.

Prychid, C. J., Rudall, P.J., Gregory, M. (2004). Systematics and Biology of Silica Bodies in Monocotyledons. *The Botanical Review* 69 (4): 377-440.

Pyankov, V. I., Gunin, P.D., Tsoog, S., Black, C.C. (2000). C4 plants in the vegetation of Mongolia: their natural occurrence and geographical distribution in relation to climate. *Oecologia* 123: 15-31.

Quine, T. A., Walling, D. & Zhang, X. (1999). Slope and gulley response to agricultural activity in the rolling loess plateau, China. *Fluvial processes and environmental change.* A.G. Brown, T.A. Quine. New York, Wiley and Sons: 71-90.

Rao, K. E., De Wet, J.M.J., Brink, B.E., Mengesha, M.H. (1987). Intraspecific variation and systematics of cultivated *Setaria italica*, foxtail millet. *Economic Botany* 41 (1): 108-116.

Raven, J. A. (1983). The transport and function of Silica in plants. *Biological Reviews of the Cambridge Philosophical Society* 58: 179-207.

Reddy, S. N. (1997). If the threshing floor could talk: integration of agriculture and pastoralism during the Late Harappan in Gujurat, India. *Journal of Anthropological Archaeology* 16: 162-187.

Reddy, S. N. (1994). *Plant usage and subsistence modeling: an ethnoarcheological approach to the late Harappan of northwest India,* University of Wisconsin-Madison.

Reissig, W. H., Heinrichs, E.A., Litsinger, J.A., Moody, K., Fiedler, L., Mew, T.W., Barrion, A.T., Ed. (1986). *Illustrated guide to integrated pest management in rice in Tropical Asia.* Los Banos, Philippines, International Rice Research Institute.

Reynolds, P. J. (1979). *Iron Age Farm: The Butser experiment.* London, British Museum Publications.

Roder, W., Phengchanh, S., Keoboulapha, B. (1997). Weeds in slash and burn rice fields in northern Laos. *Weed Research* 37: 111-119.

Roder, W., Phengchanh, S., Keoboulapha, B. (1995). Relationships between soil, fallow period, weeds and rice yield in slash and burn systems of Laos. *Plant and Soil* 176: 27-36.

Rosen, A. M. (2008). The impact of environmental change and human land use on alluvial valleys in the Loess Plateau of China during the Middle Holocene. *Geomorphology* 101: 298-307.

Rosen, A. M. (2007). *Civilizing climate: Social responses to climate change in the Near East.* Lanham, MD, Altamira.

Rosen, A. M. (2007). The role of environmental change in the development of complex societies in China: a study from the Huizui site. *Indo-Pacific Prehistory Association Bulletin* 27: 39-48.

Rosen, A. M. (1992). Preliminary identification of silica skeletons from Near Eastern archaeological sites: an anatomical approach. *Phytolith systematics: Emerging issues: Advances in Archaeological and Museum Science 1.* G. R. Rapp, Mulholland S.C, Plenum Press: 129-147.

Rosen, A. M. (unpublished). *Phytolith extraction protocol.*

Rosen, A. M. (2000). Phytolith evidence for agro-pastoral economies in the Scythian period of southern Kazakhstan. *Phytoliths: Applications in earth sciences and human history.* Lisse, Balkema: 183-198

Rovner, I. (1971). Potential of opal phytoliths for use in palaeoecological reconstruction. *Quaternary Research* 1: 343-359.

Rull, V. (1987). A note on pollen counting in palaeoecology. *Pollen et spores* (29): 471-480.

Sage, R. F. (2004). The evolution of C_4 Photosynthesis. *New Phytologist* 161(2): 341-370.

Saito, K., Linquist, B., Johnson, D.E., Phengchanh, S. Shiraiwa, T., Horie, T. (2008). Planted legume fallows reduce weeds and increase soil N and P contents but not upland rice yields. *Agroforest Systems* 74: 63-72.

Saito, K., Linquist, B., Keobualapha, B., Phanthaboon, K., Shiraiwa, T., Horie, T. (2006). Cropping intensity and rainfall effects on upland rice yields in northern Laos. *Plant Soil* 284: 175-185.

Sakamoto, S. (1987). Origin and dispersal of common millet and foxtail millet. *Japan Agricultural Research Quarterly* 21: 84-89.

Sangster, A., Hodson, M.J. (2007). Silicification of conifers and its significance to the environment. *Plants, people and places: recent studies in phytolith analysis.* M. Madella, Zurro, D. Oxford, Oxbow Books: 79-91.

Sangster, A., Hodson, M.J., Tubb, H.J. (2001). Silicon deposition in higher plants. *Silicon in Agriculture.* I. E. Datnoff, Snyder, G.H., Korndorfer, G.H.,. Amsterdam, Elsevier Science: 85-113.

Sangster, A., Wynn Parry, D. (1969). Some factors in relation to Bulliform Cell Silicification in the Grass Leaf. *Annals of Botany* 33: 315-323.

Scarry, C. M. (1986). *Change in plant procurement and production during the Moundville chiefdom.* Ph.D. dissertation. Department of Anthropology, University of Michigan, Ann Arbor: University Microfilms

Schellenberg, H. C. (1908). Wheat and Barley from the North Kurgan, Anau. *Exploration in Turkestan, Volume 3.* Pumpelly. Washington DC, Carnegie Institution. 3: 471-73.

Schiffer (1976). *Behavioral archaeology.* New York, Academic Press.

Schulze, E.-D., Ellis, R. Schulze W., Trimborn, P. (1996). Diversity, metabolic types and $\delta^{13}C$ carbon isotope ratios in grass flora of Namibia in relation to growth form, precipitation and habitat conditions. *Oecologia* 106 (352-369).

Service, E. R. (1975). *Origins of state and civilisation.* New York, Norton.

Service, E. R. (1962). *Primitive social organisation.* New York, Random House.

Shao, W. (2000). The Longshan period and incipient Chinese civilization. *Journal of East Asian Archaeology* 2(1-2): 195-226.

Shelach, G. (2000). The earliest Neolithic cultures of Northeast China: Recent discoveries and new perspectives on the beginning of agriculture. *Journal of world prehistory* 14(4): 363-413.

Shelach, G. (1999). *Leadership strategies, Economic Activity, and Interregional Interaction: Social complexity in Northeast China.* New York, Kluwer Academic/Plenum Publishers.

Shelach, G. (1994). Social complexity in North China during the early Bronze Age: A comparative study of the Erlitou and Lower Xiajiadian Cultures. *Asian Perspectives* 33(2): 261-292.

Shi, J. (2001). Archaeology in China. *Acta Archaeologica* 72(2): 55-90.

Shi, X. (1992). The discovery of the pre-Yangshao culture and its significance. *Pacific northeast Asia in prehistory: hunter-fisher-gatherers, farmers and sociopolitical elites.* C. M. Aikens, Rhee, S.N. Pullman, (WA), Washington State University Press: 125-132.

Shi, Y., Kong, Z., Wang, S., Tang, L., Yao, T., Zhao, X., Zhang, P. & Shi, S. (1993). Mid Holocene climates and environments in China. *Global and planetary change* 7: 219-233.

Sikkink, L. (1988). *From field to the house: ethnoarchaeology and ethnobotany of harvest and crop processing in Andean peasant households.* MA thesis, University of Minnesota.

Simmons, I. G. (1996). *Changing the face of the earth.* Oxford, Blackwell.

Smilauer, P. (1992). *CANODRAW version 0.3, user's guide.* Ithaca, New York, Microcomputer Power.

Smith, B. (1995). *The emergence of agriculture.* New York, Scientific American Library.

Song, J., Zhao, Z., & Fuller, D. Q. (2013). The archaeobotanical significance of immature millet grains: an experimental case study of Chinese millet crop processing. *Vegetation History and Archaeobotany* 22(2), 141-152.

Specht, R. L. (1958). An introduction to the ethnobotany of Arnhem Land. *Records of the American Australian Scientific expedition to Arnhem Land.* R. L. Specht, Mountford, B.P. Melbourne, London, New York. 3: 479-503.

Spencer, C. S. (1994). Human Agency, Biased Transmission, and the Cultural Evolution of Chiefly Authority. *Factional Competition and Political Development in the New World.* E. M. Brumfiel, Fox, J.W.,. Cambridge,, Cambridge University Press: 31-43.

Srinivas, M. N. (1976). *The remembered village.* Oxford, Oxford University Press.

Stevens, C. J. (2003). An investigation of agricultural consumption and production models for Prehistoric and Roman Britain. *Environmental Archaeology* 8: 61-76.

Stone, G. D. (2001). Theory of the square chicken: advances in agricultural intensification theory. *Asia Pacific Viewpont* 42(2/3): 163-180.

Stoskopf, N. C. (1985). *Cereal Grain Crops.* Reston, Virginia, Prentice Hall.

Stromberg, C. A. E. (2009). Methodological concerns for analysis of phytolith assemblages: Does count size matter? *Quaternary International* 193: 124-140.

Stromberg, C. A. E. (2004). Using phytolith assemblages to reconstruct the origin and spread of grass dominated habitats in the Great Plains during the Late Eocene to Early Miocene. *Palaeogeography, Palaeoclimatology, Palaeoecology* 207: 239-75.

Stromberg, C. A. E. (2002). The origin and spread of grass dominated ecosystems in the Late Tertiary of North America: Preliminary results concerning the evolution of hypsodonty. *Palaeogeography, Palaeoclimatology, Palaeoecology* 177: 59-75.

Struve, G. A. (1835). *De silicia in plantis nonnullis.* Dissertatio inauguralis Berolini

Sweeney, M., McCouch, S. (2007). The complex history if the domestication of rice. *Annals of Botany*: 1-7.

Syarifudden, K., Sundaru, A.M., Azirin, A. (1983). Farmer's weed control technology in insular Southeast Asia. *Proceedings of the Conference on weed control in rice.* I. R. R. I., Los Banos: 210-2.

ter Braak, C. J. F. (1988). *Canoco - a FORTRAN program for canonical community, version 2.1.* Ithaca, New York, Microcomputer Power.

Thompson, G. B. (1996). *The excavations of Khok Phanom Di: a prehistoric site in Central Thailand; Volume IV: Subsistence and environment: the botanical evidence, the biological remains (Part II).* London: Society of Antiquaries

Thompson, G. B. (1992). *Archaeobotanical investigations at Khok Phanom Di, Central Thailand.* Australian National University

Thorp, R. (1991). Erlitou and the search for the Xia. *Early China* 16: 1-38.

Toyama, S. (2002). The origin and spread of rice cultivation as seen from rice remains. *The origins of pottery and agriculture.* Y. Yasuda. Kyoto, International Research Center for Japanese Studies.

Trigger, B. (1999). Shang political organization: a comparative approach. *Journal of East Asian Archaeology* 1: 43-62.

Trigger, B. (1993). *Early civilisations: Ancient Egypt in context.* Cairo, American University Press.

Tsartsidou, G. L.-Y., S., Albert, R-M, Miller-Rosen, A, Efstratiou, N., Weiner, S. (2007). The phytolith archaeological record: strengths and weaknesses evaluated based on a quantitative modern reference collection from Greece. *Journal of Archaeological Science* 34(8): 1262-1275.

Twiss, P. C. (1992). Predicted World distribution of C_3 and C_4 Grass Phytoliths. *Phytolith Systematics: Emerging issues.* G. R. J. Rapp, Mulholland S.C. New York, Plenum Press.

Twiss, P. C., Suess, E., Smith, R.M. (1969). Morphological classification of grasses phytoliths. *Soil Science Society of America Proceedings* 33: 109-115.

Underhill, A. P. (2002). *Craft production and social change in Northern China.* New York, Kluwer Academic/Plenum Publishers.

Underhill, A. P. (2000). An analysis of mortuary ritual at the Dawenkou site, Shandong, China. *Journal of East Asian Archaeology* 2(1-2): 93-127.

Underhill, A. P. (1997). Current issues in Chinese Neolithic archaeology. *Journal of world prehistory* 11(2): 103-160.

Underhill, A. P. (1994). Variation in settlement during the Longshan Period of Northern China. *Asian Perspectives* 33: 197-228.

Underhill, A. P. (1992). Regional growth of cultural complexity during the Longshan Period of Northern China. *Pacific northeast Asia in prehistory: Hunter-fisher- gatherers, farmers and sociopolitical elites.* C. M. Aikens, Rhee, S.N. Washington, Washington State University Press: 173-178.

Underhill, A. P., Feinman, G. M., Nicholas, L., Bennett, G., Cai, F., Yu, H., Luan, F. Hui, F. (1998). Systematic, regional survey in South East Shandong Province, *China. Journal of field archaeology* 25(4): 453-474.

Underhill, A. P., Habu, J. (2006). Early communities in East Asia: economic and sociopolitical organisation at the local and regional levels. *Archaeology of Asia.* M. Stark. Oxford, Blackwell: 121-148.

Vaughn, D., Lu, B-R., Tomooka, N. (2008). The evolving story of rice evolution. *Plant Science* 174: 394-408.

Vaughn, D. (1994). *The wild relatives of rice: a genetic resources handbook.* Manila, International Rice Research Institute.

Vaughn, D. A. (1989). *The Genus Oryza L: Current status of taxonomy.* Manila, Philippines, International Rice Research Institute.

van de Veen, M. (2007). Formation processes of desiccated and carbonized plant remains - the identification of routine practice. *Journal of Archaeological Science* 34: 968-990.

van de Veen, M. (1999). The economic value of chaff and straw in arid and temperate zones. *Vegetation History and Archaeobotany* 8: 211-224.

Von Mohl, H. (1861). Uber das Kieselkelett lebender Pflanzenzellen. *Botanische Zeitung.* 19: 209-215, 217-221, 225-231, 305-308.

Vongsaroj, P. (n.d.). *Weeds in paddy field and their control.* Botany and Weed Science Division, Department of Agriculture. Bangkhen, Bangkok.

Waley, A. (translator). (1960). *(Shi Jing) The book of songs.* New York, Grove Press.

Wang, Q., Andrews, K.C. (2002). Technological investigation of the decorative coatings on Yangshao pottery from Henan, China. *Archaeometry* 44(2): 241-250.

Wang, T., Anderson, E.N., (1998). *Ni Tsan and his 'Cloud Forest Hall Collection of Rules for Drinking and Eating.* Totnes, Prospect Books.

Watson, L., Dallowitz, M.J. (1992). *The grass genera of the world.* Wallingford, Oxon, Centre for Agriculture and biosciences International.

Webb, E. A., Longstaffe, F.J. (2003). The relationship between phytolith and plant water $\delta^{18}O$ values in grasses. *Geochimica et Cosmochimica* Acta 67(8): 1437-1449.

Webb, E. A., Longstaffe, F.J. (2002). Climatic influences on the oxygen isotopic composition of biogenic silica in prairie grass. *Geochimica et Cosmochimica* Acta 66(11): 1891-1904.

Webb, J., Ford, A. Gorton, J. (2007). Influences on selection of lithic raw material sources at Huizui, a Neolithic/Early bronze Age site in Northern China. *Indo-Pacific Prehistory Association Bulletin* 27: 76-86.

Weiler, A. (1897). Beitrage zur Anatomie des Stockes von Saccharum. *Beitrage zur Wissenschaftlichen Botanik* 2: 141.

Weiler, A. (1893). Ueber das Vorkommen von Verstopfungen in den Gefassen mono-und dicotyler Planzen. *Medeel Proefstation Midden-Java*: 1-41.

Wiesheu, W. (1997). *China's first cities: the walled site of Wangchenggang in the Central Plain region of North China. Emergence and early change in urban societies.* L. Manzanilla. New York, Plenum Press: 87-108.

Weisner, J. (1867). *Einleitung in die technische Mikroskopie*. Vienna.

Welle, B. J. H. (1976). Silica grains in woody plants of the Neotropics, especially Surinam. Wood structure in biological and technological research. P. Baas, A. J. Bolton, D.M. Catling. Leiden, Leiden Unversity Press. Leiden Botanical Series 3: 107-142.

Werth, E. (1937). Geography and history of millets. *Angewandte Botanik* 19: 41.

de Wet, J. M. (1995). Foxtail millet. *Evolution of crop plants (2nd edition).* J. Smartt, Simmonds, N.W. London, Longman: 170-172.

de Wet, J. M., Oestry-Stidd, L.L., Cubero, J.L. (1979). Origins and evolution of foxtail millet. *Journal d'Agriculture Traditionelle et de Botanique Appliquée*. 26 Appl. 1: 53-64.

White, J. C. (1995). Modeling the development of early rice agriculture: ethnoecological perspectives from Northeast Thailand. *Asian Perspectives* 34(1): 37-68.

White, J. C. (1989). Ethnological observations on wild and cultivated rice and yams in northeastern Thailand. *Foraging and farming: the evolution of plant exploitation.* D. Harris, Hillman, G.C. London, Unwin.

Wilcox, G. (1999). Agrarian change and the beginnings of cultivation in the Near East: evidence from wild progenitors, experimental cultivation and archaeobotanical data. *The Prehistory of Food. Appetites for Change, One world Archaeology.* C. Godsen, Hather, J. London, Routleledge: 478-500.

Wilding, L. P., Drees, L.R (1968). Distribution and implications of sponge spicules in surficial deposits in Ohio. *The Ohio Journal of Science* 68(2): 92-99.

Winkler, M., Wang, P.K., (1993). The Late Quaternary vegetation and climate of China. *Global Climates since the last Glacial Maximum.* H. E. Wright, Kutzbach, J.E., Webb, T. Rudiman, W.F., Street-Perrott, F.A.,. Bartlein, P.J., Minneapolis: University of Minnesota Press: 221-264.

Woolly, L. C. (1930). *Digging up the Past.* London, Earnest Ben Ltd.

Wright, H. (1977). Recent researches on the origins of the state. *Annual Review of Anthropology* 6:379-397

Wu, W., Liu, T. (2004). Possible role of the Holocene Event 3 on the collapse of Neolithic Cultures around the Central Plain of China. *Quaternary International* 117: (1) 153-166.

Wu, Y. (1996). Prehistoric rice agriculture in Yellow River Valley. *Bulletin of Indo-Pacific Prehistory Association* 15: 31-44.

Xia, N. (1990). What is archaeology. Anthropology in China. G. E. Guldin. Armonk New York, M.E. Sharpe: 59 –67, originally published in *Kaogu* (1984).10: 931-5, 948.

Yabuno, T. (1987). Japanese barnyard millet (*Echinochloa utilis*, Poaceae) in Japan. *Economic Botany* 41(4): 484-93.

Yan, W. (1999). Neolithic settlements in China: latest finds and research. *Journal of East Asian Archaeology* 1(1-4): 131-147.

Yan, W. (1992). Origins of agriculture and animal husbandry in China. *Pacific Northeast Asia in Prehistory: Hunter-fisher- gatherers, Farmers and sociopolitical elites.* C. M. Aikens, Song Nay Rhee. Pullman, Washington State University Press: 113-24.

Yang, D. Y., Liu, Li, Chen, X., Speller, C.F. (2008). Wild or domesticated: DNA analysis of ancient water buffalo remains from north China. *Journal of Archaeological Science* 35(10): 2778-2785.

Yang, X., Wan, Z., Perry, L., Lu, H., Wang, Q., Zhao, C., Ge, Q. (2012). Early millet use in northern China. *Proceedings of the National Academy of Sciences,*109(10), 3726-3730.

Yen, D. E. (1982). Ban Chiang pottery and rice. A discussion on the inclusions in the pottery matrix. *Expedition* 24: 51-64.

Yoffee, N. (1991). Orienting collapse. *The collapse of ancient states and civilizations.* N. Yoffee, Cowgill, G.L. Tuscon, The University of Arizona Press: 1-19.

Yu, Y.-S. (1967). *Trade and expansion in Han China: a study in the structure of sino-barbarian economic relations,* Berkley: University of California Press.

Yuan, J., Flad, R.K. (2002). Pig domestication in ancient China. *Antiquity* 76: 724-32.

Zhang, C. (2002). Early pottery and rice phytolith remains from Xianrending and Diaotonghuan sites, Wannian, Jiangu province. *The Origins of Pottery and Agriculture.* Y. Yasuda. New Delhi, Roli Books: 185-192.

Zhang, H.-Q., Zhao, W.M. (2007). Mathematical modeling of the relationship between Neolithic sites and the rivers in Xian (Shaanxi province, China). *Archaeometry* 49(4): 765-773.

Zhang, J., Harbottle, Wang, C., Kong, Z. (1999). Oldest playable musical instruments found at Jiahu early Neolithic site in China. *Nature* 401: 366-368.

Zhang, J., Wang, X. (1998). Notes on recent discovery of ancient cultivated rice at Jiahu, Henan Province. *Antiquity* 72: 897-901.

Zhang, Z.-P. (1985). The social structure reflected in the Yuanjinmiao cemetery. *Journal of Anthropological Archaeology* 4: 19-23.

Zhao, S., Wu, W. (1989). Early Neolithic Hemudu Culture along the Hangzhou estuary and the origin of domestic paddy rice in China. *Asian Perspectives* 27: 29-34.

Zhao, J. (2008). Flotation techniques and their applications in Chinese Archaeology. *From the concepts of the past to practical strategies: the teaching of archaeological techniques.* P. Ucko, Qin, L., Hubert, J. London, Saffron: 233-238.

Zhao, Z. (2006). *Domestication of Millet. Palaeoethnobotanical data and ecological perspective.* Archaeology in China and Sweden. Institute of Archaeology Chinese Academy of Social Sciences and the Institute of Archaeology Swedish National Heritage Board. Institute of Archaeology. Beijing, Science Press: 97–104.

Zhao, Z. (1998). The Middle Yangtze Region in China is one place where rice was domesticated: phytolith evidence from the Diatonghuan Cave, Northern Jiangxi. *Antiquity* 72: 885-897.

Zhao, Z. (1996). *Rice domestication in the Middle Yangtze region, China: an application of phytolith analysis.* Unpublished Doctoral Dissertation. Department of Anthroplogy. Columbia, University of Missouri.

Zhao, Z., Pearsall, D. (1998). Experiments for improving phytolith extraction from soils. *Journal of Archaeological Science* 25: 587-98.

Zhao, Z., Piperno, D.R., (2000). Late Pleistocene/Holocene environments in the Middle Yangtze River Valley, China and rice (*Oryza sativa* L.) domestication: the phytolith evidence. *Geoarchaeology* 15(2): 203-222.

Zhao, Z., Pearsall, D., Benfer, R.A.Jr., Piperno, D. (1998). Distinguishing rice (*Oryza sativa* Poaceae) from wild *Oryza* species through phytolith analysis II: finalized method. *Economic Botany* 52: 134-145.

Zohary, D., &, Hopf, M (2000). *Domestication of plants in the Old World.* Oxford, Oxford University Press.

Zong, Y., Chen, Z., Innes, J.B., Chen, C., Wang, Z., Wang, H., (2007). Fire and flood management of coastal swamp enables first rice paddy cultivation in east China. *Nature* 449(27): 459-463.

Zhou, W., Yu, X., Jull, T., Burr, G.S., Xiao, J.Y. Lu, X., Xian, F. (2004). High resolution evidence from southern China of an Early Holocene optimum and a Mid Holocene dry event during the last 18,000 years. *Quaternary Research* 62: 39-48.

CHINESE AND JAPANESE LANGUAGE REFERENCES CITED

(2000). *Xia-Shang-Zhou duandai gongcheng [Xia-Shang-Zhou Chronology Project].* Beijing, Shijie Tushu Chuban Gongsi (with English summary)

(1989). *Nongye Kaogu* 1: 111.

(1987). *Nongye Kaogu* 1: 57.

(1986). *Nongye Kaogu* 1: 91.

An, Z. (1988) b. Prehistoric Agriculture in China. *Kaogu Xuebao* 4.

Anonymous (1976). *Guzi (Domesticated millet).* Beijing, Science Press.

Cai, L., *et al* (1984). Tanshisan ceding he guadai shipu yanjiu (C^{13} analysis and human diet in ancient society). *Kaogu* (Archaeology) 2: 234-250.

Cai, L., Qiu, S. (1984). Carbon-13 evidence for ancient diets in China. *Kaogu* 10: 949.

Chen, X. (2007). Lithic production of early states in China: An examination of the development of craft specialisation. *Kaogu*

Chen, X. L., L., Lee, Y-K., Wright, H., Rosen, A (2003). Zhongguo Wenming Fudi de Shehui Fuzahua Jincheng: Yiluohe Diqu de Juluo Xingtai Yanjiu (Development of social complexity in the heartland of Chinese Civilization: Yilou region settlement patterns). *Kaogu Xuebao* 2: 161-218

Chinese-American, Rizhao, Liancheng, Region, Collaborative, Archaeological, Team, (in press). Chemical analysis of the remains of alcohol at Liangchengzhen, Rizhao, Shandong. *Kaogu*

CPAM of Hebei Province and Handan Relics Preservation Station (1981). The Cishan site in Wuan, Hebei Province. *Kaogu Xuebao (Acta Archaeology Sinica)* 3

Crawford, G., Chen, X. Wang, J. (2006). Houli culture rice from the Yuezhuang site, Jinan. *East Asian Archaeology* 3: 247-251. (English abstract)

Du, J. (1992). Shilun Dawenkou wenhua Yingshui leixing) On the Yingshui Phase of the Dawenkou Culture). *Kaogu* 2: 157-169, 181.

Du, Z. (1991). Xiadai kaogu jiqi guojia fazhande tansuo (Archaeology of the Xia dynasty and the development of state). *Kaogu* 1 (43-56).

Erlitou Working Team, Institute of Archaeology, Chinese Academy of Social Sciences (2005). Henan yanshi Erlitou yizhi zhongxingqu de kaogu xinfxian (New finds from the central area of the Erlitou site in Yanshi, Henan). *Kaogu* 7: 15-20. (English abstract)

Erlitou Working Team, Institute of Archaeology, Chinese Academy of Social Sciences (2005). Henan Luoyang pendi 2001-2003 nian kaogu diaocha jianbao (Report of the 2001-2003 archaeological survey in the Luoyang basin, Henan. *Kaogu* 5: 18-37. English abstract

Erlitou Working Team, Institute of Archaeology, Chinese Academy of Social Sciences. (2004). Henan Yanshi Erlitou yizi gongcheng ji gongdianqu waiwei dalu de kanch yu fajue (Survey and excavation of the palatial town, and its outer roads at the Erlitou site in Yanshi, Henan. *Kaogu* 11: 3-13.

Erlitou Archaeological Team, Institute of Archaeology, Chinese Academy of Social Sciences (1992). 1987 nian Yanshi Erlitou yizhi muzang fajue jianbao (preliminary report on the excavation of Burials at the Erlitou site in Yanshi in 1987). *Kaogu* 4: 294-303.

Erlitou Archaeological Team, Institute of Archaeology., Chinese Academy of Social Sciences (1984). 1981 nain Henan Yanshi Erlitou muzang fajue jianbao (Brief report on the Excavation of Burials at the Erlitou Site in Yanshi, Henan, 1981). *Kaogu* 1: 37-40.

Fang., Y. (1994). Guanyu Huanghe Zhongyou Xin shiqu Shidai ZaoZhongqi Kaoguxue Wenhua de Xiangguan Wenti (Regarding the classification of archaeological cultures from the Early and middle Neolithic Period in the Middle reaches of the Yellow River). *Zhongyuan Wenwu* 2: 31-40, 80.

Fujiwara, H., Sugiyama, S. (1985)a. Fundamental studies of plant opal analysis 5:Investigation of ancient paddy fields before archaeological excavation by plant opal analysis. *Kokogaku to Shizen Kagaku* 17: 73-85. (In Japanese with English title and summary)

Fujiwara, H., Sugiyama, S. (1985)b. Fundamental studies of plant opal analysis 6: Investigation of ancient upland fields and burnt fields by plant opal analysis. *Kokogaku to Shizen Kagaku* 18: 111-26. (In Japanese with English title and summary)

First Henan Archaeological Team, I., Institute of Archaeology, Chinese Academy of Social Sciences (1984). 1979 Nian Peiligang Yizhi Fajue Baogao (Report for the excavation of the Peiligang site in 1979). *Kaogu Xuebao* 1 (1984): 23-52.

Gao, M. J., Chen, J.J. (1988). Zaipeisu Qiyuan de Tonggongmei Yanjiu (Isozymic studies on the origin of cultivated foxtail millet). *Zuowu Xuebao (Acta Agronomica Sinica)* 14(2): 131-136.

Gong, Q., Jiang, J. (1993). *Shilun Longshan Wenhua de Shehui Xingzhi (A preliminary discussion of the nature of Longshan culture in society).* Jinian Chengzia Yizhi Fajue Liushi Zhounian Guoji Xueshu Taolunhui Wenji. X. Zhang. Jinan, Qilu Press.

Gongxian County Chronicle Editorial Board. (1991). *Gongxian Zhi.* Zhengzhou, Zhongzhou Guji Press.

Henan Cultural Bureau (1961). Henan Yanshi Xian Huizui Shangdai Yizhi De Diaocha (Survey of the Shang site at Huizui in Yanshi). *Kaogu* 2: 99-100.

Henan Team, (1995). Henan Jiaxian Shuiquan Peiligang Wenhua Yizhi (The Shuiquan site of the Peiligang culture at Jia County, Henan). *Kaogu Xuebao* 1: 39-77.

Henan First Team Institute of Archaeology, Chinese Academy of Social Sciences (2002-2003). Niandu Huizui Yizhi Fajue Jianbao 2002-2003 (A brief report of the year 2002-2003 excavation at Huizui site). *Kaogu.*

Henan First Team IACASS. (2003). Henan Yanshi Huizui Yizhi Fajue de Xin Shouhuo (New Gains from the excavation at Huizui site in Yanshi, Henan). Zhongguo Shehui Kexueyuan Gudai Wenming Yanji *Zhongxin Tongxun* 5: 36-39.

Henan Provincial Institute of Cultural Relics (1999). *Wuyang Jiahu (Jiahu Site in Wuyang)*. Beijing, Kexue Press (Beijing Science Press).

Huang, S. (1982). Neolithic crops in the Yellow River Valley. *Nongye Kaogu* 2.

Institute of Archaeology, Chinese Academy of Social Sciences. (1991). *Zhongguo Kaoguxue Zhong Tan Shisi Niandai Shujuji.* Beijing, Wenwu Press

Institute of Archaeology, Chinese Academy of Social Sciences. (1999). *Yanshi Erlitou.* Beijing, Zhongguo Dabaikekquanshu Press.

Jia, T., Xinghua Jia & Xionggao Li (1992). *Lingbao Xianzhi (Lingbao County Annals).* Zhengzhou, Zhongxhou Guji Press.

Jiang, Q., Zhang, J. (1998). Phytolith evidence for rice cultivation during prehistoric periods at Baligang site of Baizhuang, Dengzhou City, Henan Province. *Acta Scientarium Naturalium Universitatis Pekinensis* 34(1): 66-71.

Jiang, X. M., Zhou, Y. (1989). SEM observation on crystals and silica in wood species of Chinese Gymnospermae. *Acta Botanica Sinensis* 31: 835-840.

Jilin University (2001). Henei Yangyuanxian Jianfjialiang yizhi zinshiqi shidai rengu DNA de yanjui. *Kaogu* 7: 74-81.

Jin, G. (2007). Phytolith analysis for the archaeological sites of Pre Qin dynasties in Shandong and some related questions. *Dongfang Kaogu* 3(25): 258-279.

Kaoguxue Jikan. (1993). Henan Anyang Huanhe Liuyu de kaogu diaocha (Archaeological survey in the Huan River valley in Anyang, Henan). *Kaoguxue Jikan* 3: 90-97.

Kaogu Yanjiushuo, Zhongguo Shehui Kexueyuan. (1999). *Yanshi Erlitou.* Beijing, Zhongguo Da Baike Quanshu Chubanshe.

Kong., S., Du, N., Xu, Q.,& Tong, G. (1992). *Zhongguo beifang quanxinshi danuanqi zhiwuqun de guqihou bodong (Palaeoclimatic fluctuations reflected in Flora of Holocene megathermal in the northern part of China).* Zhongguo Quanxinshi. Y. Shi, Kong, S.. Beijing, Haiyang Press: 48-65.

Li, S. (2004). On the date of wheat unearthed at the Donghuishan site, Gansu. *Archaeology and Cultural Relics* 6: 51-60.

Lin, H. D. (1992). *Hemudu Wenhua Chutian (A study of the Hemudu Culture).* Hangzhou, Zeijiang People's Press.

Liu, F. (1994). Cong Minzuxue Youguan Ziliao Kan Zhingguo Zaoqi Xinshiqi Shidai Yizhi de Jingji Wenhua Leixing (Economic and cultural types of Early Neolithic cultures in China considering ethnographical data). *Huaxia Kaogu* 2: 50-60.

Liu, X. (2001). Youguan Xiadai niandai he Xia wenhua cenian de jidian kanfa (Opinions on dating the Xia dynasty and Xia culture). *Zhongyuan Wenwu* 2: 32-33.

Louyang Relics Team, (2002). *Luoyang Zaojiaoshu.* Beijing, Wenwu Press.

National Bureau 1999

Pei, A. (1998)a. Pengtoushan Wenhua de Daozuo Yichun yu Zhongguo Shiqian Daozuo Nongye Zailun (Remains of rice cultivation found in the Pengtoushan assemblages and a re-discussion on the origin of rice cultivation in prehistoric China). *Nongye Kaogu* 1: 193-203.

Ren, S. (1995). Gongyuan Qian Wuqian Nian Qian Zhonguo Xinshiqi Wenhua de Jixiang Zhuyao Chenjui (Important results regarding Neolithic cultures in China Earlier than 500B.C.). *Kaogu* 8: 710-718.

Shi, X. (1986). Qian Yangshao wenhua de faxian ji qi yiyi (The discovery of pre-Yangshao cultures and their meaning). Zhongguo kaoguxue yanjiu. Beijing, *Wenwu chubanshe.* 2: 1-23.

Shi, X. (1982). *Tianye kaogu fangfa -diaocha, fajue yu zhengli. Kaogu Gongzuo Shouce.* Institute of Archaeology, Chinese Academy of Social Sciences. Beijing, Wenwu Press: 1-92.

Su, B. (1997). *Zhongguo Wenming Qiyuan Xintan.* Hong Kong, Shangwu Press.

Su, B. (1988). Zhonghua wenming de xin shuguang. Dongnan *Wenhua* 5: 1-7.

Su, B., Yin, W. (1981). Guanyu kaoguxe wenhua de quxi leixing wenti. *Wenwu* 5: 10-17.

Sun, G. (2000). Henan jingnei de Dawenkou wenhua he Qujialing wenhua (The Dawenkou culture and Qujialing Culture in Henan. Zhongyuan *Wenwu* 2: 22-28.

Teacher's College of Northwest China, (1984). Gansu Dongxiang Linjia Majiayao Wenhua Yizi Chutu di Ji Yu Dama (Hemp and Millet from the Majiayao Culture site of Linjia at Dongxiang, Gansu). *Kaogu* 7: 654-655, 633.

Tong, W. (1984). Cishan yizhi de yuanshi nongye yicun jiqi xiangguan de wenti (Remains of primitive farming at Cishan site and its related problems). *Nongye Kaogu* 7: 36-42.

Wang, J. (1997). *Shilun Dawenkou wehhua Yuchisi leixing. Kaogu Quizhiji,*Institute of Archaeology, Chinese Academy of Social Sciences. Beijing, Zhongguo Shehui Press: 213- 223.

Wang, X. (2007). Production tools from the Western Henan-Southern Shanxi region 2500-1500BC. Keji *Kaogu* 2: 116-119.

Wu, J. (1930). Pingling fanggu ji. Zhongyang Yanjiuyuan Lishi Yuyan Yanjiusuo *Jikan* 1: 471-486.

Wu, R., Xiao, H., Wei, C., Pan, Y., Huang, Y., Zhao, C., Xu, S., Orin, N. (2007). Stable isotope evidence concerning the diet of humans and pigs, the agricultural economy and pig domestication at Xinzhai, Henan. Keji *Kaogu* 2: 97-115.

Xiang, A. Q. (1991). Lun Changjiang Zhongyou Xinshiqi Shidai Jiaoqi Yichun de Nongye (The agriculture of the early Neolithic in the middle Yangzi Valley). *Nongye Kaogu* 1: 121-135.

Yan, W. (1997). Huanghe liuyu wenming de qiyuan. *Wenwu* 1: 40-49.

Yan, W. (1981). The Longshan culture and the Longshan period. *Wenwu* 6: 41-48.

Yang, Y. (1991). *Zhongguo Wenwu Dituji: Henan Juan (Cultural Relics Atlas: Henan Fence).* Beijing, Zhongguo.

Ye, W. (2000). Palaeoenvironment and ancient culture of the Zhaojiaoshu site. *Issues in environmental archaeology, vol, 2.* K. S. Zhou, Song, Y.Q. Beijing, Science Press. 2.

Yuan, J., Huang, Y., Yang,M.,Lu, p., Tao, Y., Yang, J. (2007). Research into the zooarchaeology of the Central Plains 2500-1500 BC: the cases of Taosi, Wangchenggang, Xinzhai and Erlitou. *Keji Kaogu* 2: 12-34.

Zhao, Z. (1987). Lun Erlitou yizhi wei Xiadai wanqui duyi (On the Erlitou site as a capital of the Xia dynasty). *Huaxia Kaogu* 2: 196-204,217.

Zhang, J., Kong, Z. Liu, Z. (1994). Wuyang Shiquian Daozuo Yicun Yu Huang Huai Diqu Shiqian Nongye (Paddy Rice at a prehistoric site in Wuyang and prehistoric agriculture in the Huang and Huai River areas). *Nongye Kaogu* 1: 68-77.

Zhang, W. X., Pei, A.P. (1997). Lixan Mengxi Bashidang Chutu Daogu de Yanjiu (A study on the ancient rice from Bashidang, Mengxi in Lixian County. *Wenwu* (The relics) 1: 36-41.

Zhang, X., Qiu, S., Bo, G., Wang, J., Jian, Z. (2007). Carbon-13 and Nitrogen-15 analysis of a sample of human bones at Erlitou and Taosi. *Keji Kaogu* 2: 41-48.

Zhang, Z. H. (1984). Zhongguo Yuanshi Nongye de Changsheng he Fazhan (The origins and development of indigenous agriculture in China). *Nongye Kaogu* 2: 36-42.

Zhao, C. (2001). *Zhengluo Diqu Xinshiqi Shidai Juluo de Yanbian.* Beijing, Beijing University Press.

Zhao, Z. (2007). Research into the agricultural economy of the Central Plains 2500-1500 BC. *Keji Kaogu* 2: 41-8.

Zhao, Z. (2005). Palaeoethnobotany and its new achievements in China. *Archaeology and Cultural Relics* 7: 42-49.

Zhao, Z. (2004). Study on the origin of the dry-land agriculture in North China based on the flotation result from the Xinglongou site, Inner Mongolia. *Antiquities of Eastern Asia.* Beijing, Wenwu Press. Volume A.

Zhao, Z. (2004). Zhiwu kaoguxue de tianye gongzuo fangfa - fuxuan fa (Flotation: a palaeobotanic method in field archaeology). *Kaogu* 3: 80-87.

Zhao, Z. (1987). Lun Erlitou yizhi wei Xiadai wanqi duyi [On the Erlitou site as the capital of the late Xia dynasty]. *Huaxia Kaogu* 2: 196-204.

Zhao, H., Chen, G., Xu, H. (2006). *Erlitou yizhi faxian daxing weihuan zuongqu (A large craft workshop enclosure discovered at the Erlitou site).* Zhouguo Wenwubao.

Zhao, Z., Xue, L. (2004). Analysis of the flotation results from the Zhouyuan site, Shaanxi province. *Wenwu* (10).

www.ingramcontent.com/pod-product-compliance
Lightning Source LLC
LaVergne TN
LVHW072255300926
849160LV00018B/1673